Armed Groups and the Politics of International Legitimation

Armed Groups and the Politics of International Legitimation

Edited by

Stephan Hensell
Klaus Schlichte

OXFORD
UNIVERSITY PRESS

OXFORD
UNIVERSITY PRESS

Great Clarendon Street, Oxford, OX2 6DP,
United Kingdom

Oxford University Press is a department of the University of Oxford.
It furthers the University's objective of excellence in research, scholarship,
and education by publishing worldwide. Oxford is a registered trade mark of
Oxford University Press in the UK and in certain other countries.

Published in the United States of America by Oxford University Press
198 Madison Avenue, New York, NY 10016, United States of America

British Library Cataloguing in Publication Data
Data available

Library of Congress Control Number: 2025939369

ISBN 9780198974512

DOI: 10.1093/9780198974529.001.0001

The manufacturer's authorized representative in the EU for product safety is
Oxford University Press España S.A. of Parque Empresarial San Fernando de Henares,
Avenida de Castilla, 2 – 28830 Madrid (www.oup.es/en or product.safety@oup.com).
OUP España S.A. also acts as importer into Spain of products made by the manufacturer.

Preface

Successful academic work is usually attributed to single individuals. A socialization towards 'competitiveness' as single scholars and individualized curricula vitae encourages, if not enforces, individual reputation as the main currency in academia. Established modes of organization in contemporary academic culture do not really reward truly joint work as an achievement as such. Prefaces then usually are used as occasions to thank a row of individuals for their critical comments, for their contributions in processing the manuscript, etc.

We will not fail to do this here. But first we want to stress that this book is the outcome of a truly collective endeavour, and this on various levels. The first of these collective settings was the study group on the causes of war (Arbeitsgemeinschaft Kriegsursachenforschung, AKUF) that has existed since the early 1980s as a joint teaching and research seminar at the University of Hamburg. Numerous scholars, scattered now across Europe and Africa, in universities, think tanks, and other professional fields have been socialized in that group into a critical sociological and historical approach into International Relations as a discipline. We, the two editors, have been part of this intellectual context and are grateful for this experience and all the stimuli that former and current members of it have provided to our work.

This book started out as a project entitled 'the legitimacy politics of armed groups' that was funded by the German Research Foundation (Deutsche Forschungsgemeinschaft, DFG) at the Institute of International and Intercultural Studies at the University of Bremen (2021–2024). A core group of three scholars, funded by this project, soon became much larger as we started to look out for possible project collaborators. We are deeply thankful for having found like-minded scholars with deep case knowledge and similar orientations of a historical and/or sociological approach to the subject of armed groups' politics of legitimation. This book brings together scholars from both political science and history and demonstrates that interdisciplinarity, often invoked but rather rarely practiced, is often possible without much ado. All authors of this book participated with enthusiasm in a project with uncertain outcome and the method of 'reiterative case-comparison' that we describe in the Introduction to this volume required a substantial engagement by all authors involved. In two workshops at the Université Libre de Bruxelles (2022) and the University of Bremen (2023), the authors engaged in an iterative-cyclical research process of theory-driven interpretation, empirical analysis, and comparative discussion in repeated rounds. We remember these intense discussions as perfect illustrations of what academic debate can be: mutually enriching, focused on content, going beyond each single contribution, in the spirit of critical solidarity. All authors invested time and effort into an endeavour of which the outcome cannot fully be turned into individualized symbolic capital.

This book would not have been possible without the generous funding by the DFG (project no. 424426045). We are also grateful for the financial contribution of the 'Fonds de la Recherche Scientifique' (FNRS) to our Brussels workshop. A special thanks goes to REPI at the Université Libre de Bruxelles, which hosted Stephan Hensell for almost the entire project duration as a visiting researcher. Apart from our direct project collaborators, a wider range of sympathizing but also critical colleagues from the University of Bremen, the Université Libre des Bruxelles, and other institutions contributed with remarks to our project. Without these we would not have reached a number of relevant insights. The same applies to the reviewers of this volume. We are grateful for their valuable comments. We would also like to thank Dominic Byatt and the staff of Oxford University Press for their professional assistance, recommendations, and swift work in publishing this book. Finally, we would like to thank Anne-Sophie Lassen, Maryna Kocherzhat, Sarper Gökzal, and Claudia Herold at the University of Bremen, who greatly helped in the research for our project and in producing the book manuscript.

Overall, we hope that this volume will contribute to our understanding of armed conflicts and help to come to terms with a stage of international politics in which organized violence seemingly gains more momentum. In the regions directly affected by warfare with which we deal in this volume, this challenge is nothing new, neither for scholars nor for non-scholars.

Klaus Schlichte, Bremen and Stephan Hensell, Brussels,
February 2025

Contents

List of abbreviations

AMB	al-Aqsa Martyrs' Brigades
ANC	African National Congress
AU	African Union
CPD	Political Diplomatic Commission
DFLP	Democratic Front for the Liberation of Palestine
DRV	Democratic Republic of Vietnam
EUCOCO	European Conference of Solidarity and Support for the Saharawi People
EP	European Parliament
FCO	Foreign and Commonwealth Office
FDR	Democratic Revolutionary Front
FMLN	Farabundo Martí Para la Liberación Nacional
FLN	Front de Libération Nationale
FRELIMO	Frente de Libertação de Moçambique
FRETILIN	Frente Revolucionária de Timor-Leste Independente
FSLN	Frente Sandinista de Liberación Nacional
GDR	German Democratic Republic
GNA	Government of National Accord
GNC	General National Congress
GOC	Good Offices Committee
HTS	Hayat Tahrir al-Sham
IR	International Relations
ISKP	Islamic State—Khorasan Province
LAAF	Libyan Arab Armed Forces
LNA	Libyan National Army
LTTE	Liberation Tigers of Tamil Eelam
MINURSO	United Nations Mission for the Referendum in Western Sahara
MIR	Movimiento Izquierda Revolucionaria
MPLA	Movimento Popular de Libertação de Angola
NLF	National Liberation Front
NPFL	National Patriotic Front of Liberia
NRA	National Resistance Army
NRM	National Resistance Movement
OAU	Organisation of African Unity
OAS	Organization of American States
OPT	Occupied Palestinian Territories
PA	Palestinian Authority
PAIGC	Partido Africano para a Independência da Guiné e Cabo Verde
PFLP	Popular Front for the Liberation of Palestine
PKK	Partiya Karkeren Kurdistan
PLA	Palestinian Liberation Army
PLO	Palestine Liberation Organization

POLISARIO	Frente Popular para la Liberación de Saguia el-Hamra y Rio de Oro
PRG	Provisional Revolutionary Government
PRI	Institutional Revolutionary Party
RENAMO	Resistência Nacional Moçambicana
SADR	Saharawi Arab Democratic Republic
SPLA	Sudan People's Liberation Army
SRSG	Special Representative of the Secretary-General
SWAPO	South West African People's Organisation
TTP	Tehreek-e Taliban Pakistan
UÇK	Ushtria Çlirimtare e Kosovës
UNGA	United Nations General Assembly
UNITA	União Nacional para a Independência Total de Angola
UNSMIL	United Nations Support Mission in Libya
YPG	Yekîneyên Parastina Gel
ZAPU	Zimbabwe African People's Union

Notes on Contributors

Ahmed Elsayed is a postdoctoral researcher at the Institute of Intercultural and International Studies (InIIS) at the University of Bremen. He has held research positions at various academic institutions, think tanks, and human rights organizations in Cairo, Berlin, and Washington, D.C. His research explores Middle East politics, Islamist movements, legitimacy, power, armed groups, conflict transformation, democratization, and human rights. He holds a Dr. rer. pol. in Political Science from the University of Bremen, an MPP from the Hertie School, and an MA (with distinction) in Politics from the University of East Anglia.

Stephan Hensell is a Senior Researcher at the Institute of Intercultural and International Studies (InIIS) at the University of Bremen. He has held positions as researcher and lecturer at the Institute for Peace Research and Security Policy and at the Institute of Political Science at the University of Hamburg. He was Visiting Research Fellow at the Changing Character of War Programme, University of Oxford and has done field research in Georgia, Albania, Kosovo, Morocco, Western Sahara, Algeria, and Belgium. His research interests include state theory, civil wars, armed groups, international organizations, and EU politics.

Reyko Huang is an Associate Professor of International Affairs at the Bush School of Government & Public Service at Texas A&M University. Her research examines armed conflicts and international politics, state-building, rebel politics, and international security. She is the author of *The Wartime Origins of Democratization: Civil War, Rebel Governance and Political Regimes* (Cambridge University Press), as well as articles in academic, policy, and media publications. Her research has been supported by the US Institute of Peace, the Minerva Research Initiative, the Institute for Humane Studies, and the Centre for International Security and Cooperation at Stanford University. She holds a PhD from Columbia University.

Wolfram Lacher is a Senior Associate at the German Institute for International and Security Affairs (SWP) in Berlin. His research focuses on violent conflict in Libya and the Sahel, and relies on frequent fieldwork. He has published articles in *Violence: An International Journal*, *Survival*, *Mediterranean Politics*, *Security Dialogue*, and *Foreign Affairs*, amongst others. He is the author of *Libya's Fragmentation: Structure and Process in Violent Conflict* (I.B. Tauris, 2020), and co-editor of *Violence and Social Transformation in Libya* (Hurst/Oxford University Press, 2023). He studied Arabic and African languages and history, conflict, and development, and holds a PhD in political science from Humboldt University in Berlin.

Robert J. McMahon is the Ralph Mershon Professor Emeritus at Ohio State University. He is the author of several books, including *The Cold War: A Very Short Introduction* (2nd rev. edn., 2021); *Dean Acheson and the Creation of an American World Order* (2009); and *The Limits of Empire: The United States and Southeast Asia Since World War II* (1999). He has held visiting professorships in Ireland, Germany, and Indonesia and, in

2001, served as president of the Society for Historians of American Foreign Relations.

Christian Olsson is Professor in Political Science and International Relations at Université libre de Bruxelles (ULB), director of its research centre in international relations (Recherche et Enseignement en Politique Internationale, REPI), and member of its Observatory of Muslim and Arab Worlds (OMAM). Drawing in particular on (critical) security studies and sociological approaches, his research focuses on the colonial genealogies of contemporary military doctrine, the historical sociology of security practices, and the role of private military companies in recent overseas military interventions, especially in Afghanistan and Iraq where he has carried out field research. He has published, amongst others, in *Critical Military Studies*, *Millenium*, *Cultures & Conflicts*, and *PArtecipazione e COnflitto*.

Eline van Ommen is a Lecturer in Contemporary History at the University of Leeds. She is a historian of Latin America in the twentieth century, particularly interested in revolutions, transnational and grassroots activism, and foreign policy during the Cold War. She has published on the international and transnational history of Nicaragua's revolutionary decade, European solidarity activism, and Central American-European relations. Her recent book *Nicaragua Must Survive: Sandinista Revolutionary Diplomacy in the Global Cold War* was published in 2024 with the University of California Press. Eline van Ommen obtained a BA at the University of Groningen before moving to the London School of Economics (LSE) to do an MSc and PhD in International History. Prior to being appointed as a lecturer in the School of History at the University of Leeds in 2021, she was a lecturer at the University of Utrecht. Before that, she taught courses on Latin America and the Cold War at the LSE.

Andrea Oñate-Madrazo is Assistant Professor of Modern Latin American History at California Polytechnic State University. She received her PhD in History from Princeton University in 2016. Her forthcoming book, *Insurgent Diplomacy: The Internationalized Revolution of El Salvador and the Transition to a post-Cold War Order, 1970–1992*, examines the Salvadoran Civil War from a transnational perspective. Based on multi-archival research, this work uncovers the role that transnational state and non-state actors played in shaping the course of the Salvadoran war, and showcases how Salvadoran guerrillas cum diplomats influenced foreign actors and helped forge the post-Cold War order. Oñate-Madrazo teaches various undergraduate and graduate courses on Modern Latin America, comparative social movements, revolutions and insurgency in the global south, and the Cold War.

Klaus Schlichte is Professor of International Relations and World Society, University of Bremen. With an interest in global political sociology, Schlichte carried out research in Senegal, Mali, Serbia, France, and Uganda and taught at Science-Po, Paris, the University of Washington, Seattle, the OECD Academy in Bishkek, and at several German universities. Apart from numerous works in German, he is the author of *In the Shadow of Violence: The Politics of Armed Groups* (Campus, Frankfurt/Main, 2009), and the co-editor of *The Historicity of International Politics: Imperialism and the Presence of the Past* (Cambridge University Press, 2023).

Leila Seurat is a researcher at the Centre Arabe de Recherches et d'Études Politiques (CAREP) in Paris. She has completed her PhD in political science at Sciences Po. Focusing on Hamas' foreign policy, her thesis has been published under the title *The Foreign Policy of Hamas: Ideology, Decision Making and Political Supremacy* (IB Tauris, Bloomsbury, 2022). Since 2015 she has been working on order maintenance and policing in the Arab countries.

Friedrich Kratochwil is a scholar in the field of International Relations (IR), known for his contributions to constructivism and practice theory and his focus on norms, rules, and the processes of social construction in international politics. His work emphasizes the role of language, meaning, and social practices in shaping international relations, offering a counterpoint to traditional rationalist and materialist perspectives in IR. Kratochwil taught at Princeton, Columbia, and Pennsylvania University and at the European University Institute in Florence. Until 2003, he was Professor of International Politics at the Geschwister Scholl Institute of the Ludwig-Maximilians-University of Munich.

1

Fighting for Legitimacy

Armed Groups and the Politics of International Legitimation

Stephan Hensell and Klaus Schlichte

Introduction

A countless number of armed groups—rebels, guerrillas, militias, liberation movements—have taken up arms since 1945 in order to topple governments, to overturn entire social-political systems by revolutionary struggle, or to secede and create new states.[1] Most of them failed. However, a considerable number of them have been successful. Today's governments of Nicaragua, Algeria, Eritrea, Zimbabwe, Uganda, Rwanda, or Timor-Leste, for example, all fought their way to power with violent means. They are political actors that initially started out as non-state armed groups. What began with armed rebellion ended with full recognition by other states as subjects of international law. Rebels successfully turned the power of the gun into legitimate authority and became recognized as legitimate rulers by other governments. In various other instances, however, this recognition has been withheld—armed groups tried to create new states but achieved only partial recognition, as in the cases of Kosovo or Western Sahara. Still other groups have only rudimentary international legitimacy, even though they are in full control of a recognized state, as in the case of the Taliban in Afghanistan.

The remarkable trajectory of armed groups that are initially often denounced as 'criminals', 'bandits' or 'terrorists' but later become recognized governments has not yet attracted much scholarly attention. The aim of our book is to address this gap and to advance a sociological and historically informed understanding of the international legitimacy of armed groups that seek state power, self-rule, or secession in civil wars since 1945. Our volume puts the struggle of armed groups to become legitimate actors in international politics centre stage and our theme—'Fighting for

[1] As authors we are ultimately responsible for everything stated in this introduction. But we would like to express our deep gratitude to all the authors of this volume with whom we discussed our approach intensely. All their critical comments have helped tremendously to improve our ideas. Research for this chapter has been made possible by a grant from the German Research Foundation (Deutsche Forschungsgemeinschaft, DFG), project no. 424426045.

Stephan Hensell and Klaus Schlichte, *Fighting for Legitimacy*. In: *Armed Groups and the Politics of International Legitimation*. Edited by: Stephan Hensell and Klaus Schlichte, Oxford University Press. © Stephan Hensell and Klaus Schlichte (2025). DOI: 10.1093/9780198974529.003.0001

Legitimacy'—denotes their use of all available means, violent and peaceful, in order to acquire international legitimacy. Why do some armed groups gain international legitimacy whereas others fail? How does the legitimation of armed groups work and under what conditions are their claims to legitimacy more likely to be accepted or rejected? This book suggests that the answer to these questions is to be found in the politics of international legitimation that evolve around armed groups. Based on political sociology in the Weberian tradition, our book offers a new theoretical approach for understanding this process and for analysing the politics of legitimation as it unfolds in the system of states. In doing so, this volume seeks to make four major contributions to conflict studies and to the discipline of International Relations (IR):

First, we seek to combine the study of armed groups in civil wars and studies on legitimacy in international politics. Drawing from these fields, we build on two vibrant but so far unconnected strands of research. In conflict studies, scholars have produced a wealth of case studies in the last two decades on the international politics of armed groups including their quest for international recognition and legitimacy. However, the process of legitimation has not been unpacked in a systematic fashion. In contrast, in IR, the debate on the various dimensions of international legitimacy in global governance has blossomed in recent years, gaining conceptual and theoretical depth. Yet, armed groups as a category of actors have not received much attention. We are building on both strands of the discussion by applying theoretical vocabulary from legitimation research in IR to conflict studies while at the same time suggesting that armed groups should be part of any comprehensive approach to international legitimacy. We are, thus, enriching the debate on armed groups within the field of conflict studies as well as the discussion on international legitimacy in IR while also seeking cross-fertilization with practice theory and global history approaches in IR, which we employ in order to develop our conceptual framework.

Second, we approach the issue of international legitimacy from an angle which has been neglected so far. The dominant approach in IR is to analyse the legitimacy of international institutions such as the WTO, the IMF, or the EU, whose authority is accepted in principle while their legitimacy might vary depending on the exercise of that authority (Zürn 2018, 65). In the mainstream understanding, institutions of global governance might gain or lose some legitimacy here and there but ultimately, their worthiness of being recognized as legitimate is not questioned. Their legitimacy is a question of degree that varies across issue areas and audiences but is not one of principle. From armed groups we can learn how legitimacy comes into being in the first place. They are usually seen as illegitimate outlaws and, accordingly, they are criminalized or repressed. Armed groups that seek international legitimacy aim in the first instance to leave the 'illegal', the realm of the underground. Becoming legitimate means becoming accepted in principle as a regular political actor that has 'the right to have rights' in Hannah Arendt's term or that is allowed 'rightful membership' in international society to use the language of the English School of IR (Clark 2005, 26–28).

Third, our volume offers a new and unique approach to the study of armed groups as international actors, based on political sociology and global history. This approach, which we also see as the core theoretical output of this book, consists of two elements: legitimation as a process and 'historical times'. Legitimation means constructing an individual, group, or institution as legitimate. We understand legitimation as a relational process which involves practices and publics. We identify two sets of practices: making and validating a claim to legitimacy. Claim-making is done primarily by armed groups. They put forward legitimacy claims regarding their actions and agendas. Such claims are characterized by recourse to social values such as human rights, freedom, or democracy. With claim-making, armed groups want to create the belief in the eyes of others that they are legitimate actors. Claim validation, in contrast, is typically a practice of governments, international organizations and NGOs or the 'international community', as it calls itself. By claim validation, we mean accepting the validity of a claim to legitimacy often through acts of publicly demonstrated recognition that validate armed groups' claims. We also identify two kinds of publics that are addressed by armed groups: audiences and constituencies. We use the term audience for a broader social grouping of possible supporters, undecided observers and opponents of legitimacy claims. A constituency, in contrast, is a more circumscribed group that validates these claims. Armed groups direct their claims simultaneously towards both. Whether the interplay of practices and publics brings about international legitimacy or fails to do so, we argue, is co-determined by historical context. We map four subsequent but overlapping historical times since 1945 which we understand as conditions or structural contexts of armed groups' legitimation. These are the age of decolonization, the Cold War, the time of humanitarianism, and more recently the time of sovereigntism. We argue that these historical times have set different incentives and constraints for armed groups' quest for international legitimacy, thus making their legitimation more or less likely. By relating practices and historical contexts, we argue that the politics of legitimation is socially embedded and subject to historical change. Our book offers a novel theoretical understanding of the politics of legitimation of armed groups, and it proposes analytical instruments that allow us to study such politics over time. In doing so, we also seek to propose future research directions for the study of armed groups and for international legitimacy in IR.

Fourth, our book makes a contribution by assembling eleven new in-depth crucial case studies of armed groups' legitimation covering different world regions and historical contexts since 1945. The case studies are also the core empirical research output of this book. The volume includes cases of armed groups in Indonesia, Vietnam, Western Sahara, Nicaragua, El Salvador, Uganda, Angola, Palestine (West Bank and Gaza), Afghanistan and Libya. With our cases studies we seek to explore patterns of legitimation of armed groups and explain how and why they become legitimate or fail to do so. The case studies also probe the analytical utility and empirical plausibility of our concepts. They provide thick descriptions and context-specific detailed knowledge on armed groups' legitimation politics. Using our analytical tools, we reconstruct these politics by analysing how practices of claim-making and claim validation have unfolded internationally. The case studies deliver new insights into how armed groups internationalize their cause and shed light on the norms and ideas evoked by these actors, their narratives of justification, and the responsiveness of their respective audiences and constituencies. Moreover, we spell out the embeddedness of armed groups in their historical times and how changing historical conditions affected their fates. The case studies include new and unique case material. All authors have expertise on the cases and have based their studies on primary sources including interviews, and archival material in different languages, mostly collected in the field.

In making these contributions our book seeks to advance two core arguments: First, we argue that non-state armed groups are not just central actors in contemporary civil wars, but also in international politics. As forces that can co-shape international politics and with a genuine quest for international legitimacy, armed groups are politically relevant in a double sense. They are able to manipulate, overrule, or outmanoeuvre the power of states or entire groups of states. But still more importantly, as they fight for being recognized as legitimate actors and conquer legitimate positions in the state system, they also invoke and confirm the international order. We thus take armed groups seriously as what they are: political actors in their own right, equipped with ideas, beliefs, and an agency that allows them to co-shape international politics.

Second, we argue that paying attention to the legitimation of armed groups teaches us something about the hidden structures and mechanisms of international politics. Studying how a non-state violent group leaves the realm of the informal and irregular and turns into a formal and legitimate actor, we maintain, allows us to ask pertinent questions about the rules of the game we call 'international politics'. What contradictions of international politics come to the fore in the ups and downs of legitimizing an armed group? How do inclusion and exclusion work? Where is the power located? How are domination and order challenged and reproduced? Thus, we suggest studying IR from below by focusing bottom-up on what armed groups do in order to achieve legitimacy. We argue that we can learn something about the functioning of the international system by studying the trajectories of fringe players that seek legitimate membership in that system.

The remainder of this introduction is organized into four sections, inviting readers to use it selectively: the first section briefly clarifies how we use the two core notions in our book—'armed groups' and 'legitimacy'—and it will describe our methodological approach which we term 'reiterative case comparison'. The second section focuses on the state of the art and the lacunae which we want to address. In the third section we develop our conceptual approach in three steps. We start with the notions of power, domination, and international legitimacy. Next, we clarify our understanding of legitimation and its practices and publics. Finally, we situate the politics of legitimation within four historical times which we outline in the form of brief historical sketches. In the fourth section we briefly outline the structure of the volume and present the various contributions in this book.

Definitions and methodology

Our two core terms 'armed groups' and 'legitimacy' and closely related concepts are part of a semantic field that is not only open to misunderstandings but also to political instrumentalization, often with far-reaching implications. It is therefore important to clarify right from the start how we use these terms. We will then also elucidate our methodology, which is the collective effort of 'reiterative case comparison'.

Armed groups are organized political actors that use violence in civil wars for political ends. These actors are usually described with a wide range of powerful and expressive terms such as rebels, revolutionaries, insurgents, guerrillas, terrorists, or freedom fighters. Often these terms are driven by the attempt to identify the nature of these groups on the basis of their objectives or strategies (Thompson 2014). Most of these labels, however, are ambiguous as they carry positive and negative connotations. Naming is a political act and some labels such as 'liberation movement' or 'terrorist organization' are normatively loaded and easily instrumentalized by these actors themselves or their adversaries (Bhatia 2005). Other seemingly more neutral terms such as 'non-state violent actor' are no less problematic. The prefix 'non-state' neglects the fact that armed groups are often not quite as 'non-state' as the term suggests. These groups regularly present themselves as a state-like actor and feature quasi-state elements by administering territories or conducting diplomacy (Bakonyi et al. 2006; Mampilly 2011; Huang 2016). Also, in most of these groups former state officials and militaries play leading roles (Schlichte 2009, 35). We prefer the term 'armed groups', which we see as less normatively coded and conceptually problematic as the available alternatives. The politics of armed groups cannot be reduced to their violent actions. In their repertoires they usually combine armed and civilian action, military combat and diplomatic outreach as the cases in this book show. The universe of armed groups is huge, so their trajectories vary a lot. Some groups, like the African National Congress (ANC) in South Africa, started out as political parties that only later took up arms. Others, like the Serbian Volunteer Guard, also known as Arkan's Tigers in the wars of Yugoslavia, are rather ad-hoc creations or spin-offs from state agencies. Armed groups also vary in their internal structure. Some rather resemble a loose warlord faction such as Charles Taylor's National Patriotic Front of Liberia (NPFL) while others are closer to a disciplined bureaucratic organization as in the case of the Hezbollah in Lebanon (Schlichte 2009, 30–56). While some groups have established distinct political and military wings, others have a more unified command structure while still others are only a federation of various different groups such as the Farabundo Martí Para la Liberación Nacional (FMLN) in El Salvador. Armed groups might also be part of, or associated with, a governmental structure or state such as in the case of the Frente Popular para la Liberación de Saguia el-Hamra y Río de Oro (Frente Polisario) and the Saharawi Arab Democratic Republic or the Palestine Liberation Organization (PLO) and the Palestinian Authority (PA). The notion of 'armed group' as an umbrella term accommodates for this variation and it allows for comparative insights and generalizations.

Legitimacy is a term that is not only open to different understandings but also especially controversial when used in conjunction with describing non-state armed violence. According to the classical understanding in political theory, the acquisition and exercise of authority always has a moral dimension, which is expressed through the notion of 'legitimacy'. Two main understandings of legitimacy can be distinguished (Beetham 2013, 3–15). The first one is a normative account and prominent in political philosophy. Legitimacy is seen as a quality of government which satisfies ethical values and normative principles. Normative approaches to legitimacy are interested in the justifiability of rules and criteria of acceptability of political authority according to moral principles. They seek to evaluate existing political orders against such standards, expose deficiencies, and propose remedies. The second understanding is to be found in the social sciences and uses legitimacy as an empirical concept in the tradition of Max Weber. Here, legitimacy is seen as a descriptive category about normative judgements. According to Weber (1978, 213), legitimacy is to be understood as a belief that the actions of a political entity are justified and rightful. Legitimacy, in other words, is not the right to rule but the belief in the right to rule. It is the perceived quality of an individual or institution and attributed to its power, decisions, interests, actions, and rule. This ascription is norm-referential as judgements of legitimacy are always made with reference to values which define the parameters of rightful agency and action within a given political realm. Legitimacy is therefore an inherently social phenomenon. It emerges in social relations through the collective construction of a social reality, in which a political entity is seen as consonant with values and beliefs that are presumed to be shared by others. The empirical understanding does not deny the relevance of normative principles for legitimacy but it considers the recourse to these norms itself as observable social facts. What matters for the social scientist is the legitimacy claims of rulers and the acceptance or rejection of the validity of these claims by their subjects. It is this empirical understanding of legitimacy that is used in this book. We therefore do not seek to evaluate armed groups against normative criteria in order to justify or condemn their behaviour. Instead, and in line with other empirical approaches in much of current social sciences research, we are interested in how and why armed groups come to be seen as legitimate in the eyes of others.

Reiterative case comparison: Our volume brings together contributions from IR and history in order to explore the legitimation politics of armed groups through changing world historical contexts. With the assembled authors we deepen the exchange between IR and the discipline of history, being part of a move that has intensified over the last decade with the growing interest in the historicity of international politics and historical sociology as an approach to IR.[2] Bringing together an interdisciplinary group of scholars, we also want to make a plea for a methodology that bridges the gap between disciplines of social sciences and their dominant methodological schools. In political science and IR, the dominant orientation is often the nomothetic ideal of theory-building and higher levels of causal

[2] See Go and Lawson (2017); Schlichte and Stetter (2023); Bukovansky et al. (2023).

generalizations which tends to ignore possible features and conditions that might be critically important in individual cases. In historiography, the situation is inverse. Few historians believe in far-reaching claims, instead preferring an ideographic approach in which a high value is placed on the deep study of cases or specific historical periods. With this volume we suggest a method that we call 'reiterative case comparison' in order to bridge these ideals of approaches that at first glance seem incompatible. We thus try to reconcile the contradiction between nomothetic and ideographic efforts, between the aim of making general theoretical claims and the necessity of being attentive to the contexts and particularities of cases. In doing so, we neither seek to generate theory by deduction from a priori assumptions nor do we advocate atheoretical empiricism and the inference of propositions merely from observations through induction. We seek to seize the middle ground, that is, reasoning at an intermediate level in a field of research where we can observe a certain class of phenomena but lack applicable theories to explain them. This approach is broadly related to a methodological position that has gained increased prominence over the last years in the social sciences and is labelled as 'pragmatism'.[3]

Our starting point is the subject matter itself, armed groups. We maintain that substantial research on the legitimation politics of armed groups requires an explorative approach given the state of knowledge and the difficult field access to non-state violent actors. Studying unexplored cases here usually means discovering the unknown. Social science research is all too often built not on uncharted territory but on empirical data that is more or less easy to access if not readily available and can be managed without much difficulty. While this might be understandable for reasons of efficiency, it risks the danger of reproducing an observational bias. Kaplan has described this with the metaphor of the drunk who is only looking under a lamppost for his lost key because it is easiest to look where the light is (Kaplan 1964, 11). In contrast, our book aims to shed light on a phenomenon that has not yet been sufficiently explored and on which empirical sources are often difficult to obtain. The empirical base of our case studies consists of a wide array of primary sources including field work and interviews, archival work, and grey literature. We see the in-depth study of cases, that is, delving into a complex and multifaceted historical reality, as necessary in order to identify new features and potentially powerful explanations of the phenomena observed. With our appreciation of qualitative case studies, we highlight the constructive role of empirical research in shaping the development of theory.

From our acknowledgement of the importance of well-researched case studies follows an emphasis on theory that connects fruitfully with the empirical data gathered through explorative research. We sympathize with grounded theory and its core argument that a theory must fit the situation being researched, and that its categories must be applicable to the empirical data under study and be meaningful to the behaviour that is studied (Glaser and Strauss 1967, 3). Theory is malleable and it

[3] On the philosophical underpinnings of pragmatism see Bacon (2012) and Shook (2023). On the use and application of pragmatist principles in the social sciences see Swedberg (2014); Wills and Lake (2020) and in IR more specifically Friedrichs and Kratochwil (2009); Bauer and Brighi (2009); Hellmann and Steffek (2023).

needs to take into account the specifics of the particular subfield to which it is applied. Especially in exploratory research, stipulating a fixed and limited number of definite concepts or theoretical templates at the beginning of the research process in order to structure the investigation is often not helpful. Predetermined concepts that do not pay enough attention to the characteristics of particular phenomenon are difficult to apply at best and empirically irrelevant at worst.

Building on these assumptions, our book propagates an incremental, iterative-cyclical research process which is explorative and at the same time theory-guided. The core rationale of this approach is the mutual adaptation of theoretical concepts and ever more fine-tuned empirical analysis. Reiterative case comparison, as we call it, proceeds by applying the circle of empirical analysis, comparative discussion and theory-driven interpretation in repeated rounds. The study of a single case always comes first. Insights from comparative discussion allow for further steps in the study of single cases and will then show how far such mid-level observations reach. A host of theoretical concepts is instrumental in making shared features of cases visible, and the analysis of case material will render clear which concepts are more fitting than others. Proceeding in this manner we ultimately aim at building a theory that is grounded and anchored in empirical research. Reiterative case comparison puts the open comparative discussion of cases centre stage in order to facilitate theory-building and explanation.

Our research thus started with collecting observations on the international politics of rebel groups and armed movements and their trajectories in becoming legitimate actors over time. All scholars involved had prior expertise in this field of research and on their cases. In this initial stage, we drew on secondary literature and case studies including our own data from earlier research. In addition, we reflected on our conceptual language. Drawing on literature in political sociology and IR, we adopted a few categories and concepts that allowed us to translate observations into a theoretical vocabulary and to frame the investigation. Moreover, we sought 'sensitizing concepts' (Blumer 1954, 7) that could sharpen our senses for observation and serve as a heuristic instrument in order to further explore the international politics of armed groups. On the basis of initial case sketches and mapping a class of phenomena, a reiterative comparison and cross-case analysis followed. In our collective comparative discussion, we moved from case to case, using our concepts to analyse the cases while our observations allowed for the modification and adjustment of our concepts. After these early stages of the research process, more fine-tuned case analyses were produced for a second round of reiterative case comparison. Here, we combined cross-case analysis with within-case analysis while again adjusting our concepts. On the basis of the comparison, we then established similarities and differences and identified regularities, thus increasing our understanding and moving towards an explanation and tentative theory (Swedberg 2014, 111; Friedrichs and Kratochwil 2009, 716). The outcome is, we hope, a coherent theoretical framework discussed on the background of all contributions and which, at the same time, is addressed and guides the analysis in the case studies, promising to be applicable to other cases as well. We will present our framework in the following sections of this introduction.

Reiterative case comparison as we have practised here is less formalized than other comparative methods like comparative historical analysis or qualitative comparative analysis (cf. Mahoney and Rueschemeyer 2003; Mello 2021). Most if not all comparative methods in political science are about the identification of co-variance of variables, aiming at determining causal relationships. Reiterative case comparison is not an alternative to these methods but complements them. With a reiterative comparative discussion of cases more general hypotheses can be generated, but it has also proven helpful to gain case-related insights and sharpen judgements by contrasting cases in order to see the contours of single features and causal relations more clearly. Causal reasoning, at least in this book, was however not in the foreground. Reiterative case comparison's main purpose is rather to identify patterns which might convey causality or not. It is equally relevant, in our view, to identify particularities which become visible on the basis of extended case expertise and which allow for descriptions generated in the light of constant comparative questions. What we observed in our comparative discussions were often shared features of a few cases. We found these findings more telling for our understanding of historical change as highly abstract and all-covering general propositions. Reiterative case comparison as we suggest it hinges upon the quality of the case studies and the scope of theoretical concepts that are invoked in order to 'see' what is in the single cases or what their cross-cutting features are. With this book, we would like to highlight the innovative potential of dense case descriptions and contextualized knowledge for mid-level generalization as well as the powerful role of a concept-driven approach that uses the wealth of political sociology and its theoretical vocabulary.

Literature review

Our book contributes to two major ongoing discussions. The first one is the debate on non-state violent actors in conflict studies and the second is on the role of legitimacy in international politics. We will address each of these strands in the following and then briefly also relate to practice theory and global history approaches to which we connect in developing our conceptual framework.

Armed groups and conflict studies

Non-state armed groups are central actors in contemporary civil wars, and the academic debate has focused on various aspects of these groups such as their strategies of violence (Kalyvas 2006; Weinstein 2006) or their practices of economic accumulation (Collier and Hoeffler 2004; Berdal and Malone 2000). Other works have focused on the forms of rule armed groups establish within their territories (Bakonyi et al. 2006; Mampilly 2011; Arjona et al. 2015). Still others have analysed how civil wars spill across borders and examined the role of sanctuaries or refugees in the transnationalization of 'internal' wars (Salehyan 2009; Checkel 2013). Scholars have also investigated the post-war transformations of armed groups into political parties and governmental elites (De Zeeuw 2008; Hensell and Gerdes 2017).

This book builds on a recent strand of the discussion, that is, the growing research interest in the legitimacy politics of armed groups in civil wars (Schlichte and Schneckener 2015; Duyvesteyn 2019). That insurgents in civil wars create and maintain international relations is an old insight (cf. Modelski 1964). During recent years, however, scholars have paid more attention to the international strategies of armed groups and shown that these groups try to project their struggle on the international stage. Studies have pointed out that armed groups go global in order to mobilize support, isolate their opponent, facilitate mediation, and often also to gain recognition or legitimacy. To achieve these aims they communicate their agendas to an international public, lobby foreign governments, or run media campaigns. The internationalization of conflicts through strategies of extraversion or 'rebel diplomacy' (Coggins 2015; Huang 2016) is a remarkably consistent feature of civil wars that we can observe across different types of groups and conflicts, from the era of decolonization until the present. The war of decolonization in Algeria (1954–1962) can be seen as a paradigmatic case here. The Front de Libération Nationale (FLN) fully internationalized their independence struggle that was fought in the Algerian mountains as much as in television studios or in the corridors of the UN in New York. The FLN set up offices in foreign capitals from Cairo to Jakarta, cultivated foreign news media contacts and employed propaganda films which were shown across the Eastern Bloc and in the Arab World. Their ultimate success in becoming the legitimate government of an independent Algeria, however, depended not solely on the rebels' strategies but also on a wide range of domestic and international conditions from US Cold War imperatives to the dilemmas of the French Fourth Republic (Connelly 2002). Many more armed groups employed very similar strategies and likewise faced unique circumstances marked by governmental interests, domestic politics, regional constellations and international power shifts that facilitated or obstructed the success of these groups. This includes a wide range of cases, from the ANC in South Africa (Thomas 1996), the PLO (Norton and Greenberg 1989) or the Frente Sandinista de Liberación Nacional (FSLN) in Nicaragua (van Ommen 2023) to more contemporary groups such as Afghan warlords (Malejacq 2017), the Mexican Zapatistas (Bob 2005), the Hamas (Seurat 2015), rebels in Darfur (Jumbert and Lanz 2013), Mali (Bos and Melissen 2019), or Libya (Jones and Mattiacci 2019). As these works show, armed groups make great efforts to internationalize their fights. They do so in order to accomplish their military and political objectives and this regularly also includes justifying their armed struggle and trying to generate a belief that their violent actions are rightful. Scholars have also shown how armed groups' concern for international legitimacy or recognition shapes their behaviour in different ways. Jo has argued that rebel groups are likely to comply with international law, as they want to promote their legitimacy for instrumental reasons in the eyes of audiences that care about human rights (Jo 2015: 19; see also Fazal 2018). Others have argued that social recognition matters for armed groups, and granting or withholding that recognition may have positive but also counterproductive effects on engaging these groups in conflict management (Biene and Daase 2015; Geis et al. 2021).

In several cases, the struggle for international legitimacy continues long after the fighting has ended. This is true for the so-called de facto states born out of civil wars and secessions.[4] In these cases, armed groups have transformed themselves into governments of new states that are not or only partially recognized. The fate of polities such as Kosovo, Somaliland, Transnistria, Abkhazia, South Ossetia, Nagorno-Karabakh, or Western Sahara serve as examples here. These states have their own administrations, organize multiparty elections, and some of them even print their own money. They make a claim to independent statehood but none of them is fully recognized or at least recognized by a significant majority of states, no matter how effective and established their authority is within the territory they control. A burgeoning literature has analysed the constraints of these contested polities in getting formal entry into the club of sovereign states and explored their state-building strategies and interactions with the outside world.[5] Nina Caspersen (2012) has shown that de facto states put an emphasis on successes in democratization and state-building in order to showcase their viability and, thus, to become more acceptable for the international community. Others have argued that de facto states seek to accumulate social recognition in the absence of legal recognition by mimicking traditional state diplomacy (Isachenko 2012; Bouris and Fernández-Molina 2018). Coggins argues that the decision to grant recognition to de facto states is mostly dependent on the self-interest of great powers (Coggins 2014, 8–10). Yet, as other scholars have argued, it is not only states and their executives that matter in the politics of recognition but also other actors such as courts, parliaments or regional and international organizations (Visoka et al. 2020).

As we can see, there is a rich body of literature that highlights the agency of armed groups in seeking international legitimacy during ongoing civil wars but also in post-war times when armed groups have acquired de facto statehood. International legitimacy is often an implicit or explicit aim of these groups and actively pursued through all kinds of strategies in early but also in later phases of conflicts. However, there are also gaps in the literature. The notion of international legitimacy is seldom elaborated or conceptually distinguished from related categories such as recognition.[6] The process of legitimation of armed groups is also usually not explicitly theorized, apart from a few notable exceptions (e.g., Duyvesteyn 2019a). How exactly legitimacy emerges and why some groups become legitimate and others fail

[4] The comparative literature on secessionism deals with highly diverse actors and settings. Secessionist movements can employ violent means but often also resort to peaceful tactics such as competing in elections or civil resistance. However, in seeking recognition, they all face the same structural constraints or 'recognition regime' in international politics (Griffiths 2021). We elaborate on these constraints in our theoretical section with respect to armed groups that may or may not be secessionist.

[5] Pegg (1998); Geldenhuys (2009); Caspersen and Stansfield (2011); Caspersen (2012); Isachenko (2012); Coggins (2014); Bouris and Fernández-Molina (2018); Visoka et al. (2020).

[6] Legitimacy and recognition are often used synonymously. We use the notion of recognition, too (cf. Hensell and Schlichte 2021), but we adopt a more limited and descriptive understanding of the term as we see legitimacy as the more fundamental category. We explicate this in the theoretical section below.

has not been systematically addressed in the literature. Building on these vibrant debates and including our own works on armed groups and legitimacy (Schlichte 2009; Schlichte and Schneckener 2015; Hensell and Schlichte 2021), we seek to advance the state of the art by making a contribution that is original in two ways. First, we are putting the politics of legitimation of armed groups centre stage. In doing so, we explicitly theorize international legitimacy and the process of legitimation with its gradual steps by analysing the interplay of practices and publics that bring about international legitimacy. Second, we are relating the politics of legitimation to broader and changing historical contexts within our macro-level framework of historical times.

Legitimacy in international politics

The other stream of research to which we contribute is centred in IR around the theme of legitimacy in international politics. As a concept that is commonly understood to mean 'the right to rule', legitimacy was long seen as irrelevant for understanding the international system, which is defined by the absence of a central ruling institution. Early accounts approached the issue of international legitimacy from a state-centric perspective. Inis Claude, for example, suggested that the major political function of the UN is 'collective legitimization' by approving or disapproving the actions of states (Claude 1966), and Thomas Franck argued that nations tend to obey international rules because they are seen to be legitimate (Franck 1990). Ian Clark invoked legitimacy in order to denote the existence of an international society of states and argued that principles of legitimacy express social agreement about who is entitled to participate in that society and about appropriate forms of conduct therein (Clark 2005). Others have demonstrated how states in the UN Security Council seek to use legitimacy strategically as a tool in their pursuit of political advantage (Hurd 2007). Yet others have studied the legal dimensions of international legitimacy (Charlesworth and Coicaud 2010) or explored crises of legitimacy in international politics (Reus-Smit 2007). Scholars have also shown that legitimacy requirements evolve over time in the interaction of a community of actors and institutionalized norms that prevail in an issue area such as the Kyoto Protocol on climate change (Bernstein 2011).

Building on conceptual elaborations of legitimacy in political theory (Barker 2001; Beetham 2013), scholars have also underlined the need for new methodological approaches as well as mixed and context-sensitive research designs that take the multidimensional nature of legitimacy into account including legitimation as the procedural dimension of legitimacy (Hurrelmann et al. 2007). A focus on legitimation, understood as the activity of claiming and generating legitimacy (Barker 2001, 1–29), has subsequently become a major theme in IR research. Scholars have explored the legitimacy of international organizations such as the UN, the WHO, or the World Bank from the angle of legitimation practices. An evolving literature has shown that the active claiming of legitimacy has become of central importance for these organizations (Zaum 2013; Gronau and Schmidtke 2016; Dingwerth et al. 2019). This encompasses discursive and symbolic practices such as acts of communication, rituals, ceremonies, or internal institutional reforms that aim at presenting an organization as committed to normative ideals such as welfare, human rights, democracy, or peace. Moreover, their legitimation strategies address various audiences with different and changing interests and expectations such as member states, the wider public, civil society, or the organizations' own administrative staff (Gronau and Schmidtke 2016, 542–546).

Recent studies have theorized legitimacy within the architecture of global governance. In the most comprehensive and ambitious treatment of international legitimacy so far, a group of 16 scholars explored the questions how, why, and with what consequences global governance institutions are regarded as legitimate. The 'Legitimacy in Global Governance' (LegGov) project analysed how legitimacy varies across institutions, issue areas, audiences, countries, and social groupings. The scholars of this project have developed a theoretical framework that builds on a process-oriented understanding of legitimacy and highlights the role of practices, justifications, and audiences in the legitimation or de-legitimation of international institutions (Tallberg et al. 2018; Bexell et al. 2022). According to the LegGov project, today's global governance institutions are legitimated or de-legitimated by a host of actors ranging from governments to NGOs, business associations, political parties, media channels, and ordinary citizens in highly dynamic and interactive processes (Bexell et al. 2022). The group of scholars has also shown that a notable elite–citizen gap exists in legitimacy beliefs towards global governance (Dellmuth et al. 2022) and demonstrated that major public challenges to the legitimacy of international organizations are rather rare events (Sommerer et al. 2022).

We share with these works the general understanding that legitimacy is an evolving quality of actors and institutions. Likewise, we share the emphasis on processes in the study of international legitimacy that involve the constant generation of legitimacy beliefs. With this foundation, we are also building on the conceptual vocabulary of this strand of research which includes an attention to practices, discourses, narratives, and audiences. However, we also go beyond the debate in several important ways. First, our book enlarges the scope of subjects for which legitimacy has been invoked as a relevant category. Armed actors have largely been overlooked in the boom of research on legitimacy in international politics. By focusing on wars and armed groups, we introduce a category of actors that has so far attracted almost no attention in the literature. Second, we approach the question of international legitimacy from a different and yet unexplored angle. Current research investigates the legitimacy of international institutions on the assumption that these institutions as such are undisputed. Their existence is recognized and their authority is usually regarded as desirable in principle so that certain goals and common goods can be achieved. In this strand of literature, it is only in the judgement about the appropriate exercise of that authority where the issue of legitimacy comes to the fore (Zürn 2018, 65). Our approach contrasts with this mainstream understanding in IR. At first, non-state violent actors are usually criminalized and repressed. They are generally from the outset not accepted as authorities in international politics. Thus, our starting point is the transition from the illegitimate to the legitimate in international politics. We are interested in how and why non-state actors such as armed groups attain legitimacy in a state-dominated system that discriminates against such actors. In line with other works, we problematize the issue of legitimacy not from the angle of appropriate exercise of authority but from the perspective of inclusion and membership as an authority in international politics (Clark 2005, 155–189; Griffiths 2017; McConnell 2017; Viola 2020). By analysing how and why non-state actors acquire legitimacy as actors, our book makes a core contribution to the wider literature in IR that has so far neglected the fundamental question of legitimate membership in the international state system. We will elaborate on this point below. Finally, and in contrast to most of the works in IR, we take the history of international legitimacy seriously. As our case studies show, historical context matters a lot in analysing international legitimacy. We therefore highlight the role of changing historical times for the legitimation of armed non-state actors.[7] Apart from the debates mentioned above, our book also speaks to two other strands of the literature which we will address very briefly, namely practice theory and global history.

Practice theory: Starting out with contributions based on the work of Pierre Bourdieu and Theodore Schatzki, practice theory has made its way into comparative politics and IR over the last two decades.[8] Practice theory puts an emphasis on agency, on what actors do. According to this school of thought, practices are ways of doing things that are embodied, shared by a collective and that occur regularly as habituated action. Moreover, practices imply practical knowledge, that is, the acquisition of learned 'rules of the game'. Practice theory is well-suited for our subject, as it puts an emphasis on process orientation and on the emergent character of practices. As the IR debate has shown, the social construction of objects, norms, and structures is always open to reenactment depending on the performativity of practices (Bueger and Gadinger 2018, 26–29). We pick these ideas up with our approach to international legitimation that emphasizes legitimating practices and publics invoked by these practices. And we spell out what this means in the politics of armed groups. With the use of the term 'practice', we stick to the core idea of practice theory, namely that social action is embedded in a symbolic world of which the actors are part and that they constantly reproduce by patterned action.

Global history: The last decade has seen a growing interest in the historicity of international politics and historical sociology as an approach to IR.[9] We are building on this debate, and our book is also firmly rooted in historical approaches to IR. Instead of presuming a homogeneous time-space in which international politics take place, we suggest a historically nuanced and differentiated understanding of international politics. 'History matters' is a broadly supported claim in the booming research

[7] For a similar historical approach which relates practices or regimes of legal recognition to distinct time periods, see also Fabry (2010) and Griffiths (2017).

[8] See, e.g., Adler and Pouliot (2011); Bueger and Gadinger (2018); Drieschova et al. (2022).

[9] See, for example, Buzan and Lawson (2015); Go and Lawson (2017); de Cavalho et al. (2021); Schlichte and Stetter (2023); Bukovansky et al. (2023).

on the history of international politics, but it is less clear what this means. By following arguments on the historicity of international politics (Schlichte and Stetter 2023), we suggest 'historicizing' international politics with regard to the politics of armed groups and intra-state warfare. With our argument about historical times, we understand armed groups as being part and parcel of historical changes, which are not just external to their actions. Instead, as the contributions of this volume show, the politics of armed groups are themselves a historical expression of power relations, political beliefs, known practices and attempts at innovation. As political actors, they also shape and co-constitute the historical political spaces within which they act.

Armed groups and the politics of legitimation: a theoretical approach

In this section we present our theoretical framework in order to unpack the politics of legitimation revolving around armed groups. In the first part, we clarify some basic terminology of Weberian political sociology, namely power, domination, and legitimacy with respect to armed groups. All three are foundational concepts in political science that are embedded in centuries of political thought and each of them could be the subject of entire books. For practical reasons we cannot go into the debates, and we limit ourselves to the established Weberian understanding which we see as the most fruitful one. Building on Max Weber, we also outline what we mean when we speak of the international legitimacy of armed groups. In the second part, we turn to our core notion of 'legitimation'. Armed groups seek international legitimacy, and in order to understand how and why they become legitimate, we argue that it is necessary to look at the process of legitimation that leads to the construction of armed groups as legitimate. We propose an analytical approach that differentiates between practices of claim-making and claim validation, and we also highlight the role of audiences and constituencies that are invoked and addressed by these practices. We also briefly clarify our understanding of the notion of 'recognition' in the overall process of legitimation. Finally, in the third part we outline and specify four historical times after 1945 that have set different structural conditions for armed groups' quest for international legitimacy.

Armed groups: power, domination, and international legitimacy

Our starting point is the general assumption that force and coercion are powerful forms of social action (Popitz 2017). Armed groups that employ physical violence in order to impose their will, even against resistance to quote Weber's famous definition (Weber 1978, 53), thus exert power in social relations. We understand armed groups here as outright political actors and their actions as a form of power. However, simply exerting power through the use of violence is not the only goal of armed groups fighting in civil wars. They also attempt to turn the power of the gun into stable institutionalized rule. Weber understands such continuous rule as domination, that is, 'the probability that a command with a given specific content will be obeyed by a given group of persons' (Weber 1978, 53). For armed groups, the ultimate form of such domination is usually the state. Their actions aim at conquering positions either in an existing state or in a newly established one (Schlichte 2009, 178–202). Even during war, they may emerge as alternative governments often in control of territories and state-like apparatuses (Bakonyi et al. 2006; Mampilly 2011; Arjona et al. 2015). At the latest when a war ends, be it through a rebel victory or a negotiated settlement, armed movements become new political elites as regular political parties or governments within a state (De Zeeuw 2008; Hensell and Gerdes 2017).

What armed groups, thus, do is powerful action, but in order to consolidate their positions over time they aim at continuous rule or domination. And every form of domination, as Weber has pointed out, 'attempts to establish and to cultivate the belief in its legitimacy' (Weber 1978, 213). Armed groups are no exception here. Much as any other political actor, they need to justify their actions and agendas and they actively pursue legitimacy. Be it the overthrow of an authoritarian regime, the secession of a territory or a quest for power-sharing: armed groups seek to justify the use of violence by referring to identities, institutions, interests, and political aims.[10] While this is certainly true for the domestic level including local and national arenas, we argue that gaining legitimacy at the international level is also of fundamental importance for armed groups (Schlichte 2009, 65–72, 85–115; Schlichte and Schneckener 2015).[11]

But what does it mean to have 'international legitimacy'? In classical accounts, from Weber and Arendt to Habermas and Easton, legitimacy has been associated with systems of rule-making and the hierarchical relation between a power holder and those subject to this power. Legitimacy is generally understood as the right to rule, that is, the entitlement to issue authoritative commands that require compliance from those subject to them within the confines of a political community (Beetham 2013). In this classical sense, legitimacy is tied to the authority of a domestic ruler within a nation-state. Recent IR debates likewise start from the traditional nexus between a rule-making institution and legitimacy (Tallberg et al. 2018; Bexell et al. 2022). The authority of international institutions and their right to make decisions within a particular area is taken as given while international legitimacy refers to the question whether these rights are appropriately exercised (Zürn 2018, 65). Or more simply: the existence of a political authority as such is not a problem, only the exercise of this authority is.

[10] We do not assume that this is true for all armed actors fighting in civil wars since 1945 but for a significant number of them. We also do not make an assumption here about the underlying motives of armed actors. Armed groups might be true believers in their claim to legitimacy but they might also pursue legitimacy for instrumental reasons.

[11] While the domestic dimension is not our core analytical concern in this book, we take the interactions between domestic and international legitimacy into account in our conceptualization as well as in our case studies.

Our book approaches the question of international legitimacy from a different angle. We propose an understanding of legitimacy which is more compatible with approaches that problematize membership in the international system and that differentiates between legitimate and illegitimate actors in international politics. Who is a legitimate political authority in the first place? We see this as the more fundamental question. It is an uncontested given in IR scholarship that the sovereign state is the building block and primary actor of international politics. The state is the natural inhabitant of the international political space and its major institutions. In the eyes of most observers, it is the only actor that really counts and the only subject with substantial rights and the entitlement to participate in the international system. Against the background of this state-centric orthodoxy, the issue of who is a member in the international system has received not much attention. However, a few works have problematized the question of entry barriers, belonging, and liminality in international politics and put a spotlight on practices of inclusion and exclusion.

The English School of International Relations has introduced the idea of 'rightful membership of the family of nations' (Wight 1972, 1). The collective of nation-states is seen as an international society, and international legitimacy delimits who is and who is not a member of that society and entitled to participate in it. This determination takes place as a collective judgement on the basis of what is perceived as a legitimate form of political organization or desirable character of statehood. Ian Clark has argued that membership in the international society has become more restrictive in the post-Cold War decade. New boundaries have been drawn between legitimate members that conform to the ideals of liberal democracy and good governance and illegitimate members who do not conform to these standards such as rogue or failed states (Clark 2005, 155–189). Other scholars have pointed out that the state system actively discriminates against all types of non-state international actors in order to limit access to resources and decision-making. The legitimacy of NGOs, corporations, or philanthropic foundations has been and continues to be a contested issue (Erman and Uhlin 2010; George 2015). This also applies to de facto states that are not or only partially recognized such as Kosovo, Somaliland, Transnistria, or Northern Cyprus. These polities are continuously excluded from legitimate membership status in the international system (Pegg 1998; Geldenhuys 2009; Caspersen 2012; Visoka et al. 2020). The same is true for stateless nations, indigenous communities, and national minorities that likewise remain liminal and fringe players in international politics (McConnell 2017). Ultimately, a broad range of possible members of the international system from NGOs to non-sovereign polities, sub-national regions or citizens remain marginalized and excluded despite all attempts to democratize and open up international institutions. Scholars have argued that the international system is best understood as a club of sovereign states that applies rules for admittance and closure to delimit legitimate from illegitimate members (Griffiths 2017; Viola 2020). We share with these widely different approaches the critical stance towards the international system as a closed collective of states and the idea that international legitimacy can be seen as a principle giving admission to and recognition by that collective.

Our understanding of international legitimacy thus starts from the state. Given the dominance of the sovereign nation-state, we would argue that it can also be seen as the most legitimate form of domination in international politics. As an established model of political legitimacy, the state is a deep structure that is not questioned, but taken for granted and perceived as almost natural by most practitioners and observers of international politics, be they presidents, diplomats, rebels, or IR scholars. One could argue that the state as a political form has become so normal that the question of its legitimacy does not even arise. Following Pierre Bourdieu, we could also call this the 'doxa' of the state system. The doxa is a set of fundamental, deep-founded beliefs that are unquestioned and do not need to be asserted. Within a given field, they inform the actions and thoughts of agents who perceive the social world as taken for granted and self-evident (Bourdieu 1977, 159–171). Doxa means misrecognizing the historical contingency and arbitrariness of the established current state order while recognizing the legitimacy of that order with its distinctions such as state/non-state or legal/criminal. In the Bourdieusian sense, international legitimacy can be seen as the power to naturalize the given state order as self-evident. Gaining international legitimacy often simply means becoming a state that is becoming part of the landscape of international politics, which is constituted mainly by states as the undisputed political form.

Legitimacy beyond the state, however, is contested. This is above all true for armed groups, who are almost always first treated as 'criminals' or 'terrorists'. What is at stake for armed groups is, thus, acceptance as a political force and ultimately as the authority of a state. Armed groups seek formal state power because the state is the established legitimate model of political domination, the ultimate symbolic capital in Bourdieu's sense. Becoming legitimate by appropriating what is legitimate—this could be described as the core rationale of armed groups' strategy of legitimation. In order to be successful, armed groups need to adopt state-like practices and attain state-like qualities in the perception of other political actors. Therefore, armed groups need to be seen as a likely future governmental actor or collective entity that embodies governmental qualities. Being considered a serious contender for power, the sole representative of a population, a government-in-waiting, or credible alternative to an incumbent regime is, therefore, of vital importance for the legitimacy of armed groups. However, 'declaring yourself a state and even functioning like a state does not make you a state unless the international community approves' (Griffiths 2017, 180). This means that the ultimate goal for armed groups is joining the sovereignty club through acquiring formal recognition as a state that has a legal identity, evidenced by full membership in the UN and access to all kinds of sovereign privileges.

Building on this understanding, we use the term international legitimacy as a descriptive category that refers to the shared belief in the justifiability of the agenda and struggle of an armed group within a given international constituency. It has often been noted that it is difficult or almost impossible to observe or measure legitimacy beliefs directly (Dogan 1992, 120). We subscribe to this view and take international legitimacy in the sense defined above as an end point. It is the goal of armed groups that aspire for state power, and it often coincides with their official international recognition as a political party or government. In our book, however, we are not so much interested in this end state, which is when an armed group ceases to be a violent organization. We are interested in the process during which they come to be seen as legitimate and legitimacy beliefs emerge (Duyvesteyn 2019a, 6). The various gradual and observable steps of acceptance, normalization, and incorporation which ultimately lead to legitimacy are the core focus of this book. We thus put the notion of 'legitimation' centre stage.

The legitimation of armed groups: practices and publics

By legitimation we mean the process of making someone or something legitimate. Rodney Barker has defined legitimation as the activity of insisting or demonstrating that a political actor and his actions are justified (Barker 2001, 30). It is a justification that often intentionally but also unintentionally generates, shapes, and cultivates legitimacy beliefs. The politics of legitimation comprise all those efforts and activities that lead to the construction of an individual, group, order, or institution as legitimate.[12] Analysing the politics of legitimation means disentangling a deeply relational and interactive process with different arenas, sets of actors, and practices (Geis et al. 2012; Duyvesteyn 2019a, 6). We differentiate in the following between practices and publics as heuristic devices. We see two sets of practices at the heart of the politics of legitimation: making and validating a claim to legitimacy. Moreover, we differentiate between two kind of publics, namely audiences and constituencies.

Practices: claim-making and claim validation

Armed groups seek international legitimacy for themselves or the institutions they govern. In doing so they make legitimacy claims regarding their actions, orders, and agendas. Such claims are usually characterized by recourse to social values. They aim at generating conformity with established norms and at presenting an actor as committed to ideals such as human rights, freedom, democracy, welfare, or peace. We suggest differentiating between ideation-based and performance-based legitimacy claims or, more simply, between words and deeds, that is, what armed groups say and what they do (Schlichte and Schneckener 2015, 417). Ideation-based claims refer to ideas and ideologies such as nationalism, human rights, Marxism, Maoism, or Islamism as well as to local traditions and cultures which armed groups employ in their discourses and symbolic representations. Performance-based claims are linked to their actual behaviour and outcomes of their policies such as combat performance,

[12] We often also see de-legitimation in violent conflicts. Armed groups use it as a tool against their opponents and they are themselves often de-legitimated as the long history of criminalization, repression, and counter-insurgency shows. As we are interested in the legitimation of actors that are initially considered as illegitimate, we do not explicitly theorize de-legitimation as other approaches do (e.g., Tallberg et al. 2018; Bexell et al. 2022). We rather understand the legitimation of an armed group as a process which usually also includes de-legitimating their opponents in order to cast the actions of the group in a better light. We account for such practices in our case studies.

defensive capacities, the provision of security or elements of welfare, the use of democratic procedures or successes in diplomatic negotiations. With 'claim-making' we refer to a set of practices. Claim-making is basically about communicating who you are and what you do. Armed groups repeatedly engage in claim-making in order to create the belief in the eyes of observers but often also their followers that they are not only powerful but also legitimate actors. We see three broad single practices here:

Fighting: Armed action and employing physical violence is at the core of what armed groups do and can be seen as perhaps their basic repertoire, a form of ultimate 'action power' (Popitz 2017). They strike and carry out attacks, they conquer territory and inflict losses on their opponents in order to force or threaten their enemies. Fighting, however, is not only a form of violent action but can also be seen as an act of claim-making. Armed groups demonstrate their military might and their capacity to impose their will with violent means. As violence is particularly likely to draw the attention of global publics (Koloma Beck and Werron 2018, 288–289), armed groups' practices of fighting often aim at gaining visibility and raising awareness of their cause. Armed groups need to be dealt with as a result, by taking steps against them or by establishing relations and seeking negotiated solutions. In either case, one can then no longer ignore the existence of the armed group and their claims to legitimacy.

Voicing: Claim-making is about conveying information to a public, and voicing is the practice of articulating and expressing a claim by putting it into spoken or written words. In order to make a claim to legitimacy, armed groups need to give an account of themselves, their cause and their aims. And in doing so they need to employ all kinds of communicative means in order to broadcast and disseminate information and to frame their actions in a positive way. Armed groups need first of all to generate and cultivate a narrative about their struggle. Usually, this includes positive public self-presentations of their leaders, fighters, and political orders including the violent actions of the group. However, armed groups also need to evoke a particular understanding of the overall conflict in order to bolster their claim to legitimacy. Thus, practices of voicing also include portraying the plight of civilians, staging victimhood, denouncing and delegitimizing the enemy, exposing the misconduct and failure of third parties, or highlighting international norms.

Mimicking: Imitation or mimicking can be seen as a universal social practice constitutive of society (Tarde 1903 [1890]). In the politics of legitimation, mimicking plays a performative role in making a claim to legitimacy. Armed groups need to gain a state-like position in the perception of others and in order to do so, they seek to inscribe themselves into the symbolic world of the state. Presenting themselves as a future governmental actor, adopting the typical insignia of the state and acting like a state through formal procedures or ceremonies matters for armed groups. Oftentimes this happens out of necessity—armed groups need to act as functional authorities after they have acquired a position of power. Mimicking the state also means exercising authority over a territory and its population through coercion and wielding violence but also through administering public institutions such as hospitals and schools. Moreover, armed groups adapt to the rules and habits of interstate diplomacy by opening embassies abroad, setting up external missions or lobbying other governments in foreign capitals (Huang 2016; see McConnell et al. 2012). Performing the state in order to be perceived as a state is the rationale of armed groups' practices of mimicking that ultimately aim at accommodation in the international state order.

When we speak of 'claim validation', we refer to another set of practices. By claim validation we mean accepting or approving the validity of a claim to legitimacy and establishing its worthiness. Practices such as popular acclamation, taking part in a ceremony, concluding an agreement or entering into a contract can be understood as performative acts of public recognition and acknowledgement that confer legitimacy (Beetham 2013, 18). All kinds of international actors regularly employ such practices when engaging with armed groups, and in doing so they often either intentionally or unintentionally also endorse a claim to legitimacy made by that group. International organizations, NGOs, regional organizations and single governments are the most prominent examples for such practitioners. Practices of claim validation can be motivated by a host of reasons and do not provide direct evidence of people's belief in legitimacy. They may or may not entail a subjective belief in the legitimacy of an armed group. However, the practices usually entail acts with a symbolic or declaratory force so that they constitute an acknowledgement or recognition that is publicly demonstrated (Beetham 2013, 90–97, 266–268). We would thus argue that such acts also go some way towards validating agents' claims by directly or indirectly confirming them. Validating a claim at a certain point in time can be regarded as a 'legitimacy episode' that may or may not add up to international legitimacy in the long run. It provides provisional or proto-legitimacy that, however, remains open to revision. We identify three single practices of claim validation.

Acknowledging: At the most basic level, acknowledging means taking note of the reality that an armed group exists as a political force with a claim to legitimacy. This is by no means trivial, as the politics of using names for violent actors shows (Bhatia 2005). Armed groups are often portrayed as 'terrorists', 'extremists' or 'bandits' that are morally deficient and have no political agenda. This depoliticizes them right from the start as actors that cannot claim any legitimacy for their actions. Acknowledging, in contrast, means mentioning an armed group as part of a political problem or as a party to a conflict. And in doing so, the issue of legitimacy also enters the picture. We would argue that acknowledging is not only an act of recognizing a group as a political actor. It also means admitting that—in principle—a claim to legitimacy exists. Acknowledging the existence of a claim can be seen as the very first step in the practices of claim validation. Whether the claim as such is also validated depends on how the group, its actions, and its aims are framed through media coverage and political discourse. Naming armed groups 'freedom fighters' or 'liberation movements' or describing their actions as 'defence' or 'protection' are positive judgements and can be seen as indicators of validating a claim to legitimacy.

Supporting: Often, armed groups would not be viable without external support. The history of civil wars shows that these groups have profited from a wide array of forms of public support, sponsoring, and cooperation. Armed groups are often supported by international organizations, foreign governments, diaspora groups, NGOs, or sympathizing activists with all kinds of assistance. The kinds of external resources armed groups have received are numerous, ranging from arms, money, safe haven, logistics and aid to diplomatic backing and advocacy. We would argue that the practices of openly providing these resources also to some extent validate a claim to legitimacy. While covert support is a more limited practice as it only signals to the group that its actions are justified, overt support implies a wider audience. We see especially military support as a crucial form of claim validation. Sponsoring an armed group through weapons procurement and delivery, the provision of training in guerrilla warfare, or access to military intelligence can be seen as a strong indicator here, as it signifies that not only the armed group, but also their struggle with violent means, is perceived as justified.

Officializing: By officializing we mean making something or someone official and public. According to Pierre Bourdieu, officialization is the passage from the clandestine and semi-official to the open and official. Allowing someone to present themselves in front of an attentive audience can be seen as an act of validation, and the practice of officialization aims at making this recognition public and official through acts and gestures of verbalization and demonstration (Bourdieu 2015, 156–160; 2016, 371–374). This logic can also be observed in international politics. In the case of armed groups, this includes all kinds of formalized interactions or more simply 'red carpet treatment' such as meetings and negotiations within a framework of diplomatic protocol. Officialization often happens through negotiations and is done by governments. Attending talks, sitting at the negotiating table, signing agreements, gaining formal status in an international organization and other such formal practices can be seen as validating a claim to legitimacy.

Given these diverse practices of claim validation, how do we know then that armed groups have attained international legitimacy? We argue that the diplomatic practice of recognition serves as an indicator here. In the classical understanding of international law, recognition is a legal act (Lauterpacht 1948). The recognition conferred by a state to a political entity, usually another state, means the formal acknowledgement and acceptance of that entity as a bearer of rights and subject of international law. While recognition is usually seen as a classic interstate practice (Fabry 2010), de facto or de jure recognition has historically been extended to all kinds of authorities including rebellious parties during civil wars, from 19th century insurgencies in Latin America to the Spanish civil war and national liberation movements of the age of decolonization (Lauterpacht 1948, 270–310; Freudenschuss 1982). In the 1970s, for instance, the UN formally recognized the South West African People's Organisation (SWAPO), the PLO, the ANC, and the Polisario each as representative of a political community. Many other armed groups have been recognized in this way as political authorities, that is, as the representative of a people, as a government-in-exile or future government of a state. We argue that the legal or political recognition of an armed group in international diplomacy is an indicator or proxy of its international legitimacy. In line with a newer literature, we see international recognition not only as a formal but also as a social act that includes recognizing the agency, status, and agenda of a political actor.[13] Recognition, in this understanding, also implies the confirmation of a claim to legitimacy through its formal validation. The common phrase 'recognized as legitimate', which can be found in countless statements and documents by international organizations and states, hints at what we mean here. In our case studies we are tracing these explicit and identifiable acts of formal recognition that confer authorized agency or even legal personality onto an armed group.

Publics: audience and constituency

Legitimation is relational involving not only agents of legitimacy claims but also a public that is addressed by these claims. In line with practice theory, we understand such a public not as a defined and bounded space but rather as a constructed sphere that is the outcome of practices (Best and Gheciu 2014, 33–35). Making a claim is a goal-oriented activity. It is directed to a public which is assumed to see its interests being implicated or at stake in a given claim and which is supposed to engage with the claim, to validate it or to endorse the values on which it relies. Validating a claim through publicly demonstrated and performative acts of recognition and acknowledgement also imply the presence of a public. We therefore understand claim-making and claim validation as public-defining acts. They are practices 'that seek to claim particular problems, actors, or processes *as* public—or of common concern—and in doing so, that effectively work to constitute those issues as public' (Best and Gheciu 2014, 33, emphasis in original). As such, practices of claim-making and claim validation are also public-producing. They have constitutive effects as they invoke a public and an understanding that something is of public concern. Violent non-state actors that engage in claim-making generally have to cope with divergent views and divided publics that hold or withhold legitimacy beliefs vis-à-vis the armed group. Picking up on works on political representation (Saward 2010, 35–56) and legitimation research (Bexell et al. 2022), we differentiate between two types of publics here, namely *audiences* and *constituencies*. Both of these groupings are spoken to and addressed by a claim to legitimacy.

We use the term 'audience' for a broader social grouping. Armed groups that make a claim to legitimacy generally have to address national and transnational audiences. On the national level, the audience of armed groups is basically the domestic population within the territory of a state. Seeking international legitimacy, however, armed groups also address a transnational audience that is formed by states, governments, and international organizations but that also includes all kinds of other actors such as parliaments, political parties, NGOs, advocacy groups, diaspora populations, and the media (Schlichte 2009, 86–87; Schlichte and Schneckener 2015, 419–421). We assume that audiences are heterogeneous groupings whose members hold widely different views regarding the claims made by armed groups. The audience may contain supporters who validate or reiterate the claim but also undecided observers, passive

[13] For a broader understanding of recognition in IR that relates the concept to the semantics of social goods and similar terms such as esteem or rights see Lindemann and Ringmar (2012); Daase et al. (2015), and with respect to armed groups Geis et al. (2021).

listeners, and bystanders who prefer a neutral standing or who ignore the claim, but also the opponents who contest and reject the claim altogether or make counter-claims. In the age of social media and sophisticated technologies which potentially address a global public, it is often difficult to empirically delineate who exactly the audience is (see Bos and Melissen 2019; Jones and Mattiacci 2019).

By 'constituency' we mean a more circumscribed and limited public which is usually part of a larger audience. The constituency is the group that recognizes a given claim as being made for it and that accepts and validates the claim or is likely to do so. Thus, the constituency is the group in which legitimacy is ordained and reproduced. It is the group whose legitimacy beliefs count. Like audiences, we assume that constituencies are national and transnational. Generally, armed groups that seek domestic legitimacy need to have a domestic constituency on the national level, where they control territory and exercise authority over a part of the population and where they mobilize resources and followers. Armed groups need to address a national constituency which the group seeks to represent and on whose support it depends. However, when armed groups strive for international legitimacy, they also need to build a transnational constituency. Sovereign states and international organizations that are constituted by and for those states are most important here. They form the core transnational constituency for all non-state violent actors who ultimately seek legitimate state power for themselves. Yet, as our case studies show, legitimacy also emanates from transnational constituencies that are formed by civil society organizations or political parties.

We thus use the notion of audiences to indicate who exactly is addressed and spoken to by an armed group, and we assess the groups' legitimacy by the extent of their constituencies (Beetham 2013, 265). To illustrate our argument: The Polisario and the Taliban each run a website in five different languages, and they address various audiences on the domestic, regional, and international level. Their actual constituencies, however, are more circumscribed. The Polisario has a local constituency in the displaced Sahrawi population exiled in Algeria but it also has an international constituency among states in Latin America and Africa as well as among a solidarity movement in Europe. Both constituencies vigorously validate the Polisario's claim to legitimacy. The Taliban, in contrast, have a constituency among a significant part of the Afghan population, but their international constituency was short-lived and very small, when Saudi-Arabia, the United Arab Emirates, and Pakistan recognized the Islamic Emirate of the Taliban government between 1996 and 2001. Today, the Taliban's international constituency is almost non-existent (see the chapters by Stephan Hensell and Ahmed Elsayed).

Armed groups direct their claims simultaneously towards multiple and diverse types of audiences and constituencies and we assume shifting and flexible boundaries between these groupings as well as overlap. Importantly, neither constituencies nor audiences can be perceived as passive target groups that are simply the receivers of claims to legitimacy. They are often agents in their own right, active in legitimation or de-legitimation through reiterating and disseminating or challenging and criticizing legitimacy claims in order to shape the legitimacy beliefs of others (Saward 2010,

35–56; Bexell et al. 2022). Winning over the audience and turning it into a legitimating constituency is a key challenge for an armed group, and we assume interacting local and international dynamics here. The local standing of an armed group often feeds into the groups' international support and vice versa, but both could be also at odds with each other as the case of the PLO in Palestine makes clear (see the chapter by Christian Olsson). The international legitimacy of an armed group is a matter of degree across various constituencies.

Armed groups have to master conflicting expectations and messages when addressing different publics. Most importantly, they have to be aware of the ambiguous role of violence. Armed groups employ violence in order to achieve their military and political goals. Calculating the effects of violence with respect to audiences and constituencies, however, is difficult for them. The central paradox of the politics of armed groups is that violence can bolster but also undermine the legitimacy of these groups (Schlichte 2009, 57–84). On the one hand, armed groups make deliberate use of violence in order to attract attention to their cause. Employing violence, as we have argued above, is a form of making a claim to legitimacy. On the other hand, employing violence always has destructive effects and tends to derail. If it is seen as excessive or inappropriate, it ultimately endangers the claim to legitimacy (Koloma Beck and Werron 2018). Finding a practical way out of this dilemma is a core challenge for armed groups.

Summary

Armed groups seek international legitimacy, and in order to understand how and why they become legitimate we need to look at the process of legitimation. We understand legitimation as a transformative process but also as the result of that process. Its core dynamic lies in the interaction of practices and publics: the reinforcement of claim-making and claim validation in the interplay of armed groups and their audiences and constituencies. The legitimation of armed groups is thus a deeply relational matter. It is a collective endeavour that involves communication, perception, accommodation, and acceptance. The emergence of international legitimacy through legitimation, however, is by no means a linear development. Unpacking the politics of legitimation means tracing a gradual and incremental process that may or may not bring about the legitimacy of an armed group. Historical context matters here. In the next section we suggest a conceptualization to take this into account.

Historical times

The politics of legitimation by armed groups allows us to see features of international politics that are less visible or even rendered invisible in the other fields of research and theoretical schools. It is not just our case studies but a broader reading of case literature on non-state war actors that shows how deeply intertwined globalized discourses, legitimating ideas, and the politics of armed groups have been through various periods of world politics since 1945. The international legitimacy of armed groups is a time-bound phenomenon. It can hardly be explained without a recourse to global history. Building on our earlier work (Schlichte 2009, 107–113; Hensell and Schlichte 2021) and partly modifying earlier categorizations, we differentiate four historical times that have set different incentives and structural conditions for armed groups' politics of legitimation. These are the time of decolonization, the time of the Cold War, the humanitarian time, and the time of the new sovereigntism. This final section seeks to show how they mattered for the politics of legitimation of armed groups. We will first explain our understanding of what a historical time is, and we will then present these times in more detail including an outlook on contemporary developments.

Let us start with elucidating what we aim to achieve with these new notions by delineating them from established theoretical paradigms in IR. In a realist or liberal tradition, what we mean by historical times would simply be explained as 'contexts' that influence the way states act (Goertz 1994). Realists or Neorealists would point to international power structures as the overarching context. The bipolar or multipolar mode of that system would determine in large part how states act (Waltz 1979). While we do not deny that such power structures or polarities exist, we find this way of thinking under-complex and ahistoric. It cannot take into account the manifold ideas, convictions, and political cultures that all play into the limits and chances of the politics of legitimation since 1945. It is blind, for example, to the imperial structure of international relations up until the 1960s, it is silent about colonialism and its aftermath. And, in its state-centredness, it does not account for the agency and power of non-state actors like armed groups in the making of international politics. In contrast, a liberal view on international politics would highlight norms and rules as a context of state behaviour, and it would accept the change of content in international politics with regard to normative values. The debate about international regimes (Krasner 1983) basically is such an early tale of progress and qualitative change in international relations. But again, imperial politics in this paradigm, and the pathologization of all non-state violent actors, is in our view as unacceptable as the tendency to see the Western world as the avantgarde of moral betterment. A more self-reflexive view on international politics, to which we subscribe in this introduction, takes a rather sober view on the role of the West in warfare after 1945 including Cold War militarization and the use of force in Western interventions, often without sufficient legal justification. And it would take seriously what post-colonial studies have reminded us: to take non-Western actors into account, not treating them just as objects of Western governmental strategies. In our conception of historical times, however, we include elements of both realism and liberalism. Yet, we argue that these times are not just about power politics or norms and values, and we account for the tenets of critical IR approaches, in particular from historical and post-colonial perspectives.

So what do we mean by historical time? Our conceptualization starts from the differentiation between a quantitative and qualitative understanding of time, also known as the classic notions of 'chronos' and 'kairos' (Hutchings 2008, 4–9). First, time can be understood in a chronological sense as a certain duration of events or conditions that can be measured as a sequence of years or decades. Second, however, it can also be understood in the kairotic sense as a significant time opening up possibilities or proper time for action. In a related sense, Reinhart Koselleck has differentiated between natural, measurable time and historical time. The latter arises out of the former but is not reducible to it. According to Koselleck, time is not simply the medium in which all histories take place, but it can gain also a historical quality and become a dynamic force in its own right. 'Historical time ... is bound up with social and political actions, with concretely acting and suffering human beings and their institutions and organizations. All these actions have definite, internalized forms of conduct, each with a peculiar temporal rhythm' (Koselleck 2004, 2). The two seminal categories he employs here are the 'space of experience' (*Erfahrungsraum*) and the 'horizon of expectation' (*Erwartungshorizont*). Every human community has, on the one hand, a space of experience out of which people act, in which past things are present or can be remembered, and on the other, people always act with reference to a specific horizon of expectation which includes hopes, wishes, and fears. These experiences and expectations of active human agents constitute not merely subjective contingencies but enduring, historically accessible structures (Koselleck 2004, 255–275).

We use the term 'historical time' in this qualitative sense. In our understanding, a historical time describes a period of global politics that is marked by the combination of a particular set of interests and ideas, of international power relations with prevailing understandings of international legitimacy. Historical times include relations and major cleavages between states or groups of states such as between 'East' and 'West' during the Cold War. They also comprise transboundary political figurations of non-state actors and forms of transnational activism such as the 'Global Left' (Wallerstein 2022) in the 1960s. Historical times also include time-bound modes and technologies of warfare as well as organizational templates of rebellion and ways of representing conflicts in an emerging global public sphere. The sequence of historical times since the end of World War II can also be recounted as shifts in prevailing conceptions of international legitimacy or of different time-bound 'legitimate orders' that are 'held by at least part of the actors to define a model or to be binding' (Weber 1978, 31; see also Clark 2003, 81–89). A historical time gives rise to particular legitimating ideas, norms, and moral principles against which it is possible to make judgements about the legitimacy of actors and institutions. It favours specific claims to legitimacy that are more normatively justifiable and acceptable than others. Hence, it also makes the validation of these claims more likely. International legitimacy thus does not denote the existence of one coherent global value system or normative regime. Rather, this notion embraces normative indeterminacy, tensions, and contradictions as well as compromises and adjustments (Clark 2003, 93). In temporalizing international legitimacy, we understand a historical time as representing a zeitgeist that is a period of global politics in which certain understandings of legitimacy prevailed, meaning that they took shape over a number of years, were in full swing during a core period, and slowly waned in validity without ever losing all their pertinence in subsequent times.

The various historical times since 1945 are not just external conditions for armed groups. Just like any other political actor, these groups are part and parcel of their times. Armed groups have to insert themselves within the legitimate order of their time and in making a claim to legitimacy, they need to capture the imagination of an international public. In order to get validated, their claims need to resonate with the dominating legitimating ideas of the time, be they liberation, social revolution, or human rights. This ideational resonance is more often than not an effect of shared beliefs. The history of leftist internationalism and Islamism show that the simultaneity of such beliefs is more an expression of the very same understanding of legitimacy rather than the result of strategic choice. Historical change matters for the legitimation politics of armed groups. With changing power relations, discourses and attention, older ideas and agendas might lose pertinence, even though their validity might not be openly questioned. World historical timing, we argue, is crucial for armed groups in order to receive attention and influence international political agendas.

Historical times differ not only in terms of political power constellations and valid norms. They are each also marked by major political events such as the 1954 defeat of the French in Dien Bien Phu in Indochina, the 1995 massacre of Srebrenica, or the terrorist attacks of 11 September 2001, creating at times 'communicating' relations between different wars and allowing for the simultaneity of armed groups' actions. Wars in which armed groups act are not isolated events, but they are part and parcel of global tendencies, often contradictory ones, in which what happens in one country can have an effect on perceptions and political opportunities on other continents. These times also overlap and so can the struggle of armed groups stretch across different times. It is in particular those armed groups that have fought over decades for their cause like the PLO or Polisario (see the chapters by Christian Olsson and Stephan Hensell) where we can observe these long-term changes in notions of legitimacy.

The following section is neither a history of international politics since 1945 nor a global history of armed groups. We rather understand our historical times as condensed historical sketches. They are in part based on the insights from the case contributions in this book as well as on secondary sources on wars, armed groups and international politics. The literature here is almost endless. We cannot engage in all relevant discussions related to these times, and we will only selectively refer to scholarly work in order to substantiate our arguments. We outline the historical times of decolonization, Cold War, humanitarianism, and new sovereigntism. During the historical time of decolonization, self-rule and national self-determination became universally valid norms. It was a period in which rebellions against colonial rule quickly gained almost undisputed international legitimacy. In the second historical time, the Cold War, armed groups either connected their claims to legitimacy with one camp of the superpowers or they tried to avoid the polarization of international politics in order to get their claims validated. During the third historical time, the age of humanitarianism, human rights became the dominant legitimizing discursive anchor. Armed groups acted by denouncing the human rights records of their opponents and presenting themselves as credible alternatives. The fourth historical time, the new sovereigntism, is marked by a renewed emphasis on sovereignty and national self-assertion coupled with selective interventions in civil wars. While legitimating ideas like decolonization and human rights still linger on and have some leverage for gaining the attention of an international audience, the basis for the international legitimacy of today's armed groups is much less clear.

The time of decolonization: undisputed legitimacy

The time of decolonization with its core period between the 1940s and the end of the 1960s was marked by the dissolution of European colonial empires. Compared to other global times that we sketch out below, it was a period in which rebellions against colonial rule quickly gained almost undisputed international legitimacy. The fight of armed groups and the success of decolonization movements more generally can be seen as an overlooked watershed in international relations. It is in this period that the effect of armed groups on the international system had the perhaps strongest effects: Decolonization meant nothing less than the transformation of an international system of empires into an international system of nation-states. The politics of armed groups is an essential part of this major transformation.[14] What becomes apparent here is the role that armed groups played as actors in international politics, as real subjects, not as mere objects of subjugation and treatment by others. The historical time of decolonization shows particularly well that determined political action by the 'subaltern' shaped political developments to a much larger extent than the realist narrative of great powers want to make us believe. Decolonization revolutionized the international system. Empires vanished and dozens of new states emerged.

At the end of World War II, empire was still considered a legitimate order by most European powers. However, the claim to imperial possessions lost validity very quickly while the newly founded UN affirmed the right of national self-determination, which had gained prominence after World War I and the political reordering of the European continent (Fisch 2010; Manela 2007). Colonialism increasingly became superseded by a new legitimate order of post-colonial states created through decolonization. While this process was overall rather peaceful, conflicts between colonial powers and indigenous populations escalated into wars of independence in many cases as, for instance, in Indonesia, French Indochina, Kenya, Algeria, Guinea-Bissau, Cameroon, or Mozambique. Anti-colonial movements started to employ militant strategies of resistance and empowerment in an international system seen as hostile and created by empires. They turned into armed groups striving for an independent state as the manifestation of national liberation. Fighting became an increasingly legitimate thing to do in the 1950s. As a consequence, the recognition of governments that were formed by armed groups after or even before gaining independence was much more readily granted than in later decades. For instance, during

[14] Self-determination, to be sure, was originally a more radical project that envisioned national independence only as the first step in transforming the international order in a more egalitarian post-imperial world, free of domination and economic injustice (Getachew 2019).

the war of decolonization in Algeria, the FLN declared Algeria as independent in 1958, four years before the end of the war, and several Arab states as well as states from the Non-Aligned Movement immediately recognized the FLN as the legitimate representative of the Algerian people (Connelly 2002). The independent Algeria, in turn, quickly developed into a hub for other rebel groups. Once it became the formal government, the FLN multiplied and strengthened relationships with freedom fighters, militants, and revolutionary forces across the continent. It offered money, arms, training, and safe haven, encouraged liberation movements to coordinate with one another, facilitated diplomatic contacts across continents, and helped with travels to potential sponsors (Byrne 2016, 201–208, 248–256). Other decolonized or socialist states like Cuba, Ghana, Egypt, North Vietnam, China, and Yugoslavia with an 'anti-imperialist' agenda likewise supported various liberation movements. Anti-colonial solidarity with the disenfranchized facilitated instant mutual recognition of armed groups willing to fight colonialism with violent means. Especially with the delegitimizing violence that characterized decolonization wars in Malaysia, Vietnam, and Algeria, with India and Pakistan gaining independence in 1948 and with the Chinese Revolution in 1949, a considerable part of the world was politically not only outspokenly anti-colonial, but also ready to validate the claim to legitimacy of armed movements fighting against colonial powers that were reluctant to withdraw.

While the imperial figuration was the main political frame in which each single rebellion resonated with others, the Cold War also had an accelerating effect on decolonization and made recognition of liberation movements much more likely than in any other global time after 1945. Both superpowers vied for alliances, and none of them could afford to ignore the pleas for self-determination, not the least because their own historical self-understanding was related to emancipation from imperial rule. The Soviet Union had always denounced colonial possessions as Western imperialism and perceived the oppressed colonial people as natural allies of the first proletarian state (Kanet 1975). The US equally supported the 'new nations' in the wake of Wilsonian ideas of self-determination and Roosevelt's decolonization policy with its aim of exporting US ideas of political and economic liberty (Ryan and Pungong 2000). However, Cold War imperatives usually had a higher priority than anti-colonialism. The US, for instance, supported decolonization movements often half-heartedly or forced European powers to decolonize out of strategic calculus as in the case of the Netherlands and Indonesia (see the chapter by Robert McMahon).

Decolonization gained its own momentum as ever more former colonies became independent states and then members of international organizations, or they formed their own regional organizations. These organizations subsequently also became legitimating constituencies for armed groups. Throughout the 1960s and 1970s, the UN not only criticized all remaining colonial possessions but also explicitly justified the struggle of liberation movements with violent means and officialized some of these movements by granting them observer status at the UN (Herr 2017, 118–121). In the case of the Organisation of African Unity (OAU), support for violent decolonization movements was even institutionalized. When it was founded in 1963 by thirty African states, the OAU was dedicated to the eradication of all forms of colonialism and white minority rule that existed in Africa. To that end, the OAU supported various independence movements active at that time. The OAU's 'Liberation Committee' was tasked with mobilizing international solidarity for the liberation struggle as well as with assisting armed groups financially and materially including weapons procurement and training. Moreover, the OAU recognised various armed groups as the sole representation of oppressed people, such as the ANC, SWAPO, the Zimbabwe African People's Union (ZAPU), the Frente de Libertação de Moçambique (FRELIMO), the Movimento Popular de Libertação de Angola (MPLA), and the Partido Africano para a Independência da Guiné e Cabo Verde (PAIGC). The OAU formally recognized all these groups, sometimes even before they made substantial military progress or effectively controlled territory. The OAU also campaigned for the recognition of these groups internationally, which eventually led the UN to accord representatives of the ANC and SWAPO observer status in the UN General Assembly (Reno 2011, 25, 80). The OAU policies expressed a consensus on the righteousness of violent anti-colonial struggle in Africa and established the legitimacy of a broad range of liberation movements and freedom fighters.

Decolonization was a unique time in global history. Fighting for the liberation from empire was legitimate, but challenging the post-colonial state was not. In the late 1970s the UN started to reemphasize norms of state sovereignty and territorial integrity while self-determination was interpreted in a more restrictive manner. The term 'liberation movement' slowly vanished from UN resolutions and a less empathetic wording for armed groups entered the UN vocabulary. Armed groups were viewed less favourably (Herr 2017, 121–124). In retrospect, the amount of international support for armed decolonization movements is remarkable. It indicates, as we argue, a different historical time in which armed groups could position and legitimate themselves along a fundamental line of conflict between the remnants of empire and the norm of self-determination. In the context of decolonization and the political climate of anti-imperialism, the use of violence for political ends as such was not deemed illegitimate. With the intensifying Cold War, however, decolonization wars became increasingly reframed as embodiments of the conflict between the superpowers.

The time of the Cold War: divided legitimacy

As a historical time, the Cold War overlapped considerably with decolonization, but it had its own legitimating logics. The competition between the US and the Soviet Union that was so central for actors in the Third World can also be interpreted as a competition between two legitimate orders. The rivalry of capitalism and communism was essentially also a contest between the ideologies of liberty versus justice, a competition between two models of modernity over the universalism of its values and the rightness of its cause (Westad 2007, 8–72). As the Cold War unfolded globally, the lines between the two blocks became less clear, though. De-Stalinization and the Chinese Revolution had shattered the ideological unity of the 'Socialist Camp', while the so-called 'Free World' remained a politically heterogeneous group of democratic and authoritarian regimes. Situated between East and West was the

'Third World', constituted by the large group of new nations in Africa, Asia, and Latin America. The Third World was seen as an arena of contestation between the superpowers (McMahon 2013). However, far from being only passive observers of the contest between two blocs, many middle and small powers critically shaped the course of the Cold War. Regional actors had significant agency and pursued their own agendas including the formulation of ideological alternatives to the superpower conflict such as Asian–African Internationalism, Non-Alignment, pan-Islamism, or Eurocommunism (Lüthi 2020).

In the Cold War, non-state violence was interpreted less as violence against foreign rule such as had been the case in the age of decolonization. It was rather seen as an expression of the conflict between the superpowers. Civil wars in Africa, Latin America, and Asia were allegedly 'proxy wars', and actors in these regions were seen as 'fifth columns' of the Soviet Union, the US, or the People's Republic of China. Although these generalizations were untrue—all these conflicts originated from local cleavages (Jung et al. 2003)—this politicization did not spare the actors themselves. Entire regional war complexes, such as in southern Africa, the Horn of Africa, or Southeast Asia as well as civil wars in Latin America, were coded along the binary semantics of the East–West divide. Many armed groups oriented themselves either to the East or to the West and so became part of the global constellation of the Cold War, even though local actors hardly ever followed the directions suggested by superpowers or their allies. The autonomy of armed groups, their ability to mobilize quite diverse sources of support and to play on the register of international politics like formal governments becomes evident in several chapters of this volume. Yet, it is true: the US, the Soviet Union but also the People's Republic of China became sponsors of liberation and resistance movements in Africa, Asia, and Latin America and often openly supported them with military training, financial resources, and weapons (Westad 2007; Lovell 2019). Great power support was a demonstrative practice of claim validation. Overall, the Cold War had a galvanising role for violent conflicts. It could escalate, prolong, and internationalize but also shorten them (McMahon 2013). This opened up opportunities but also set restrictions. Armed groups in ongoing anti-colonial struggles but also in new civil wars had to adapt to the legitimating or de-legitimating narratives of the two camps in the Cold War.

Groups with leftist beliefs usually gravitated towards the Soviet camp. National revolutionary movements such as the MPLA, FRELIMO, SWAPO, and ZAPU, which fought against foreign rulers or indigenous regimes, emphasized the idea of liberation, freedom from external control, social revolution, and cultural regeneration. With their blend of Marxist ideas, nationalist particularism, and a liberal conception of self-determination, they had an ideological affinity with the Soviet Union or China (MacFarlane 1985, 5–6, 213–214). The same applies to social-revolutionary guerrilla movements in Latin America but also to various other groups, such as the ANC in South Africa, the Sudan People's Liberation Arm' (SPLA), or the Partiya Karkeren Kurdistan (PKK) or the New People's Army in the Philippines. For these insurgencies, versions of Maoism and Marxism became in many cases not only dominant legitimating ideas but also organizational templates, while cities such as Algiers, London, Moscow, Beijing, and Dar-es-Salaam developed into transnational hubs for rebel leaders. Mao Zedong's 'little red book' and Frantz Fanons 'The Wretched of the Earth' but also the writings of Amílcar Cabral or Che Guevara became extremely popular and influential. They circulated on various continents among intellectuals, leftist activists, and would-be revolutionaries who engaged in a transnational exchange of ideas, strategies, and techniques (Cook 2014; Lovell 2019; Byrne 2016). Elective affinities emerged between progressive forces in the North and militant groups in the South sharing a radical anti-imperialism that was part of the global zeitgeist (Parrott and Lawrence 2022). Rebel groups and liberation movements were an integral part of the 'Global Left', and its leaders and activists sought political mobilization in the West just as in the Third World (Hobsbawm 1994, 436–460). Leftist political parties, progressive political circles, and solidarity groups formed the transnational constituencies of armed groups. Activists became the 'grassroots ambassadors' for the revolutionary cause and tailored the rebels' message to domestic audiences, as the case of the Nicaraguan Sandinistas and their diplomacy in Western Europe demonstrates (see the chapter by Eline van Ommen). This constituency occasionally also included governmental actors. Progressive Scandinavian governments played a particularly interesting role here, as they were the first to recognize armed groups as legitimate actors in world politics that were still seen as enemies by other Western governments. The ANC in Africa, SWAPO in Namibia, FRELIMO in Mozambique, and the Frente Revolucionária de Timor-Leste Independente (FRETILIN) in Indonesia were cases in point, but also the FMLN. Various other movements in the Cold War had more conservative and revisionist orientations such as the Resistência Nacional Moçambicana (RENAMO), the União Nacional para a Independência Total de Angola (UNITA), the Mujaheddin in Afghanistan, or the Contras in Nicaragua. The agenda of these groups was more compatible with US support and the 'Reagan Doctrine' that sought to promote anti-communist guerrillas in order to weaken Soviet-backed pro-communist governments in Africa, Asia, and Latin America (Westad 2007, 331–363). These groups were part of an 'anti-communist international' in which right-wing leaders sought to build alliances with kindred guerrilla movements, often with the help of US covert actions (Burke 2018). For groups such as the Angolan UNITA, conservatives in the US were a core international constituency (see the chapter by Reyko Huang).

The dynamics of legitimation in the Cold War could not be reduced to relations between governments in the East and West on the one hand and non-state violent actors on the other. Armed groups were embedded in large transnational social movements and networks with a leftist or rightist orientation. Thus, they became part of two competing legitimate orders that either legitimated social-revolutionary or anti-communist warfare. A binary logic was the result of this division of the world: actors could roughly calculate who would validate their legitimacy claims, depending on the side of the Cold War with which they aligned themselves. Only in a few exceptional cases such as in the case of the National Resistance Army (NRA) in Uganda (see the chapter by Klaus Schlichte) could armed groups avoid alignment with either the Soviet Union or the US and gain international recognition by becoming de facto rulers.

In the contributions in this volume that cover cases from the 1980s, three other features of armed groups' politics of legitimation become very clear: First, the 1980s was a period in which the discourse of human rights gained traction. To denounce their opponents as gross human rights violators, to delegitimize governments by taking recourse to a US foreign policy discourse born in the late 1970s, has become a major discursive strategy of armed groups since then. This applied not only to the NRA in Uganda or the ANC in South Africa but also to more outright leftist groups such as FLMN in El Salvador and the Sandinistas in Nicaragua (see the chapters by Andrea Oñjate Madrazo and Eline van Ommen). Second, in all these cases, regional constellations were more decisive than the superpowers' approval in getting support and recognition. In retrospect it seems that in the case of Fatah as well as those of the Polisario, the FMLN, the NRA, and others, it mattered more for armed groups to insert themselves into regional political landscapes and to gain approval from neighbouring governments than to lobby in Washington or Moscow. This was possible as third, the camps of the Cold War, NATO on the one side and Warsaw Pact countries on the other, were not as homogeneous as it is often portrayed. Armed groups knew this and at times exploited these rifts adeptly. The FMLN was able to mobilize the support of France and Mexico, in open contradiction to the US position on the war in El Salvador (see the chapter by Andrea Oñjate Madrazo). In Uganda, British support for the old regime continued even though its war crimes had already been discussed in the US Congress and had led to a halt of US support (see the chapter by Klaus Schlichte). The last two features show again what leverage armed groups actually had, even during the Cold War, a period that is often imagined as a stable and rigid international order in which few if any actors had any choices. The cases of the FMLN, the FSLN, and the NRA also show that during the Cold War, armed actors could gain legitimacy by de-legitimizing their opponent's violent actions. Legitimacy in the international system was divided among two competing legitimate orders, and we assume a normative effect of the discourses of both camps that constantly hinted at the immoderate violence of the respective opponent.

The end of the Cold War is often associated with the fall of the Berlin Wall and the end of the communist regimes in Eastern Europe. While '1989' has been rightly questioned as marking an epochal shift in world order given its uneven global impact (Lawson et al. 2010), it certainly marked the end of the ideological rivalry that was so characteristic for the Cold War. That also changed the policy priorities of governments. Some armed groups such as UNITA, that have long been seen as a loyal ally by the US government, quickly fell from grace in Washington and almost overnight became a criminal organization punished with sanctions (see the chapter by Reyko Huang). In the dawning new historical time, freedom, democracy, and the market became central legitimating ideas for the restructuring of societies and political systems. Yet, in the evaluation of armed conflicts, at least in Western publics, human rights and humanitarian issues became paramount with marked consequences for armed groups.

The time of humanitarianism: ambiguous legitimacy

With the end of Cold War, the discourse of human rights and humanitarianism as a new set of ideas and values became a central political theme in the politics of legitimation. It is for that reason that we call this period the humanitarian time. It marks the quarter-century period from the end of the Cold War till the mid-2010 years when the wars in Syria and Libya unfolded. Humanitarianism discourse, the aspiration of a transnational morality, had accompanied warfare since the late 19th century (Barnett 2011). Yet it became widespread only in the 1980s, and then dominant after 1995, to the degree that humanitarianism was considered by many as a morally better substitute for the Westphalian order of states. Western activists and governments such as the Carter administration in the US had started to embrace human rights during the 'human rights revolution' in the 1970s, even though its foundations had been laid in the decolonization of the 1960s (Moses et al. 2020) and at the end of World War II (Iriye et al. 2012). The time of humanitarianism was in a way prepared during the late phases of the Cold War. A prominent case here was the ANC, which had tried to de-legitimize apartheid in South Africa, but those claims had fallen largely on deaf ears for decades (Thomas 1996). It was only in the 1980s that the anti-apartheid movement gained more international traction. Acknowledgement as a practice of claim validation was increasingly based on human rights, leading to a peaceful regime change and to the legitimation of the ANC as South Africa's new government in 1994. In retrospect, however, the massacres of Rwanda (1994) and Srebrenica (1995) and the failure to protect civilians from massive violence seem to have been the major turning points. The legitimate order of humanitarianism put the rights of individual physical integrity centre stage. It was marked by a general de-legitimization of violence committed by states and armed groups against citizens, but also by a selective legitimation of counter-violence through external interventions. These interventions were increasingly justified with humanitarian arguments.

The US intervention in Somalia in 1992 was one of the earliest to be legitimated on humanitarian grounds, but the NATO intervention in the Kosovo war in 1999 was the first clear-cut case. The apex can be seen in the Libyan civil war in 2011. Here, the UN Security Council for the first time authorized the use of force for the purpose of human protection against the will of the acting government (Hehir 2013). At the heyday of the humanitarian time, human rights tended to override the legitimacy of the state as the norm of sovereignty was relativized in favour of 'R2P'—the 'responsibility to protect' (Bellamy 2011). It allowed the international community to engage in war-affected regions with 'robust mandates' and the use of force. The terminology of 'humanitarian intervention' (Wheeler 2000) was the perhaps clearest sign of a normative change in the justification of war and military intervention. However, the actual decisions by Western governments to intervene in humanitarian crises were selective and driven by domestic factors such as competing foreign policy elites and media pressure, which led the US to intervene in Somalia in 1992 but not in Bosnia that same year (Western 2002). The failure of the intervention in Somalia slowed down the engagement of the US and its NATO allies in other civil wars, as their tragic role during the genocide in Rwanda 1994 showed. With this failure in mind, the next globally publicized massacre in the Bosnian town of Srebrenica triggered the wave of new interventionism that continued unabated till the mid-2010 years. This was in part conditioned by 'the unilateral moment' (Krauthammer 1990), a period in which the US outcompeted any other major force and was able to establish a military hegemony, using a language that mirrored core liberal values, but employing means that were part of the traditional set of foreign policy tools, including open warfare. Yet the new interventionism would not have been possible without widespread agreement in Western publics and beyond that civil wars and massive human rights violations could no longer be tolerated. The US facilitated not only the logistics of military interventions, but its hegemony also enabled the UN to become a functional arena and an overall legitimizing force in this new setting. Also, the reluctance of Russia and China to put stricter limits to Western-funded interventions allowed humanitarianism to become the new legitimate order.

We follow Didier Fassin in his diagnosis that the 'humanitarian reason' is a distinct 'intelligibility regarding world politics and a particular form of collective experience' (Fassin 2011, 223). As many observers have noted, it followed liberal traditions of putting individual rights above political purposes, and moral sentiments began to overrule the requirements of 'realpolitik' (Fassin 2011, 1–18). Conflicts became events that urged action, with roles of victims and forfeiters, of saviours and saved (Ticktin 2014), judged by a globalized public that watched 'suffering at a distance' (Boltanski 1993) with a much more widespread attention, covering many arenas at the same time. Changes in global media and the televising of the world had a significant effect here. Even in the 1980s, little information was available on distant wars in Africa, Latin America, or Asia, and in Western media they were often reported sparsely or not at all. In the 1990s, however, new technologies had transformed the news media into a permanent flow of global real-time news and the speed and ubiquity of information, also known as the CNN effect, increased the attention on civil wars, humanitarian crises, and suffering people (Robinson 2002). Media coverage has remained very selective but has massively impacted the politics of legitimation. The potential transnational audiences and constituencies of armed groups became much wider. In covering violent conflicts, the media also disseminated and debated claims to legitimacy and as a consequence, armed groups started to make use of a broad set of tools to cultivate their public image (Bob 2005). Armed groups made deliberate efforts to draw international attention to their cause by making human rights violations globally known. But they themselves also became subjected to the communicative effects of exerted violence (Koloma Beck and Werron 2018, 276).

The 'politics of life' (Fassin 2007) in the historical time of humanitarianism delegitimized government repression as much as armed insurrections with obvious consequences for the legitimation of non-state violent actors. The human rights discourse helped armed groups to de-legitimate their adversaries by scandalizing and denouncing state violence and justifying their own action by framing it as 'defence' or 'protection' (Fassin 2011). However, for armed groups the international reactions to their claims to legitimacy became more difficult to calculate overall. Using violence while at the same time portraying themselves as victims of state violence and defenders of endangered populations became a balancing act for armed groups whose legitimacy remained ambiguous. A lot depended on the international framing of the governments that were being attacked by armed groups. As long as these governments could be blamed for atrocities and aggressions against civilians, armed groups had good chances of getting their claims validated. The Ushtria Çlirimtare e Kosovës (UÇK) in Kosovo, for instance, was especially successful with internationalizing the conflict and portraying itself as the defender of Albanian civilians against massive human rights violations by the Serbian state. Its strategy of claim-making and victimization not only led to the recognition of the UÇK as an official representative of the Kosovo-Albanians but also triggered an international military intervention that was the main goal of the Albanian guerrillas (Perritt Jr. 2008, 130–151; Phillips 2012). Likewise, the Sudanese rebels in the Darfur crisis of the 2000s succeeded in framing the conflict as 'genocide' and a 'new Rwanda', thus leading to their recognition as legitimate negotiating partners (Jumbert and Lanz 2013, 198). Libya can even be seen as a case of claim validation before claim-making. When the regime of Muammar al-Gaddafi initiated a violent crackdown on a popular uprising, the international reaction followed immediately. While the UN Security Council authorized the use of force for the purpose of human protection, only five days after the rebels founded a 'National Transitional Council', France recognized it as the legitimate Libyan government. Several other European states followed within the next months (Hehir 2013; Fernández-Molina 2023). The SPLA in South Sudan, the Liberation Tigers of Tamil Eelam (LTTE) in Sri Lanka, and Hamas in Palestine, however, ran into the problem of delegitimizing their cause when their violent practices attracted international attention. For the politics of armed groups and governments alike, the change towards humanitarianism meant that they entered an incalculable competition about who would be blamed more for human rights violations. In fact, armed groups proved to be much more responsive to international humanitarian law than governments, which began to avoid making official war declarations so as not to be subject to international norms of warfare (Jo 2015; Fazal 2018).

The age of humanitarianism is not entirely over, but in the current situation, strong and numerous critiques have diminished the pertinence of the humanitarian discourse: First, the R2P could not be fulfilled by the interveners, in particular as Western governments were reluctant to engage with their own troops on the ground. Kosovo and Libya are the prominent examples here. Second, interventions were accompanied by the problem of exit options which were usually connected with an agenda of 'state-building' (Paris and Sisk 2009; Bliesemann de Guevara 2012). In most cases, the results of these efforts were disappointing. The 'security predicament' (Ayoob 1995), namely the dilemma to form a central authority in war-torn societies with a minimum of democratic standards but without any major use of force, could not be solved. Third, humanitarian interventions remained selective and arbitrary.

Conflicts in Sub-Saharan Africa became primary objects, while insurrections in the former Soviet Union such as in Chechnya or Tajikistan did not spark multilateral engagement. While the case of Libya has been widely seen as putting R2P into practice, it paradoxically also heralded its decline. The NATO intervention in 2011 was supposed to be limited to protecting civilians from state violence but it evolved into a regime-change operation to topple Muammar al-Gaddafi, thus shattering international consensus on the R2P doctrine. Since then, China and Russia have opposed other such interventions in the UN Security Council. With the retreat of the US forces from Afghanistan in 2021, the interventionist policy of Western powers finally seems to have come to an end. After twenty years, humanitarianism has lost its appeal, even though it is still an important register of evaluating political dynamics, not only in liberal and leftist Western circles. Humanitarianism has always remained contested in non-Western regions with recourse to claims of national sovereignty, a motive that has seemed to gain the upper hand in our contemporary times.

Contemporary developments: the new sovereigntism

It is more difficult to identify the present day's historical time and its legitimate order. History is always much better written in retrospect. For the time being, the label of 'new sovereigntism' seems to cover best what is marking our present times.[15] Like in earlier historical times, older normative frames still linger on and can be invoked in political arguments. New sovereigntism is still confronted with powerful humanitarian objections, but can also draw on a resurgence of claims from the decolonization period. Sovereignty is the foundational idea of the modern age according to which a state has supreme authority domestically over its subjects and independent authority internationally in relation to other states (Jackson 2007). As a symbolic form (Bartelson 2014), sovereignty has organized the world since the early modern period and, thus, always played a major role as the defining principle of international politics. By 'sovereigntism', we mean the tendency of stressing the notion of sovereignty to the detriment of international obligations. Sovereigntism can be seen as a paradigm that follows old realist patterns of international politics: sovereign states compete with each other in an international system in which might is more important than right. Different sovereigntist claims can be traced throughout the history of international politics (Alles and Badie 2016).[16] The last two decades have seen their forceful re-emergence.

Peter Spiro has described the US policy elite at the beginning of the new century as the 'new sovereigntists'. Driven by anti-internationalist thinking, the new sovereigntists dismiss international law and multilateral regimes in fields such as human rights, war crimes, or arms control and instead emphasize sovereignty and the autonomy of

[15] We see this phrasing as rather hypothetical. Keeping in mind the Hegelian proverb about the owl of Minerva spreading its wings only at dusk, over time perhaps better and more concise characterizations will appear.

[16] To be sure, this is also true for the period since 1945, often in contradiction to the idea of solidarity and cooperation within the Third World. The conference in Bandung 1955 and later the founding conference for the Non-Aligned Movement in Belgrade 1961 as well as the Helsinki final act in 1975 had already emphasized the norms of sovereign equality, sovereign rights, non-interference, and independence.

the nation-state (Spiro 2000). This tendency has probably found its clearest expression in what is referred to as the 'Bush revolution in US foreign policy'. Shedding constraints imposed by allies and international institutions and using the strength of the US, including military force, to increase America's security but also to make the rest of the world more stable and prosperous became a hallmark of the Bush presidency (Daalder and Lindsay 2003). Following 9/11, the preference for the unilateral exercise of power or ad hoc coalitions of the willing guided especially the US wars against Iraq and Afghanistan that aimed at regime change in states deemed to be rogue or suspected of teaming up with terrorists. That a sovereigntist view of international politics has gained the upper hand vis-à-vis liberal views of normative diffusion and global governance can also be gauged from the self-assertive politics of other states. Great power contestation of US hegemony on the part of China and Russia signals the end of the 'unipolar moment' that had allowed for a US-led liberal order in the first two decades after the Cold War (Cooley and Nexon 2020). Sovereigntism thus goes hand in hand with an increasing multipolarity of the world. However, domestic political polarizations also matter. In various parts of the world, in the US, Russia, and Western Europe but also in India, Venezuela, or Brazil, a discourse of people's sovereignty has taken hold in the guise of populism, nationalism, and a revisionist foreign policy (Minakov 2021; Jenne 2021; Basile and Mazzoleni 2022). New sovereigntism seems to be on the rise in Africa, too. In the Sahel, France has always played a major role as a former colonial power and intervening force (Idrissa 2021). However, civil wars and military coup d'états in Mali, Burkina Faso, and Niger have led to an open confrontation with France and its European allies, signalling the rise of an anti-French 'néosouverainisme' in the Sahel, as well (Mbembe 2023). Sovereigntism, we would argue, seems to be the new historical time of our present. It is marked by a renewed emphasis of the legitimate order of the sovereign nation-state which liberates itself from external authority seen as encroaching and illegitimate.

In many regards, the global 'War on Terror' paved the way towards the new sovereigntism. Proclaimed by the US in the aftermath of the attacks of 9/11, the War on Terror had major consequences for a variety of armed groups and their quest for legitimacy. Those who were listed by Western governments as 'terrorist groups' like Hamas in Gaza, Hizbollah in Lebanon, the LTTE in Sri Lanka, the PKK in Turkey, or al-Shabaab in Somalia remained excluded from gaining legitimate status, even if they were nationally legitimate actors or controlled large territories in state-like manners. The case of Hamas who won a landslide electoral victory in 2006 and subsequently represented the Palestinian Authority (PA) in the Gaza strip is exemplary here (see the chapter by Leila Seurat). During the early 2000s, when the War on Terror brought about concerted action of all major powers, the main danger for armed groups was to be labelled as being part of the 'global terror networks', mostly of an imagined global Islam (Mamdani 2004; Keen 2006). The chance of connecting with a major power in international politics via ideological identification was largely gone. The War on Terror and post-9/11 securitization has not only unified Western governments but also facilitated international cooperation with authoritarian regimes such as Russia and China. However, the multilateralism in security affairs did not last long. Some observers see the US-led invasion into Iraq (2003–2011) as the turning point towards a 'Hobbesian moment' in international order (Battistella 2006, 153–155). The US-led war against the regimes of Saddam Hussein and the Taliban in Iraq and Afghanistan, both deemed as supporters of terrorism, not only heralded the return of the state as unilateral actor but also as a possible target of externally supported regime change. As part of the global War on Terror, the wars in Iraq and Afghanistan can thus be seen as a catalyst in the shift from humanitarianism towards a new historical time in which the idea of sovereign state power has gained renewed prominence.

The wars in Iraq and Afghanistan, however, also pushed the spread of radical movements and jihadi groups that were part of an Islamist resurgence that had already started in the 1980s. Both wars affected the political landscapes in the Muslim world of the Middle East, Sub-Saharan Africa, Central Asia, and East Asia. Being part of a broader transnational Islamic movement, jihadi groups fighting in various civil wars anchored their claims to legitimacy in an ideal Islamic order and imagined global Ummah (Roy 2004). In their ideological commitments and international embeddedness, these armed groups were not unlike the former Marxist revolutionary insurgents in the Cold War (Kalyvas 2018). The violent conflicts in Iraq, Syria, Yemen, and Libya illustrate how the legitimation of Islamist movements unfolded in the time of new sovereigntism. In these four cases, old-standing regimes began to crumble due to domestic opposition and foreign interference, ranging from full-blown invasion in Iraq in 2003 to military intervention in the case of Libya in 2011. In all four countries, radical Islamist groups posed a major challenge for Western governments who supported domestic protest and regime change but also wanted to halt undemocratic Islamist movements from taking over. The attempt to contain the spread of radical Islamist movements, in particular Daesh/Islamic State, contrasted with the growing reluctance of Western governments to engage directly with their own troops in these conflicts. As a result, a new range of ambitious mid-sized or great powers stepped in.

New sovereigntism means a new landscape of self-assertive states such as Turkey, Russia, Iran, Saudi-Arabia, Qatar, or the United Arab Emirates. As intervening powers, they have become increasingly important in the last decade for the course of civil wars. These regional powers not only provided armaments to various non-state groups, but also mediated between the conflict parties or brokered cease-fires, thus validating their claims to legitimacy through supporting and officializing. For armed groups, the new sovereigntism means new opportunities for legitimation. They can in fact pick and choose from an array of potential external supporters or play allies off against each other as the case of Haftar in Libya clearly demonstrates (see the chapter by Wolfram Lacher) and as can also be observed in other civil wars such as Iraq, Yemen, Syria, Somalia, or Sudan (Ero and Atwood 2023). Armed groups, thus, do not simply react to what states do nor are they mere objects in a great power game. Equipped with their own agency, they rather reproduce the logic of sovereigntism once in power. The case of the Taliban that are in control of Afghanistan and seek international recognition but only 'on their own terms' (see the chapter by Ahmed Elsayed) demonstrates this. Finding ways to validate their claims to legitimacy seems to be easier for many armed groups in the present age of sovereigntism. Social media and new forms of digital self-representation allow armed groups to reach broader publics, to demonstrate their power and to disseminate their claims to legitimacy (Bos and Melissen 2019; Jones and Mattiacci 2019). However, as a less active role for the UN seems to be the order of the day, regional and Western governments pursue their own national interests through selective interventions and proxy warfare. The recent experience also holds the lesson for armed groups that these alliances with external actors can be quite opportunistic and unreliable as the case of the Kurdish Yekîneyên Parastina Gel (YPG) demonstrates (Elsayed 2023).

It is interesting to note, though, that all the previous historical times with their normative references have not ceased to matter. The 'presence of the past' (Schlichte/Stetter 2023) applies to the field of legitimation as well and we can see that the historical times are still 'around' in a way. The reference to decolonization matters for armed groups and their supporters in long-standing conflicts. The Polisario as well as Palestinian groups such as Fatah and Hamas see themselves as liberation movements and still frame their fight as 'anti-colonial'. These references ring a bell and sound plausible to the constituencies and wider audiences of these groups and yet, they seem to have lost traction with the change of historical times. Also, the arguments and normative claims of humanitarianism have not entirely vanished. They have survived as references within public discourse and remain the official yardstick for the evaluation of warring parties and their actions, at least in Western audiences. As the most recent Gaza war between Israel and Hamas since 2023 makes clear, human suffering, in particular the suffering of civilian victims, is still a powerful trope for mobilizing international audiences and blaming Western governments and their allies. The current Gaza war, however, also shows that humanitarian norms and international law do no longer have the same normative leverage as they used to have. In present times, with the rise of sovereigntism, foreign policy takes more recourse to national power politics. The frequently invoked Israeli 'sovereign right of self-defence' seems to be above all other concerns while international institutions are sidelined in the conflict management. At the same time, they have not become entirely irrelevant—the constant appeal to international courts and organizations then can also be seen as an indication that not only governments but armed groups and a wider public expect some elucidation or at least a gain in legitimacy by how international institutions judge and frame ongoing conflicts.

The contributions in this book

This volume is roughly structured by the chronological order of our eleven case studies. Following the sequence of our historical times, we start with older historical cases and then move on towards more contemporary ones. Given that some of our cases are long-standing and ongoing conflicts that stretch from the age of decolonization until the present, this is not a strict systematic though. In our conclusions, we summarize the empirical findings of the cases and draw some lessons for IR. The book closes with an afterword and some further general thoughts on what the view on history as suggested in this volume might imply for theory-building in IR.

Robert McMahon examines in his chapter two of the most significant anti-colonial rebellions, those in Indonesia and Vietnam. Doing so, he also explores the intersection between two historical times. The chapter shows that, although wholly distinct from the Cold War in its origins and consequences, decolonization coincided temporally with the superpower rivalry and was profoundly shaped by it. The central argument of this chapter holds that in Indonesia and Vietnam aspiring national liberation movements deployed an effective combination of armed strength and diplomatic advocacy in their quest for independence and international legitimacy. The insurgencies not only employed military means but also presented themselves as states-in waiting and future governments. They relied on fighting and mimicking as practices of claim-making while reaching out to international audiences and potential constituencies in the West and East. Cold War priorities shaped the US and Soviet responses in each case. The superpowers sought to shape the outcome of the decolonization struggle, offering or withholding diplomatic support and military aid as a means of influence. For their part, the independence movements tried to leverage the bipolarity of world politics to enhance their prospects for recognition in the state-based international system. The specific historical constellation presented by the overlap between the historical times of decolonization and the Cold War thus created unique historical opportunities for nationalist insurgencies who skilfully seized them by combining armed resistance with diplomatic overtures in their bid for international legitimacy.

Stephan Hensell explores in his contribution the politics of legitimation of the Polisario that seeks to establish an independent state in Western Sahara. It argues that the Polisario has put forward a widely validated claim to legitimacy that, however, has been trumped over time by realism and disinterest. The historical time of late decolonization was favourable for the liberation movement, who claims to have legitimacy according to the right of self-determination and independence that has been granted to all colonial people. The legitimating idea of the anti-colonial struggle based on principles of international law resonated within the UN and a constituency formed by a leftist European solidarity movement and states from the Third World that shared the historical experience of colonialism. Being part of an international anti-imperial movement in the late 1970s to mid-1980s, the group could count on attention, solidarity, and support from like-minded allies. However, the Polisario has rather lost out as the conflict has dragged on after the end of the Cold War. The chapter makes two core observations in that regard. First, realist considerations have always been a powerful counterweight to legitimacy claims in the case of Western Sahara. Morocco as the main opponent of the Polisario has since the early days of the conflict rejected the movement's claims. Doing so, it has been tacitly supported by the US, France, and Spain in the name of stability and realpolitik. Today, Morocco is more successful than ever in pursuing its irredentist politics and compelling others to accept its sovereignty over Western Sahara. Second, claims to legitimacy that have been convincing in the past do not necessarily capture the imagination of future audiences. In the third decade of the 21st century, the Polisario continues to frame the issue of Western Sahara as a matter of decolonization, but it struggles to break the wall of silence surrounding the conflict. As a long-term case, the conflict over Western Sahara exemplifies the impact of changing historical times on armed groups. International legitimacy is time bound. It emerges in a particular historical time but cannot necessarily be reproduced in another.

In her contribution, *Eline van Ommen* examines the revolutionary diplomacy of the FSLN in Nicaragua and their campaign for international legitimacy in the late 1970s. The chapter draws particular attention to the role of solidarity activists in the politics of legitimation of the guerrilla movement. While the Sandinistas developed a truly global campaign in order to convince foreign audiences of the legitimacy of their armed struggle, it was up to local activists to ensure these diplomatic efforts fell onto fertile ground. Solidarity activists in the Americas and Europe translated and tailored the Sandinista message to appeal to a broad international audience. These 'grassroots ambassadors' presented the FSLN as a moderate group with clever arguments regarding the Cold War, human rights, democracy, and social justice. Moreover, they introduced Sandinistas to journalists, politicians, and civil society groups, and actively lobbied governments in their own countries to recognize the FSLN as the legitimate representative of the Nicaraguan people. Voicing as a practice of making a claim to legitimacy was thus at the heart of the FSLN's politics of legitimation. The Sandinistas' strategy proved ultimately successful. Political leaders and parties in Latin America and Western Europe with a social democratic or progressive background validated these claims by acknowledging the FSLN as a legitimate political actor and by supporting the group. The case of the FSLN thus highlights the key role of solidarity activists in the politics of legitimation. As early members of the FSLN's transnational constituency they paved the road towards the Sandinistas' transformation into a legitimate political actor on the global stage.

The chapter of *Andrea Oñate-Madrazo* examines the politics of legitimation in the case of the FMLN in the Salvadoran Civil War. In its bid for political power, the insurgent movement sought to secure international legitimacy as a rightful representative of the Salvadoran people mainly through diplomatic efforts. A key role here played the guerrillas relations with the Mexican government and the willingness of the latter to provide a safe haven for the FMLN's diplomatic endeavours. Based in Mexico's capital, the FMLN's Diplomatic Commission made a claim to international legitimacy by mimicking a state's foreign ministry and by voicing its political messages to an international audience. The Mexican government, in turn, validated these claims through supporting the Salvadoran guerrillas with refuge and through acknowledging and officializing the FMLN via the Franco–Mexican Declaration of 1981. This bilateral declaration had far-reaching implications as it formally and publicly recognized the FMLN as a legitimate party to the conflict and representative force of opposition in El Salvador by the governments of Mexico and France. It constituted the Salvadoran revolutionaries' first decisive political-diplomatic breakthrough in their quest for legitimacy on the international stage. Mexico's agency in this process highlights not only the significance of regional actors in the FMLN's efforts at international legitimation but also evidences how regional legitimacy can expand into broader international recognition for armed groups. The Mexican–FMLN nexus had profound implications both for the Salvadoran insurgents in their quest to secure international recognition, and also for the Mexican government's bid to demonstrate that it was a Cold War player of consequence.

The chapter of *Klaus Schlichte* argues that NRA's pathway to legitimate state power in Uganda in the early 1980s was a result of different factors, ranging from combat performance to favourable global historical timing and strategic engagement with neighbouring governments. Military might but also the reputation of being a much more disciplined force than the slowly degrading army of the old regime allowed for the NRA's successful legitimation. The gross human rights violations by government forces and the erosion of legitimate rule in Uganda greatly facilitated the NRA's success, as the rebel group systematically publicized the human rights violations of its opponent in order to attain international legitimacy for itself. While the conflict remained under the radar of superpowers, its significance lies in the fact that it can be located at the intersection of different historical times. The war led by the NRA took place at the threshold between the Cold War and the time of humanitarianism: the immediate post-colonial role of Great Britain waned in the course of the war, and with the changes in the Soviet Union, humanitarian issues became more prominent. Both developments worked to the advantage of the NRA. The case also shows that it was the claim validation by regional neighbouring governments that advanced the officialization of rebel groups, followed by the former colonial power, the US, and the two German governments.

The chapter by *Reyko Huang* uses the case of UNITA's rebel diplomacy in the Angolan Civil War to examine how armed groups go about the task of seeking and gaining legitimacy abroad. Focusing on Washington DC as a dominant site of UNITA's politics of legitimation, the chapter shows how UNITA lobbied the US government by opening a representation office, making public appearances, and working with professional lobbyists for its strategic outreach to the US government, media, and civil society. The case study shows that UNITA's rebel diplomacy involved ample political showmanship and diplomatic professionalism combined with a strategic narrative to reassure its ally of its commitment to shared values. UNITA representatives easily intermingled with those in Washington's highest political circles, delivered compelling speeches and touted the ideology of 'anti-Marxism', 'freedom', and 'democracy'. These qualities made UNITA appealing and acceptable for its conservative constituency in Washington. In making a claim to legitimacy, UNITA engaged in mimicking by adopting state-like formal behaviour through diplomatic customs. The group also practised voicing and employed communications professionals who helped frame UNITA's vision and actions in ways that resonated with American officials and lawmakers. The US government, in turn, validated these claims by officializing UNITA in the framework of diplomatic protocol, acknowledging its leader, Jonas Savimbi, as a 'patriot' and 'loyal friend', and supporting UNITA with money and weapons. As a result, UNITA and its conservative constituency in the US developed a tight symbiotic relationship in which claim-making and claim-validation became a collusive effort. Reflecting the historical time of the Cold War, the politics of legitimation in the case of UNITA was of considerable mutual benefit for both sides. UNITA received political backing while the US could count on a loyal ally.

Christian Olsson in his chapter focuses on Fatah, historically the main embodiment of the Palestinian struggle for statehood and also the de facto caretaker of PLO, the PA, and the state of Palestine. This contribution adopts a long-term perspective considering Fatah's trajectory from its foundational engagement in the PLO under the leadership of Yasser Arafat to the Oslo Accords and the stalled peace process until today. The central argument is that Fatah long maintained a delicate balance between armed struggle and the quest for diplomatic recognition. While fighting as a practice of claim-making earned Fatah domestic legitimacy from its Palestinian constituencies, an increasing number of states also validated Fatah's legitimacy claims by acknowledging the PLO as the representative of the Palestinian people and recognizing the state of Palestine. However, in the 1990s outright contradictions between external and internal legitimacies took centre stage in Fatah's trajectory. The group became increasingly caught between, on the one hand, the responsibilities imposed by their formal international recognition and the Oslo Accords and, on the other hand, demands from its Palestinian constituencies for taking decisive action in the face of obstacles on the way towards Palestinian independence. In the absence of concrete steps to end Israel's military occupation, Fatah progressively lost its domestic legitimacy in favour of more radical groups like Hamas which claimed to embody the struggle for national liberation. The case study on Fatah highlights how the pursuit of political legitimacy by an armed group affected its use of armed force and vice versa. While armed struggle initially stood not in a direct contradiction to international legitimacy, after Oslo and Arafat's renouncement of violent action any direct engagement in armed struggle became costly for Fatah in terms of international legitimacy. The case also shows the trade-off armed groups can face when they simultaneously address domestic constituencies and international audiences in their quest for political legitimacy.

The Palestinian quest for international legitimacy is continued with the chapter of *Leila Seurat* on Hamas. While Hamas is usually perceived as a terrorist organization, this case study highlights the group's quest for political legitimacy and it puts Hamas' politics of self-legitimation centre stage. Having won a landslide victory in the 2006 Palestinian legislative elections, the group has de facto been recognized as a legitimate actor by its Palestinian constituency. However, its victory at the ballot box was not followed by international recognition. The chapter argues that since then Hamas has tried to make a claim to legitimacy. Hamas initially was active in state mimicking as it duplicated the institutions of the PA in the Gaza Strip. By representing the PA in diplomatic encounters, Hamas also sought international legitimacy. Later, however, the group counted more on fighting as a practice of claim-making. In doing so, Hamas has gradually favoured domestic over international legitimacy. Winning the public support of the Palestinians by confronting Israel and presenting itself as the embodiment of Palestinian resistance became the core of Hamas' politics of legitimation. This trajectory from government to armed struggle led to various violent clashes with Israel and ultimately culminated in the 7th October attacks which provoked the most recent devastating Gaza war. As in the case of Fatah, this chapter shows the contradictions between domestic and international legitimacy armed groups can face. It also makes clear, however, the communicative function of violence. Hamas' surprise violent attacks can be seen as a typical communication strategy aimed at various audiences. While the group has presented itself as the embodiment of the Palestinian struggle for independence to the Palestinians in the various occupied territories, it has also sent a message to Arab governments that sought a normalization of diplomatic relations with Israel.

Ahmed Elsayed examines in his contribution the legitimation politics of the Taliban since the 1990s until today by focusing on how the Taliban legitimized themselves and de-legitimized their rivals. The chapter analyses how the Taliban leveraged fighting, voicing, and mimicking as practices to claim legitimacy and explores the extent to which acknowledging, supporting, and officializing by regional and international actors have directly or indirectly served to validate these claims. It argues that although the Taliban actively seek external recognition of their Islamic Emirate, they simultaneously enact policies—especially those related to human rights, inclusive governance, and foreign militants on Afghan soil—that undermine this very objective. His chapter explores this apparent paradox, arguing that the Taliban desire international legitimacy but only 'on their own terms', unwilling to compromise their doctrine, which is rooted in their distinct religious beliefs. A key observation of this text is that, unlike most other armed groups which struggle to adapt their legitimacy claims when addressing disparate domestic and international audiences, the Taliban seem to have relieved themselves of this burden. Since their inception, they have anchored their legitimating ideas in what they perceive as true Islam and Afghan national tradition and they shun all requirements for international recognition if these are deemed detrimental to their worldview. The group's ideational and performance-based claims to legitimacy have remained remarkably consistent over the years. As the new rulers of Afghanistan since 2021, the Taliban continue to be condemned by the West whereas a silent diplomatic normalization with China, Russia, and neighbouring states takes place on the regional level.

In his contribution, *Wolfram Lacher* focuses on the Libyan Arab Armed Forces (LAAF) and its leader Khalifa Haftar in the Libyan civil war. This chapter argues that a particular set of domestic and international circumstances enabled Haftar and his LAAF in rapidly gaining international legitimacy. First, Haftar benefited from the legitimacy crisis of the national unity government that erupted along with the civil war Haftar himself helped to provoke. Second, the political and military support of regional powers such as Egypt or the United Arab Emirates proved crucial to the advance of the Haftar and his LAAF. In the politics of legitimation, mimicking as a practice of claim-making played a key role. Mimicking the state by establishing territorial control allowed Haftar to claim that he was an indispensable interlocutor for foreign governments, and possibly the future ruler of Libya. Foreign support both enabled the LAAF's territorial expansion and was premised on its continuation. The more Haftar's LAAF expanded, the more convincing another facet of mimicking became, namely the imitation of state security forces and portraying the LAAF as a disciplined, regular army. Foreign governments, in turn, validated Haftar's legitimacy claims by acknowledging and officializing him and the LAAF through red carpet treatment and high-profile official meetings. Haftar's rise to international respectability owed much to the particular international moment in which it occurred. In the historical time of sovereigntism with its increasing multipolarity, the role of newly assertive regional powers proved crucial for Haftar's accreditation on the international stage. Regardless of his dubious human rights record, foreign states continuously pursued the legitimation of the LAAF and Haftar's authoritarian warlord rule.

In their conclusion, *Stephan Hensell* and *Klaus Schlichte* discuss a range of observations based on the conceptual framework, the case studies and the reiterative case comparison of this volume. The conclusion first points to some more general lesson for IR as a discipline by arguing that armed groups should be taken seriously as what they are: actors that co-shape and reproduce international politics. Second, the conclusion presents a number of empirical findings and insights from the case analyses and discusses their conceptual implications as well as open questions. While there is no simple equation that would explain why some armed groups are becoming legitimate but not others, the conclusion points to some typical features, recurring dilemmas and strategies in the politics of legitimation of armed groups. A first set of our observations is related to the politics of legitimation as such and highlights its relational and multilevel character. A second set of observations relates to the strategies of armed groups, namely the use of military and diplomatic means as well as the role of coupling and timing in the politics of armed groups. A final set of observations discusses the role of legitimation in conflict resolution. Lastly, the conclusion gives short retrospect and prospect on armed groups and the social origins of armed rebellion.

In his afterword, *Friedrich Kratochwil*, known for his contribution to practice theory in IR, relates the volume to bigger theoretical issues discussed in IR. He reflects upon the understanding of history in IR in general and in this book, trying to steer a middle-way between the Scylla of outdated teleological ideas and the Charybdis of sacrificing the idea of progress altogether. Much in the vein of this book, he argues for an understanding of theory-building that resembles the game of Scrabble more than sticking to the illusions of grand theory.

References

Adler, Emanuel and Vincent Pouliot, eds. 2011. *International Practices*. Cambridge: Cambridge University Press.

Alles, Delphine and Bertrand Badie. 2016. 'Sovereigntism in the International System: From Change to Split'. *European Review of International Studies* 3 (2): pp. 5–19.

Arjona, Ana, Nelson Kasfir, and Zachariah Mampilly, eds. 2015. *Rebel Governance in Civil War.* Cambridge: Cambridge University Press.

Ayoob, Mohammed. 1995. *The Third World Security Predicament: State Making, Regional Conflict, and the International System.* Boulder, CO: Lynne Rienner Publishers.

Bacon, Michael. 2012. *Pragmatism: An Introduction.* Cambridge: Polity Press.

Bakonyi, Jutta, Stephan Hensell, and Jens Siegelberg, eds. 2006. *Gewaltordnungen bewaffneter Gruppen. Ökonomie und Herrschaft nichtstaatlicher Akteure in den Kriegen der Gegenwart.* Baden-Baden: Nomos.

Barker, Rodney. 2001. *Legitimating Identities. The Self-Presentation of Rulers and Subjects.* Cambridge: Cambridge University Press.

Barnett, Michael. 2011. *Empire of Humanity: A History of Humanitarianism.* Ithaca, NY: Cornell University Press.

Bartelson, Jens. 2014. *Sovereignty as Symbolic Form.* London: Routledge.

Basile, Linda and Oscar Mazzoleni, eds. 2022. *Sovereignism and Populism: Citizens, Voters and Parties in Western European Democracies.* London: Routledge.

Battistella, Dario. 2006. *Retour de l'etat de guerre.* Paris: Armand Colin.

Bauer, Harry and Elisabetta Brighi, eds. 2009. *Pragmatism in International Relations.* Abingdon: Routledge.

Beetham, David. 2013. *The Legitimation of Power.* 2nd edn. Houndmills: Palgrave Macmillan.

Bellamy, Alex J. 2011. *Global Politics and the Responsibility to Protect: From Words to Deeds.* Abingdon: Routledge.

Berdal, Mats and David M. Malone, eds. 2000. *Greed and Grievance: Economic Agendas in Civil Wars.* Boulder, CO: Lynne Rienner Publishers.

Bernstein, Steven. 2011. 'Legitimacy in Intergovernmental and Non-State Global Governance'. *Review of International Political Economy* 18 (1): pp. 17–51.

Best, Jacqueline and Alexandra Gheciu. 2014. 'Theorizing the Public as Practices: Transformations of the Public in Historical Context'. In *The Return of the Public in Global Governance*, edited by Jacqueline Best and Alexandra Gheciu, pp. 15–44. Cambridge: Cambridge University Press.

Bexell, Magdalena, Kristina Jönsson, and Anders Uhlin, eds. 2022. *Legitimation and Delegitimation in Global Governance: Practices, Justifications, and Audiences.* Oxford: Oxford University Press.

Bhatia, Michael V. 2005. 'Fighting Words: Naming Terrorists, Bandits, Rebels and Other Violent Actors'. *Third World Quarterly* 26 (1): pp. 5–22.

Biene, Janusz and Christopher Daase. 2015. 'Gradual Recognition: Curbing Non-State Violence in Asymmetric Conflict'. In *Recognition in International Relations: Rethinking a Political Concept in a Global Context*, edited by Christopher Daase, Caroline Fehl, Anna Geis, and Georgios Kolliarakis, pp. 220–236. Houndmills: Palgrave Macmillan.

Bliesemann de Guevara, Berit, ed. 2012. *Statebuilding and State-Formation: The Political Sociology of Intervention*. Abingdon: Routledge.

Blumer, Herbert. 1954. 'What is Wrong with Social Theory?'. *American Sociological Review* 19 (1): pp. 3–10.

Bob, Clifford. 2005. *The Marketing of Rebellion: Insurgents, Media, and International Activism*. Cambridge: Cambridge University Press.

Boltanski, Luc. 1993. *La souffrance à distance. Morale humanitaire, medias et politique*. Paris: Métailié.

Bos, Michèle and Jan Melissen. 2019. 'Rebel Diplomacy and Digital Communication: Public Diplomacy in the Sahel'. *International Affairs* 95 (6): pp. 1331–1348.

Bourdieu, Pierre. 1977. *Outline of a Theory of Practice*. Cambridge: Cambridge University Press.

Bourdieu, Pierre. 2015. *Sociologie générale, Vol. 1, Cours au Collège de France (1981–1983)*. Paris: Seuil.

Bourdieu, Pierre. 2016. *Sociologie générale, Vol. 2, Cours au Collège de France (1983–1986)*. Paris: Seuil.

Bouris, Dimitris and Irene Fernández-Molina. 2018. 'Contested States, Hybrid Diplomatic Practices, and the Everyday Quest for Recognition'. *International Political Sociology* 12 (3): pp. 306–324.

Bueger, Christian and Frank Gadinger. 2018. *International Practice Theory*. 2nd edn. London: Palgrave Macmillan.

Bukovansky, Mlada, Edward Keene, Christian Reus-Smit, and Maja Spanu, eds. 2023. *The Oxford Handbook of History and International Relations*. Oxford: Oxford University Press.

Burke, Kyle. 2018. *Revolutionaries for the Right: Anticommunist Internationalism and Paramilitary Warfare in the Cold War*. Chapel Hill, NC: The University of North Carolina Press.

Buzan, Barry and George Lawson. 2015. *The Global Transformation: History, Modernity and the Making of International Relations*. Cambridge: Cambridge University Press.

Byrne, Jeffrey J. 2016. *Mecca of Revolution: Algeria, Decolonization, and the Third World Order*. Oxford: Oxford University Press.

Caspersen, Nina. 2012. *Unrecognized States: The Struggle for Sovereignty in the Modern International System*. Cambridge: Polity Press.

Caspersen, Nina and Gareth Stansfield, eds. 2011. *Unrecognised States in the International System*. Abingdon: Routledge.

Charlesworth, Hilary and Jean-Marc Coicaud, eds. 2010. *Fault Lines of International Legitimacy*. Cambridge: Cambridge University Press.

Checkel, Jeffrey T., ed. 2013. *Transnational Dynamics of Civil War*. Cambridge: Cambridge University Press.

Clark, Ian. 2003. 'Legitimacy in a Global Order'. *Review of International Studies* 29: pp. 75–95.

Clark, Ian. 2005. *Legitimacy in International Society*. Oxford: Oxford University Press.

Claude, Inis L. 1966. 'Collective Legitimization as a Political Function of the United Nations'. *International Organization* 20 (3): pp. 367–379.

Coggins, Bridget. 2014. *Power Politics and State Formation in the Twentieth Century: The Dynamics of Recognition*. Cambridge: Cambridge University Press.

Coggins, Bridget L. 2015. 'Rebel Diplomacy: Theorizing Violent Non-State Actors' Strategic Use of Talk'. In *Rebel Governance in Civil War*, edited by Ana Arjona, Nelson Kasfir, and Zachariah Mampilly, pp. 98–118. Cambridge: Cambridge University Press.

Collier, Paul and Anke Hoeffler. 2004. 'Greed and Grievance in Civil War'. *Oxford Economic Papers* 56 (4): pp. 563–595.

Connelly, Matthew. 2002. *A Diplomatic Revolution: Algeria's Fight for Independence and the Origins of the Post-Cold War Era*. Oxford: Oxford University Press.

Cook, Alexander C., ed. 2014. *Mao's Little Red Book: A Global History*. Cambridge: Cambridge University Press.

Cooley, Alexander and Daniel H. Nexon. 2020. 'How Hegemony Ends: The Unravelling of American Power'. *Foreign Affairs* 99 (4): pp. 143–157.

Daalder, Ivo H. and James M. Lindsay. 2003. *America Unbound: The Bush Revolution in Foreign Policy*. Washington DC: Brookings Institution Press.

Daase, Christopher, Anna Geis, Caroline Fehl, and Georgios Kolliarakis, eds. 2015. *Recognition in International Relations: Rethinking a Political Concept in a Global Context*. Houndmills: Palgrave Macmillan.

De Cavalho, Benjamin, Julia Costa Lopez, and Halvard Leira, eds. 2021. *Handbook of Historical International Relations*. London: Routledge.

De Zeeuw, Jeroen, ed. 2008. *From Soldiers to Politicians: Transforming Rebel Movements After Civil War*. Boulder, CO: Lynne Rienner Publishers.

Dellmuth, Lisa, Jan Aart Scholte, Jonas Tallberg, and Soetkin Verhaegen, eds. 2022. *Citizens, Elites, and the Legitimacy of Global Governance*. Oxford: Oxford University Press.

Dingwerth, Klaus, Antonia Witt, Ina Lehmann, Ellen Reichel, and Tobias Weise, eds. 2019. *International Organizations under Pressure: Legitimating Global Governance in Challenging Times*. Oxford: Oxford University Press.

Dogan, Mattei. 1992. 'Conceptions of Legitimacy'. In *Encyclopedia of Government and Politics*, edited by Mary Hawkesworth and Maurice Kogan, pp. 116–126. London: Routledge.

Drieschova, Alena, Christian Bueger, and Ted Hopf, eds. 2022. *Conceptualizing International Practices: Directions for the Practice Turn in International Relations*. Cambridge: Cambridge University Press.

Duyvesteyn, Isabelle, ed. 2019. *Rebels and Legitimacy: Processes and Practices*. London: Routledge.

Duyvesteyn, Isabelle. 2019a. 'Rebels and Legitimacy: An Introduction'. In *Rebels and Legitimacy: Processes and Practices*, edited by Isabelle Duyvesteyn, pp. 1–15. London: Routledge.

Elsayed, Ahmed. 2023. 'From the Mountains to the Élysée: The Precarious International Legitimacy of the Syrian Kurdish YPG and PYD'. *The Middle East Journal* 77 (1): pp. 53–77.

Erman, Eva and Anders Uhlin, eds. 2010. *Legitimacy Beyond the State? Re-examining the Democratic Credentials of Transnational Actors*. London: Palgrave Macmillan.

Ero, Comfort and Richard Atwood. 2023. 'Sudan and the New Age of Conflict: How Regional Power Politics are Fueling Deadly Wars'. *Foreign Affairs*, 26 May. https://www.foreignaffairs.com/sudan/sudan-and-new-age-conflict

Fabry, Mikulas. 2010. *Recognizing States: International Society and the Establishment of New States Since 1776*. Oxford: Oxford University Press.

Fassin, Didier. 2007. 'Humanitarianism as a Politics of Life'. *Public Culture* 19 (3): pp. 499–520.

Fassin, Didier. 2011. *Humanitarian Reason: A Moral History of the Present*. Berkeley: University of California Press.

Fazal, Tanisha M. 2018. *Wars of Law: Unintended Consequences in the Regulation of Armed Conflict*. Ithaca, NY: Cornell University Press.

Fernández-Molina, Irene. 2023. 'The International Recognition of Governments in Practice(s): Creatures, Mirages, and Dilemmas in Post-2011 Libya'. *International Studies Review* 25 (4): pp. 1–27.

Fisch, Jörg. 2010. *Das Selbstbestimmungsrecht der Völker: Die Domestizierung einer Illusion*. München: C. H. Beck.

Franck, Thomas M. 1990. *The Power of Legitimacy among Nations*. New York: Oxford University Press.

Friedrichs, Jörg and Friedrich Kratochwil. 2009. 'On Acting and Knowing: How Pragmatism Can Advance International Relations Research and Methodology'. *International Organization* 63 (4): pp. 701–731.

Freudenschuss, Helmut. 1982. 'Legal and Political Aspects of the Recognition of National Liberation Movements'. *Millennium* 11 (2): pp. 115–129.

Geis, Anna, Frank Nullmeier, and Christopher Daase, eds. 2012. *Der Aufstieg der Legitimitätspolitik: Rechtfertigung und Kritik politisch-ökonomischer Ordnungen*. Baden-Baden: Nomos.

Geis, Anna, Maéva Clément, and Hanna Pfeifer, eds. 2021. *Armed Non-State Actors and the Politics of Recognition*. Manchester: Manchester University Press.

Geldenhuys, Deon. 2009. *Contested States in World Politics*. Houndmills: Palgrave Macmillan.

George, Susan. 2015. *Shadow Sovereigns: How Global Corporations Are Seizing Power*. Cambridge: Polity Press.

Getachew, Adom. 2019. *Worldmaking after Empire: The Rise and Fall of Self-Determination*. Princeton, NJ: Princeton University Press.

Glaser, Barney G. and Anselm L. Strauss. 1967. *The Discovery of Grounded Theory: Strategies for Qualitative Research*. New Brunswick: Aldine Transaction.

Go, Julian and George Lawson, eds. 2017. *Global Historical Sociology*. Cambridge: Cambridge University Press.

Goertz, Gary. 1994. *Contexts of International Politics*. Cambridge: Cambridge University Press.

Griffiths, Ryan D. 2017. 'Admission to the Sovereignty Club: The Past, Present, and Future of the International Recognition Regime'. *Territory, Politics, Governance* 5 (2): pp. 177–189.

Griffiths, Ryan D. 2021. *Secession and the Sovereignty Game: Strategy and Tactics for Aspiring Nations*. Ithaca: Cornell University Press.

Gronau, Jennifer and Henning Schmidtke. 2016. 'The Quest for Legitimacy in World Politics: International Institutions' Legitimation Strategies'. *Review of International Studies* 42 (3): pp. 535–557.

Hehir, Aidan, ed. 2013. *Libya, the Responsibility to Protect and the Future of Humanitarian Intervention*. London: Palgrave Macmillan.

Hellmann, Gunther and Jens Steffek, eds. 2023. *Praxis as a Perspective on International Politics*. Bristol: Bristol University Press.

Hensell, Stephan and Felix Gerdes. 2017. 'Exit from War: The Transformation of Rebels into Post-War Power Elites'. *Security Dialogue* 48 (2): pp. 168–184.

Hensell, Stephan and Klaus Schlichte. 2021. 'The Historical Mapping of Armed Groups' Recognition'. In *Armed Non-State Actors and the Politics of Recognition*, edited by Maéva Clément, Anna Geis, and Hanna Pfeifer, pp. 30–46. Manchester: Manchester University Press.

Herr, Stefanie. 2017. 'National Liberation Movements and the UN: Favour Won and Lost?' In *Resistance and Change in World Politics: International Dissidence*, edited by Svenja Gertheiss, Stefanie Herr, Klaus Dieter Wolf, and Carmen Wunderlich, pp. 109–141. London: Palgrave Macmillan.

Hobsbawm, Eric. 1994. *Age of Extremes: The Short Twentieth Century, 1914–1991*. London: Abacus.

Huang, Reyko. 2016. 'Rebel Diplomacy in Civil War'. *International Security* 40 (4): pp. 89–126.

Hurd, Ian. 2007. *After Anarchy: Legitimacy and Power in the United Nations Security Council*. Princeton, NJ: Princeton University Press.

Hurrelmann, Achim, Steffen Schneider, and Jens Steffek, eds. 2007. *Legitimacy in an Age of Global Politics*. Houndmills: Palgrave Macmillan.

Hutchings, Kimberly. 2008. *Time and World Politics: Thinking the Present*. Manchester: Manchester University Press.

Idrissa, Rahmane. 2021. 'The Sahel: A Cognitive Mapping'. *New Left Review* 132: pp. 5–39.

Iriye, Akira, Petra Goedde, and William I. Hitchcock, eds.. 2012. *The Human Rights Revolution: An International History*. Oxford: Oxford University Press.

Isachenko, Daria. 2012. *The Making of Informal States: State-building in Northern Cyprus and Transdniestria*. London: Palgrave Macmillan.

Jackson, Robert. 2007. *Sovereignty: Evolution of an Idea.* Cambridge: Polity Press.

Jenne, Erin K. 2021. 'Populism, Nationalism, and Revisionist Foreign Policy'. *International Affairs* 97 (2): pp. 323–343.

Jo, Hyeran. 2015. *Compliant Rebels: Rebel Groups and International Law in World Politics.* Cambridge: Cambridge University Press.

Jones, Benjamin T. and Eleonora Mattiacci. 2019. 'A Manifesto, in 140 Characters or Fewer: Social Media as a Tool of Rebel Diplomacy'. *British Journal of Political Science* 49 (2): pp. 739–761.

Jumbert, Maria Gabrielsen and David Lanz. 2013. 'Globalised Rebellion: The Darfur Insurgents and the World'. *The Journal of Modern African Studies* 51 (2): pp. 193–217.

Jung, Dietrich, Klaus Schlichte, and Jens Siegelberg. 2003. *Kriege in der Weltgesellschaft: Strukturgeschichtliche Erklärung kriegerischer Gewalt (1945–2000).* Opladen: VS Verlag für Sozialwissenschaften.

Kalyvas, Stathis N. 2006. *The Logic of Violence in Civil War.* Cambridge: Cambridge University Press.

Kalyvas, Stathis N. 2018. 'Jihadi Rebels in Civil War'. *Daedalus* 147 (1): pp. 36–47.

Kanet, Roger, ed. 1975. *The Soviet Union and the Developing Nations.* Baltimore, ML: John Hopkins University Press.

Kaplan, Abraham. 1964. *The Conduct of Inquiry: Methodology for Behavioral Science.* San Francisco, CA: Chandler Publishing.

Keen, David. 2006. *Endless War?: Hidden Functions of the 'War on Terror'.* London: Pluto Press.

Koloma Beck, Teresa and Tobias Werron. 2018. 'Violent Conflictition: Armed Conflicts and Global Competition for Attention and Legitimacy'. *International Journal of Politics, Culture, and Society* 31 (3): pp. 275–296.

Koselleck, Reinhart. 2004. *Futures Past: On the Semantics of Historical Time.* New York: Columbia University Press.

Krasner, Stephen, ed. 1983. *International Regimes.* Ithaca, NY: Cornell University Press.

Krauthammer, Charles. 1990. 'The Unipolar Moment'. *Foreign Affairs* 70 (1): pp. 23–33.

Lauterpacht, Hersch. 1948. *Recognition in International Law.* Cambridge: Cambridge University Press.

Lawson, George, Chris Armbruster, and Michael Cox, eds. 2010. *The Global 1989: Continuity and Change in World Politics.* Cambridge: Cambridge University Press.

Lindemann, Thomas and Erik Ringmar, eds. 2012. *The International Politics of Recognition.* Boulder, CO: Paradigm Publishers.

Lovell, Julia. 2019. *Maoism: A Global History.* New York: Alfred A Knopf, Inc.

Lüthi, Lorenz M. 2020. *Cold Wars: Asia, the Middle East, Europe.* Cambridge: Cambridge University Press.

MacFarlane, S. Neil. 1985. *Superpower Rivalry and Third World Radicalism: The Idea of National Liberation.* Baltimore, CO: The John Hopkins University Press.

Mahoney James and Dietrich Rueschemeyer, eds. 2003. *Comparative Historical Analysis in the Social Sciences.* Cambridge: Cambridge University Press.

Malejacq, Romain. 2017. 'From Rebel to Quasi-State: Governance, Diplomacy and Legitimacy in the Midst of Afghanistan's Wars (1979–2001)'. *Small Wars & Insurgencies* 28 (4): pp. 867–886.

Mamdani, Mahmood. 2004. *Good Muslim, Bad Muslim: America, the Cold War, and the Roots of Terror*. New York: Pantheon.

Mampilly, Zacharia. 2011. *Rebel Rulers: Insurgent Governance and Civilian Life during War*. Ithaca, NY: Cornell University Press.

Manela, Erez. 2007. *The Wilsonian Moment: Self Determination and the International Origins of Anticolonial Nationalism*. Oxford: Oxford University Press.

Mbembe, Achille. 2023. 'Les putschs en Afrique de l'Ouest annoncent la fin d'un cycle qui aura duré près d'un siècle'. *Le Monde*, 4 August, p. 4.

McConnell, Fiona. 2017. 'Liminal Geopolitics: The Subjectivity and Spatiality of Diplomacy at the Margins'. *Transactions* 42 (1): pp. 139–152.

McConnell, Fiona, Terri Moreau, and Jason Dittmer. 2012. 'Mimicking State Diplomacy: The Legitimizing Strategies of Unofficial Diplomacies'. *Geoforum* 43 (4): pp. 804–814.

McMahon, Robert J., ed. 2013. *The Cold War in the Third World*. Oxford: Oxford University Press.

Mello, Patrick A. 2021. *Qualitative Comparative Analysis: An Introduction to Research Design and Application*. Washington, DC: Georgetown University Press.

Minakov, Mikhail. 2021. 'The Sovereigntist Turn: Sovereignty as a Contested Concept Again'. *The Ideology and Politics Journal* 17 (1): pp. 87–114.

Modelski, George. 1964. 'The International Relations of Internal War'. In *International Aspects of Civil Strife*, edited by James N. Rosenau, pp. 14–44. Princeton, NJ: Princeton University Press.

Moses, A. Dirk, Marco Duranti, and Roland Burke, eds. 2020. *Decolonization, Self-Determination, and the Rise of Global Human Rights Politics*. Cambridge: Cambridge University Press.

Norton, Agustus R. and Martin H. Greenberg. 1989. *The International Relations of the Palestine Liberation Organization*. Carbondale: Southern Illinois University Press.

Paris, Roland and Timothy D. Sisk, eds. 2009. *The Dilemmas of Statebuilding: Confronting the Contradictions of Postwar Peace Operation*. London: Routledge.

Parrott, R. Joseph and Mark A. Lawrence, eds. 2022. *The Tricontinental Revolution: Third World Radicalism and the Cold War*. Cambridge: Cambridge University Press.

Pegg, Scott. 1998. *International Society and the De Facto State*. Aldershot: Ashgate.

Perritt, Henry Jr. 2008. *Kosovo Liberation Army: The Inside Story of an Insurgency*. Urbana, IL: University of Illinois Press.

Phillips, David. 2012. *Liberating Kosovo: Coercive Diplomacy and U.S. Intervention*. Cambridge, MA: MIT Press.

Popitz, Heinrich. 2017. *Phenomena of Power: Authority, Domination and Violence*. New York: Columbia University Press.

Reno, William. 2011. *Warfare in Independent Africa*. Cambridge: Cambridge University Press.

Reus-Smit, Christian. 2007. 'International Crises of Legitimacy'. *International Politics* 44: pp. 157–174.
Robinson, Piers. 2002. *The CNN Effect: The Myth of News, Foreign Policy and Intervention*. London: Routledge.
Roy, Olivier. 2004. *Globalized Islam: The Search for a New Ummah*. New York: Columbia University Press.
Ryan, David and Victor Pungong, eds. 2000. *The United States and Decolonization: Power and Freedom*. Houndmills: Palgrave Macmillan.
Salehyan, Idean. 2009. *Rebels Without Borders: Transnational Insurgencies in World Politics*. Ithaca, NY: Cornell University Press.
Saward, Michael. 2010. *The Representative Claim*. Oxford: Oxford University Press.
Schlichte, Klaus. 2009. *In the Shadow of Violence: The Politics of Armed Groups*. Frankfurt am Main: Campus; Chicago: University of Chicago Press.
Schlichte, Klaus and Stephan Stetter, eds. 2023. *The Historicity of International Politics: Imperialism and the Presence of the Past*. Cambridge: Cambridge University Press.
Schlichte, Klaus and Ulrich Schneckener. 2015. 'Armed Groups and the Politics of Legitimacy'. *Civil Wars* 17 (4): pp. 409–424.
Shook, John R. 2023. *Pragmatism*. Cambridge, MA: MIT Press.
Seurat, Leila. 2015. *Le Hamas et le monde (2006–2015): La politique étrangère du mouvement islamiste palestinien*. Paris: CNRS Éditions.
Sommerer, Thomas, Hans Agné, Fariborz Zelli, and Bart Bes, eds. 2022. *Global Legitimacy Crises: Decline and Revival in Multilateral Governance*. Oxford: Oxford University Press.
Spiro, Peter J. 2000. 'The New Sovereigntists: American Exceptionalism and Its False Prophets'. *Foreign Affairs* 79 (6): pp. 9–15.
Swedberg, Richard. 2014. *The Art of Social Theory*. Princeton, NJ: Princeton University Press.
Tallberg, Jonas, Karin Bäckstrand, and Jan Aart Scholte, eds. 2018. *Legitimacy in Global Governance: Sources, Processes, and Consequences*. Oxford: Oxford University Press.
Tarde, Gabriel. 1903. *The Laws of Imitation*. New York: Henry Holt and Company.
Thomas, Scott. 1996. *The Diplomacy of Liberation: The Foreign Relations of the ANC since 1960*. London: IB Tauris.
Thompson, Peter G. 2014. *Armed Groups: The 21st Century Threat*. Lanham, ML: Rowman & Littlefield.
Ticktin, Miriam. 2014. 'Transnational Humanitarianism'. *Annual Review of Anthropology* 43: pp. 273–289.
van Ommen, Eline. 2023. *Nicaragua Must Survive: Sandinista Revolutionary Diplomacy in the Global Cold War*. Oakland, CA: University of California Press.
Viola, Lora Anne. 2020. *The Closure of the International System: How Institutions Create Political Equalities and Hierarchies*. Cambridge: Cambridge University Press.
Visoka, Gëzim, John Doyle, and Edward Newman, eds. 2020. *Routledge Handbook of State Recognition*. London: Routledge.

Wallerstein, Immanuel. 2022. *The Global Left: Yesterday, Today, Tomorrow*. New York: Routledge.

Waltz, Kenneth M. 1979. *Theory of International Politics*. Reading: Addison-Wesley.

Weber, Max. 1978. *Economy and Society: An Outline of Interpretive Sociology*, edited by Guenther Roth and Claus Wittich, Berkeley: University of California Press.

Weinstein, Jeremy M. 2006. *Inside Rebellion: The Politics of Insurgent Violence*. Cambridge: Cambridge University Press.

Westad, Odd A. 2007. *The Global Cold War: Third World Interventions and the Making of Our Times*. Cambridge: Cambridge University Press.

Western, Jon. 2002. 'Sources of Humanitarian Intervention: Beliefs, Information and Advocacy in the U.S. Decisions on Somalia and Bosnia'. *International Security* 26 (4): pp. 112–142.

Wheeler, Nicholas J. 2000. *Saving Strangers: Humanitarian Intervention in International Society*. Oxford: Oxford University Press.

Wight, Martin. 1972. 'International Legitimacy'. *International Relations* 4 (1): pp. 1–28.

Wills, Jane and Robert W. Lake, eds. 2020. *The Power of Pragmatism: Knowledge Production and Social Inquiry*. Manchester: Manchester University Press.

Zaum, Dominik, ed. 2013. *Legitimating International Organizations*. Oxford: Oxford University Press.

Zürn, Michael. 2018. *A Theory of Global Governance: Authority, Legitimacy, and Contestation*. Oxford: Oxford University Press.

2

Decolonization and the Cold War

The Superpowers and the Anti-colonial Insurgencies in Indonesia and Vietnam

Robert J. McMahon

During their four-and-a-half-decade-long contest 'for the soul of mankind', as President George H. W. Bush (2004, 1) so aptly described it, the United States and the Soviet Union frequently faced the dilemma of whether to recognize or to confer legitimacy upon armed insurgencies vying to overturn established civil authority.[1] That titanic struggle, at once ideological and geopolitical, came to encompass the entire globe by the 1950s, with the Third World emerging as the chief zone of competition between the two superpowers—as well as the main source and site of non-state armed groups. The potential transformations of these armed groups into governments threatened to alter the prevailing balance of regional and global power. Policymakers in Washington and Moscow, consequently, needed to calculate carefully how the granting of de jure or de facto recognition to a particular armed insurgency, or the accordance of some form of legitimacy to it, might affect broader Cold War goals.

Cold War era armed insurgencies took different forms. Many were anti-colonial rebellions that sought independence from long-time colonial powers. Others, as in Greece, Cuba, Nicaragua, Somalia, and Ethiopia, are best labelled as civil wars or insurrections against unpopular ruling governments. Still others, as in Hungary and Afghanistan, were revolts against externally imposed regimes. All raised the twinned issues of recognition and legitimacy.

Why did the United States and the Soviet Union, this chapter asks, choose to recognize some insurgencies, deny recognition to others, and actively oppose still others? How, for their part, did these various armed, non-state actors seek to gain support, recognition, and legitimacy from one, or both, of the superpowers? What factors best explain their relative successes or failures in those endeavours?

The present chapter explores those larger questions by focusing specifically on armed decolonization movements, arguably the most common and most consequential of the non-state, armed groups that emerged throughout the post-1945 period. The dismantling of the colonial empires transpired over three decades, principally

[1] This chapter has been reproduced with adaptations from Robert J. McMahon, 2023. 'Decolonization and the Cold War'. *Journal of Global Strategic Studies* 3 (2): pp. 3–21. Reprinted with permission.

Robert J. McMahon, *Decolonization and the Cold War*. In: *Armed Groups and the Politics of International Legitimation*. Edited by: Stephan Hensell and Klaus Schlichte, Oxford University Press. © Robert J. McMahon (2025).
DOI: 10.1093/9780198974529.003.0002

between 1945 and 1975, ranging from the Asian and African continents to the Middle East and the Caribbean. Many of those independence struggles played out peacefully, to be sure, as imperial powers and indigenous nationalists bargained over the precise terms and timing of self-rule. Such was the case in the Philippines, India, Burma, Ceylon, Malaya, Jamaica, and many former British and French dependencies in Africa. Other struggles devolved into open warfare between colonial military forces and indigenous armed groups. Those will be the focus here.

Although wholly distinct from the Cold War in its origins and consequences, decolonization coincided temporally with the superpower rivalry and was profoundly shaped by it.[2] I will explore the intersection between decolonization and the Cold War by focusing on two of the most significant anti-colonial rebellions, those in Indonesia and Vietnam. In each of those long-colonized territories, Washington and Moscow sought to shape the ultimate outcome of the independence struggle, wielding as diplomatic weapons the prospect of diplomatic recognition or economic-military support—or the withholding of the same. In each case, the aspiring national liberation movement deployed a combination of armed strength and diplomatic advocacy in a bid for legitimacy and recognition, seeing the support of one, or both, of the superpowers as instrumental to the overarching goal of full-fledged sovereignty and acceptance within the prevailing international state-based system. Indeed, the central argument of this chapter holds that it was that shrewd admixture of effective fighting strength *and* adroit diplomatic outreach that eventually led to the success of the independence movements in both Indonesia and Vietnam. The insurgencies not only employed military means but also presented themselves as states-in-waiting and future governments. Or in the theoretical vocabulary of this book: they relied on *fighting* and *mimicking* as practices of claim-making while reaching out to international audiences and potential constituencies in the West and East in their quest for international legitimacy.[3]

Anti-colonial rebellions erupt in Southeast Asia

Southeast Asia emerged as the first front in the anti-colonial wars of insurgency. In the days and weeks that followed World War II's fiery denouement, anti-colonial nationalists in the Dutch East Indies and French Indochina boldly acted to shake off the shackles of colonialism by unilaterally declaring independence. On 17 August 1945, Sukarno and Mohammed Hatta, nationalist leaders who had long opposed Dutch colonial rule, proclaimed an independent Republic of Indonesia. Just over two weeks later, legendary Vietnamese patriot Ho Chi Minh announced the birth of an independent Democratic Republic of Vietnam. In each country, longstanding and well-organized independence movements acted swiftly to capitalize on the power

[2] On the historical times of decolonization and the Cold War and their role for the legitimation of armed groups see the Introduction to this volume.

[3] On these practices and publics of legitimation, see the Introduction to this volume.

vacuums opened by the sudden surrender of Japan's occupying forces. Prominent, charismatic leaders and their ardent followers in both colonies sought to turn brash proclamations of independence into reality, utilizing a mix of diplomatic outreach and military pressure (McMahon 1999, 19–25).

Sukarno almost immediately appealed to US president Harry S. Truman for support, requesting American mediation in the anticipated fight with a Dutch government eager to restore imperial rule. He hoped that the persistent American anti-colonial pronouncements of the wartime years might translate into sympathy for Indonesia's aspirations. Ho, for his part, sent a series of personal letters to the new US chief executive in which he pleaded for US support and recognition. Neither man received the response he craved. Secretary of State James F. Byrnes informed Sukarno that a request for mediation could only come from the 'territorial sovereign'; and no member of the US government even responded to Ho's plaintive missives. Instead, the United States adopted a policy of formal neutrality towards the brewing colonial disputes in Indonesia and Indochina and called for negotiations between the contending parties. On 19 December 1945, the US State Department issued a formal statement that acknowledged the territorial sovereignty of the Netherlands over the East Indies islands while denying the Indonesian Republic's quest for status as an equal party to the dispute.[4]

Throughout the next year-and-a-half, as armed clashes between colonizer and colonized intensified, death tolls mounted, and negotiations stalled, the Truman administration urged the respective contending parties to pursue peaceful resolutions of these colonial disputes. At the same time, the United States declared its strict neutrality towards each of them. By so doing, Washington was actually conferring a degree of legitimacy on the two independence movements by acknowledging them as legitimate parties to the conflict.[5] One does not call for negotiations with terrorists or bandits, after all. Clearly, Washington never placed Indonesian and Vietnamese fighters and diplomats in those categories, nor did it ever explicitly consider them to be *illegitimate* aspirants to power. Indeed, US officials themselves, including former president Franklin D. Roosevelt, had long called for an end to colonialism and the eventual evolution towards indigenous self-rule across the colonial world. Acutely conscious of the burgeoning support for the fledgling Indonesian and Vietnamese republics among the peoples and governments of the emerging Afro-Asian states, the Truman administration instead sought to find a middle ground: maintaining solidarity with key Western allies while straining to avoid alienating Third World anti-colonialist sensitivities or blatantly contradicting its own professed principles. State Department Asian expert and ambassador to the Netherlands Stanley K. Hornbeck reflected that, in attempting to address the decolonization conundrum: 'We in effect attempted to support neither side, and yet favored one and hoped not unduly to offend the other' (Hornbeck 1948, 132–133).

[4] James F. Byrnes. Letter to Consul General Walter Foote, 31 October 1945. Document 856E.00/10-2845. US Department of State Records, Record Group 59. College Park, MD: US National Archives; US Department of State Bulletin, 13 (23 December 1945): 1021–1022.

[5] On acknowledging as a practice of claim-validation, see the Introduction to this volume.

Indonesia's struggle for independence

Indonesia's leaders actively, and shrewdly, courted support from the United States and from the wider international community—their principal audiences. Diplomatic outreach thus formed a core part of their strategy for achieving recognition and legitimacy—and, ultimately, full independence. From the outset of the independence struggle, republican leaders vacillated between two alternative policies: *diplomasi* or *perdjuangan* (struggle). Such spokesmen as Sukarno, Hatta, and Sutan Sjahrir believed that the military, organizational, and ideological weaknesses of the infant republic necessitated a predominantly diplomatic approach. Broadly conceived, their view was that the republic must seek to conclude a negotiated settlement with the Dutch which would be guaranteed by the great powers; everything else would have to be subordinated to that central goal. Once a settlement had been attained, they could then turn to the critical task of transforming and democratizing Indonesian society. The logic of diplomacy thus demanded that any radical or revolutionary tendencies within the nationalist movement be suspended or suppressed to appease international opinion. The republic, according to this calculation, had to demonstrate that it was a responsible, capable, and stable government. By promising to protect foreign properties and displaying a willingness to welcome foreign investments—as it did in its widely circulated political manifesto of 1 November 1945—the Republic of Indonesia vied to project an image of moderation and responsibility to the world. It was, in effect, staking a claim to international legitimacy as a respectable and trustworthy state-in-waiting.[6]

The risk inherent in a policy of *diplomasi* was that the attainment of international recognition, if the policy succeeded, might occur at the expense of internal social reform. *Perdjuangan*, the alternative strategy, based its hopes instead on the vigour and revolutionary potential of the national freedom movement. Its chief advocate, popular communist leader Tan Malaka, a decades-long veteran of the anti-colonial struggle, believed that true independence could never be bestowed as a gift from the Western powers. Only through armed struggle could independence be won, a fundamental transformation of Indonesian society effected, and a cohesive, organized sense of purpose created out of the formless national movement (McMahon 1981, 104–106).

At key junctures, those competing tendencies within the nationalist movement complemented each other. Yet the preference of senior republican representatives for diplomatic outreach to the West as the most efficacious strategy became clear before the end of 1945 and permeated all aspects of governmental policy over the next four years. On 19 December 1945, the State Department released its first official statement on the Dutch–Indonesian conflict, expressing disappointment with the breakdown of negotiations between the two parties and calling for 'a peaceful settlement recognizing alike the natural aspirations of the Indonesian peoples and

[6] 'The Political Manifesto of the Republic of Indonesia', in Wolf (1948, 172–175).

the legitimate rights and interests of the Netherlands'.[7] Although falling well short of the Indonesian leadership's hope that the United States would acknowledge the republic's status as an equal party to the dispute, Prime Minister Sutan Sjahrir heaped praise on President Truman in a personal Christmas telegram. 'The last message of good will and encouragement issued by your State Department', Sjahrir wrote, 'has given us great comfort in the struggle we are waging to establish freedom, justice and democracy in Indonesia'. He added: 'We look to you, as the head of a country that has always been in the forefront of the fight for liberty, justice, and self-determination, to use the benefit of your influence to stop the present bloodshed in Indonesia.'[8]

Over time, Cold War priorities compelled the United States to accord ever more legitimacy to the Indonesian Republic. A Dutch 'police action' of July 1947, aimed at eliminating the republic as a viable entity, failed to suppress Indonesian guerrilla resistance while bringing forth widespread international condemnation of Dutch aggression. Days after the commencement of the offensive, Prime Minister Amir Sjarifuddin broadcast a statement from the republican capital of Jogjakarta, calling for a halt to the bloodshed and appealing to the republic's friends in India, China, Europe, and the United States to intervene 'quickly and effectively'. The aim of the Dutch military action, he stated flatly, was the complete liquidation of the republic and the restoration of the old colonial rule 'under the cover of nice words and phrases'.[9]

Washington carefully monitored the international and domestic response to the renewed fighting. A State Department memorandum for Secretary of State George Marshall reported that the Dutch attack 'started a wave of sympathy for the Indonesians which has swept around the world'.[10] Popular correspondent Theodore H. White thundered in the *New Republic*: 'Men were once more being killed for the sin of seeking freedom' (White 1947, 8). Added the *New York Times*, in its first anti-Dutch editorial: 'This action cannot be interpreted otherwise, therefore, than as an effort by the Dutch government to impose by arms what it believed it was not going to gain by negotiation and which it did not choose to submit to arbitration.'[11]

President Sukarno appointed ex-prime minister Sjahrir as an ambassador-at-large to plead the Indonesian case before the UN Security Council. His first stop enroute to New York was New Delhi, where Jawaharlal Nehru, India's interim prime minister, expressed his outrage at the Dutch assault and promised unqualified Indian support for the republic. After meeting with Sjahrir on 24 July, Nehru gave a forceful indication of the growing self-consciousness and solidarity among the former colonial territories of Asia. 'What has become of the U.N. Charter', he asked? 'The spirit of the new Asia will not tolerate such things. No European country, whatever it may be, has any business to set its army in Asia against the people of Asia. When it does so, Asia will not tolerate it.' The leader of about-to-be-independent India vowed to

[7] US Department of State Bulletin, 13 (23 December 1945): 1021–1022.

[8] Sutan Sjahrir. Telegram to President Harry S. Truman, 25 December 1945. *Foreign Relations of the United States* (hereafter FRUS), 6: p. 1186.

[9] Sjarifuddin broadcast, quoted in *New York Times*, 24 July 1947, p. 3.

[10] Secretary of State's Weekly Summary, 18 August 1948, Record Group 59, US Department of State Records.

[11] Nehru statement, quoted in *New York Times*, 22 July 1947, p. 2.

call formally for Security Council consideration of the Dutch–Indonesian dispute.[12] US officials, who clung to the hope that they would not be forced to choose between their need to maintain warm relations with their European allies and their determination to forge close ties with the newly emerging nations of the Third World, grew alarmed.

In line with their broader diplomatic strategy, Indonesians kept up the pressure on what they saw as their most important audience in the West. On 25 July, Sukarno broadcast an urgent appeal to the United States, beseeching its help to bring about a peaceful settlement. 'Just as your American ancestors fought 170 years ago for your liberty and independence', he declared, 'so are we Indonesians fighting for ours. Just as you then rebelled against domination by a country far across the seas, so are we.' He implored Americans 'to stand by the principles of justice and right for which you fought so valiantly only two hundred years ago'.[13]

The urbane Sjahrir presented Indonesia's case for independence to the Security Council with an effective blend of eloquence and passion. During subsequent council debates, the Soviet Union championed the Indonesian cause while berating the Netherlands. So too did Australia and many non-Western states, thereby imposing a nettlesome dilemma on US decision makers. The debates showcased the breadth of Indonesia's support in its bid for international legitimacy and recognition. US diplomats could not afford to ignore that reality without running the risk of alienating the growing number of countries backing Indonesia's aspirations. To head off formal UN intervention, the United States offered to serve as a mediator. When Indonesian officials made known their preference for UN over US mediation, however, Washington shifted course. It acquiesced to a Security Council resolution that called for a cease-fire and the creation of a three-member UN Good Offices Committee (GOC), which came to include the United States, to help implement the cease-fire and foster renewed negotiations (McMahon 1981, 173–191).

Citizen-diplomat Frank Porter Graham, president of the University of North Carolina and Truman's appointee as US representative to the GOC, played a pivotal role in negotiating the Renville Agreement of January 1948, signed onboard a US naval vessel docked in the Java Sea. Shortly after arriving in the Dutch East Indies, Graham informed Secretary of State Marshall that he believed the Indonesian Republic to be the 'rallying center of the largest, ablest and most dedicated single group of Indonesians in this struggle for independence'. The present leadership, he emphasized, 'seems as moderate, reasonable and responsive to Western ideas of any likely to arise in the future'.[14]

When Dutch–Indonesian negotiations faltered again in December 1947, and another military action appeared imminent, the State Department drafted a compromise agreement that, while unfavourable in many respects to the republic, averted

[12] Quoted in Henry F. Grady. Telegram from US Ambassador to India to the State Department, 26 July 1947 (FRUS), 6: pp. 990–991.

[13] Sukarno broadcast, quoted in *New York Times*, 25 July 1947, p. 3.

[14] Frank Porter Graham. Telegram to Secretary of State George C. Marshall, 31 December 1947. Document 501. BC-Indonesia/2-3147. US Department of State Records.

a military clash and facilitated a tentative negotiated settlement. In a crucial meeting with Indonesian leaders, Graham urged them to sign what became the Renville Agreement because he said it would at least ensure the continued existence of the republican government. 'You are what you are,' he famously stated, 'and that is what you will remain.' In other words, he was assuring them that signing the agreement would not adversely affect their current status as the de facto government in much of Java and Sumatra—the two islands that contained approximately ninety per cent of the total population of the archipelago. That pledge won them over, despite substantial domestic resistance that soon led Prime Minister Sjarifuddin to fall from power.[15]

Officially, the US position towards the Dutch–Indonesian struggle remained one of nominal neutrality, and American participation on the GOC helped to bolster Washington's image as an impartial and even-handed mediator. In reality, though, American actions consistently tilted towards the Dutch. US policies and statements at the UN and mediation efforts on the GOC invariably worked to the advantage of The Hague, as most informed observers quickly realized. When Marshall Plan aid began to flow to the Netherlands in 1948, moreover, the United States was placed in the anomalous position of serving on a UN commission as a nominally impartial arbitrator at a time when its financial assistance was at least indirectly financing the Netherlands' aggressive policy.

Its pro-Dutch orientation remained dominant until the early months of 1949, in the immediate aftermath of a second Dutch military offensive. The rehabilitation and reintegration of Western Europe and the corresponding desire to present a solid front against the Soviet Union led the United States to support the Dutch position in virtually all negotiations between them and the Indonesians. American support was not uncritical, of course; Washington strongly advised the Dutch against resorting to military force before both military offensives and continually urged the Netherlands to reach an equitable settlement with the Indonesian nationalists. It was pressure from Washington, in fact, that led the Netherlands reluctantly to accept the Renville settlement.

But the United States never pushed the Dutch too hard. The success of the containment strategy, the European Recovery Program, and the incipient North Atlantic Treaty Organization necessitated the steadfast support of Western European nations, including, of course, the Netherlands. The intensification of the Cold War during those years underscored this need. American officials, moreover, tended to view Indonesia as an essential cog in the economic vitality of the metropole. It would contribute to the economic health of the Netherlands, which in turn would contribute to the economic health of Western Europe. Indeed, the Marshall Plan operated on

[15] United Nations. Summary Records of 60th and 61st Meetings of the GOC, 12 and 13 January 1948. Document UN/S/AC.10/SR.60. UN Library, New York: United Nations Records. Hadji A. Salim. Telegram to Frank Porter Graham, 19 June 1948. In: Indonesia Files, Frank Porter Graham Papers. Chapel Hill, NC: University of North Carolina Library.

the assumption that the European imperial powers would continue to draw strength from their colonies in Southeast Asia (McMahon 1981, 206–250).

In the wake of the second Dutch 'police action', launched in December 1948, American policy changed drastically. The nearly universal denunciation of The Hague's aggression by the international community and widespread support for Indonesian independence merged with similar sentiment in the American public and Congress to bring the issue to a head. Events within Indonesia, moreover, exposed Dutch policy as hopelessly flawed. Indonesian guerrilla forces fought the Dutch army to a standstill, while the Dutch-sponsored non-republican states on the outer islands, which they were grooming to be pliable puppets in a thinly disguised divide-and-rule tactic, summarily abandoned the Netherlands and moved to join forces with the republic. With that political shift by the outer islanders, the Republic of Indonesia demonstrated beyond any doubt that it had achieved strong domestic legitimacy across the entire archipelago. In the eyes of the overwhelming majority of Indonesians—on Sulawesi, Borneo, and Bali as well as on the far more populous islands of Java and Sumatra—it was now accepted as the sole, legitimate representative of Indonesian nationalism and the emerging Indonesian state.

Ironically, it was primarily European rather than Asian considerations that led the Truman administration to apply mounting pressure on the still-recalcitrant Dutch government to accept Indonesian independence. As a result of Dutch intransigence, Congress placed substantial pressure on the administration to move in that direction by threatening to cut off all funds to the Marshall Plan and to hold up passage of the then-pending Atlantic Pact. Those programs, which lay at the heart of the administration's Cold War strategy, were far too vital to be jeopardized by a colonial war in Indonesia—a conflict that to most senior American policy experts was an annoying sideshow (McMahon 1981, 251–294).

In the end, the Indonesians' strategy of combining military action with diplomatic overtures in their independence struggle proved highly effective. Worried that their Dutch ally now represented a greater threat to peace and order than the decidedly moderate nationalists did, and that the Soviet Union and local communists would gain influence if the fighting continued, the United States pressured the Netherlands to take immediate steps that would lead to genuine independence. Secretary of State Dean Acheson, in a climactic meeting with Dutch Foreign Minister Dirk U. Stikker, on 31 March 1949, made it clear that anything less might force the United States to withhold economic assistance that the Netherlands desperately needed.[16] That policy shift led, just days before the decade's end, to the formal transfer of authority from the Netherlands to a sovereign Indonesian state. The United States became one of the first nations to confer formal recognition on the new Indonesian government, its abrupt shift to a pro-Indonesian policy having profoundly affected the timing and scope of that outcome.

[16] Memorandum by Acheson of a conversation with Stikker, 31 March 1949, FRUS, 4: 258–261; Wiebes and Zeeman (1992, 64–65).

Vietnam's jagged path to independence

Ultimately, US policy towards the Vietnamese national liberation movement charted a strikingly different path. A crucial difference derived from the divergent ideological character of the respective insurgencies. From the first, the Indonesian independence movement was led by moderate, non-communist nationalists. The non-communist bona fides of the Sukarno-led government were proven, in the eyes of US analysts, by its bloody suppression of an internal communist revolt at Madiun, East Java, in September 1948. In Vietnam, on the other hand, the outstanding nationalist figure—Ho Chi Minh—was a decades-long communist, former resident of Moscow, and veteran Comintern agent. Worried that if Ho prevailed over the French it would lead to 'an independent Vietnam State which would be run by orders from Moscow', the State Department acknowledged, in September 1948, that 'we are all too well aware of the unpleasant fact that Communist Ho Chi Minh is the strongest and perhaps ablest figure in Indochina and that any suggested solution that excludes him is an expedient of uncertain outcome.'[17]

Ho and the Viet Minh insurgency he helmed faced the same basic challenge as their counterparts in Indonesia: How best to achieve legitimacy and recognition? What blend of armed resistance and diplomatic outreach would prove most salutary to the goal of full sovereignty and independence? After fruitless negotiations with France broke down in the summer of 1946, the Viet Minh had little choice but to pursue the path of armed resistance to an obstinate colonial overlord. In November, a brutal French naval bombardment of Haiphong claimed more than 6,000 Vietnamese lives. Ho Chi Minh and his supporters fled Hanoi, the French moved quickly to establish administrative control in the north, and the Viet Minh mobilized for guerrilla warfare. Conflict soon engulfed much of Vietnam. It would rage unabated for the next six-and-a-half years (Hess 1987, 197–204).

Yet the Vietnamese did not forsake diplomacy. Diplomatic outreach continued to serve as an indispensable adjunct to the ongoing contest of arms. Rather than immediately turning to the Soviet Union, as ideological affinity might have seemed to dictate, the Democratic Republic of Vietnam (DRV) initially hedged between the West and the East, still hoping that it might draw some degree of sympathy and support from a United States proud and protective of its anti-colonial heritage. Indeed, in 1948, the ruling Indochina Communist Party instructed cadres to refrain from criticizing the United States in public statements and to adopt a posture of neutrality towards the Cold War struggle between the superpowers. 'The foreign policy of our government', it stipulated, should not turn against the Americans 'or act in any way so as to incur their animosity. When it comes to public matters, it is formally prohibited to write, in any document, newspaper or book, one single word or one single line capable of incurring harmful repercussions on the foreign policy of our government in terms of its relations with the United States of America.'[18]

[17] US Department of State Policy Statement on Indochina, 27 September 1948, FRUS, 6: 48.
[18] Quoted in Logevall (2013, 225).

Until 1950, Ho always took care to strike a neutral pose, a balancing strategy that deepened Soviet leader Josef Stalin's suspicion that the Vietnamese leader was more of a nationalist than a communist—not unlike Yugoslavia's Josip Broz Tito. Yet Ho, for his part, reasoned that reaching out to international audiences and potential constituencies in both the West and the communist world best served the nationalist cause (Logevall 2013, 224–226; Gaiduk 1996).

The Truman administration's gravitation towards open support for the French by 1949 helped push Ho into the arms of the communist powers. With consummate cynicism, France propped up the weak and pliable ex-emperor Bao Dai as an indigenous alternative to Viet Minh rule. Despite deep misgivings, American policy makers accepted the ruse as the only available alternative to a communist-dominated Vietnam. That would have constituted an intolerable outcome for an administration that increasingly saw communism as a monolithic movement and greatly feared the geostrategic and ideological repercussions of Mao Zedong's recent triumph in the Chinese civil war (Hess 1978).

America's shift from formal neutrality to open embrace of the French colonialists drew the DRV ever closer to Stalin's Soviet Union and (after October 1949) Mao's China. In January 1950, the Vietnamese leader made a secret trip to Beijing in search of Chinese diplomatic backing and material aid. He then travelled to Moscow the next month, at the very moment that the landmark Sino–Soviet Treaty was being finalized, and personally beseeched Mao and Stalin for military assistance. Weeks earlier, first Beijing and then Moscow had extended formal recognition to the Democratic Republic of Vietnam. During the three-way meetings in Moscow between Stalin, Mao, and Ho, the former made clear his preoccupation with European issues and his lingering distrust of the Vietnamese and Chinese suitors whom he privately considered to be closet Titos. Stalin reaffirmed Moscow's diplomatic recognition of the DRV, but he ruled out any direct USSR aid. Instead, the Soviet dictator urged Mao to provide the Vietnamese insurgents with military equipment and training, a directive that Mao readily accepted ((Zhai 2000, 16–18).

Beijing's support, which commenced soon thereafter, proved indispensable to the Vietnamese anti-colonial fighters. It helped secure military victories against superior French forces, culminating in the climactic triumph at Dien Bien Phu, in March 1954. For its part, the United States abandoned any pretence to neutrality in the wake of the communist powers' recognition of Ho's government. It formally recognized the Bao Dai regime in February 1950 and a few months later began providing massive amounts of military materiel to the French. In American eyes, however skewed their vision might have been, the Viet Minh formed part of a monolithic communist threat headquartered in the Kremlin. Despite the largely ineffectual character of French military and political policies, US officials were in agreement that the French counter-insurgency campaign served US Cold War interests.

The Geneva Accords of 1954 brought an end to the First Indochina War. The settlement divided the country, temporarily, at the 17th parallel. The negotiators established two regroupment zones, one in the north for the Viet Minh and one in the south for the French and those Vietnamese who had fought with the colonialists.

Nationwide elections, slated to take place within two years, were to create a unified Vietnam. But they never took place. Rather, rival North and South Vietnamese states developed. The former, which kept as its official name the Democratic Republic of Vietnam (DRV), was formally recognized by the Soviet Union, China, and all the Eastern bloc countries; the latter, which called itself the Republic of Vietnam, received strong backing from the United States and formal recognition by Washington and its allies. A new guerrilla insurgency then erupted in the south, in 1957, aimed at overthrowing the US-supported Saigon regime of Ngo Dinh Diem. Three years later, the so-called Viet Cong guerrillas, supported and controlled by the DRV, established a politico-diplomatic arm: the National Liberation Front (NLF) (Duiker 1995, 107–137).

Vietnam's National Liberation Front

The NLF followed the script pioneered by the Republic of Indonesia and being pursued simultaneously by the Front de Libération Nationale (FLN), in Algeria, seeking international legitimacy and recognition as a government-in-waiting through a flurry of worldwide diplomatic activism to complement the ongoing military campaign. Military and diplomatic successes proved mutually reinforcing, as the Indonesians had earlier discovered. By 1961–1962, the Viet Cong/NLF insurgents controlled approximately half of the territory and half of the population of South Vietnam, helping to validate their claim that the Saigon regime lacked legitimacy in the eyes of the populace—the basic constituency that each were appealing to—and that only the NLF possessed that legitimacy (Prados 2009, 62–72).

The NLF's Foreign Relations Committee launched a diplomatic offensive, in mid-1962, aimed at winning support and recognition from both communist and non-communist states. Deliberately downplaying its intimate ties with and subservience to Hanoi, the NLF presented itself as a genuinely autonomous South Vietnamese actor, intent upon forming a 'neutralist' government independent of both Hanoi and Washington. 'We believe in a neutral solution to the Viet Nam crisis', declared NLF Secretary General Nguyen Van Hieu. 'Vietnam is capable of determining its own future without outside interference and the National Liberation Front is willing to engage in negotiations right now to produce a peace-loving and democratic government' (quoted in Brigham 1999, 17). The campaign met with considerable success; the neutralist message and image appealed to many non-aligned states and to certain liberal and pacifist groups in the West as well. Hieu travelled to Jakarta to meet with Sukarno, for example, and the Indonesian president praised the NLF while predicting that 'it will surely win'. In a subsequent press conference, on 20 September 1962, Sukarno called for immediate negotiations between the United States and the NLF to end hostilities (Brigham 1999, 24). Following talks with Prince Norodom Sihanouk, in Phnom Penh, the Cambodian ruler likewise praised the NLF while proposing that South Vietnam 'should become a neutral zone' (Brigham 1999, 25).

Throughout the mid- and late-1960s, the NLF strove to maintain the pretence of independence and continued following the path of diplomatic activism, even as the conflict in South Vietnam morphed into a major land war with more than half a million US troops deployed there along with tens of thousands of North Vietnamese regulars. To break the military stalemate on the ground—plainly evident by 1967—Hanoi, Moscow, and Beijing insisted that the NLF had to be accepted as a legitimate entity and negotiating partner in any peace talks between the United States and its South Vietnamese ally. Albeit with great reluctance, the administration of Lyndon B. Johnson, increasingly desperate to bring an end to the divisive Vietnam War, accepted that bitter pill as a condition for opening the Paris peace talks in May 1968. Winning a coveted seat at those formal negotiations and being officialized in that way as a legitimate negotiating partner constituted a supreme diplomatic triumph for the NLF.[19] It offered undeniable political legitimacy to the southern insurgency that US military forces had for years laboured fruitlessly to liquidate.

Johnson's successor, Richard M. Nixon, shifted the real negotiating action from the formal bargaining among the Americans, North Vietnamese, South Vietnamese, and NLF to secret, behind-the-scenes talks between his representative, National Security Adviser Henry A. Kissinger, and senior North Vietnamese Politburo member Le Duc Tho. Those negotiations deadlocked over Washington's refusal to allow a role for the Provisional Revolutionary Government (PRG; the NLF's new name) in South Vietnam's postwar governance. Hanoi remained adamant on that essential point. Only the Nixon administration's belated and pained reversal of course on the PRG's political legitimacy, allowing it to participate in a postwar coalition government, led to the compromises embedded in the deeply flawed Paris Peace Agreement of January 1973 (Asselin 2002).

Conclusion

In each of these decolonization upheavals, national liberation movements vied for legitimacy and recognition by combining armed resistance with diplomatic overtures. Indonesian nationalists, with their adroit comingling of *diplomasi* and *perdjuangan* set the template. They calculated, from the outset, that support from the United States could be crucial to their bid for independence; and that highlighting the moderate, non-communist character of the fledgling republic they had formed in the days that followed Japan's surrender could help sway American opinion. At the same time, Sukarno and his inner circle realized that, absent a strong military arm, Dutch forces could capsize the republican experiment, snuffing out organized nationalist resistance to the return of imperial rule. They could then present the international community with a fait accompli.

For the United States, Indonesia's diplomatic and military achievements together proved instrumental to its belated support for the freedom struggle. Not only had

[19] On officializing as a practice of claim-validation see the Introduction to this volume.

the republic effectively demonstrated its moderation, openness to Western trade and investment, and broad-based support from constituencies in the Afro-Asian world and among progressive groups within American society, but its fierce guerrilla campaign against attacking Dutch units stymied all Dutch efforts to 'pacify' the archipelago. In the end, Washington's decision to pressure the Netherlands to grant Indonesia full sovereignty proved to be surprisingly easy. Competing policy priorities shifted decisively towards the Indonesian cause in early 1949 for two essential reasons: first, the republic had plainly attained legitimacy internally and in much of the wider world; second, Indonesian armed resistance to the Dutch counter-insurgency campaign had proved remarkably effective, rendering a Dutch military victory unobtainable and continued fighting both fruitless and dangerously unstable.

The Soviet and Chinese decisions to recognize the Democratic Republic of Vietnam proved just as easy. As the Cold War heated up in the late-1940s, the search for allies in the rapidly decolonizing Third World grew in importance on both sides of the Cold War divide. The American denial of support to the DRV—before and after the Geneva Convention of 1954—seemed as natural as Moscow and Beijing's denial of the same to, first, the French-supported Bao Dai regime and, after Geneva, the Republic of (South) Vietnam. Recognizing a communist-dominated liberation movement was anathema to the United States, recognizing communist North Vietnam, after 1954, equally so.

Unquestionably, the force of arms proved much more decisive to the outcome of the First and Second Indochina Wars than the art of diplomacy. Yet diplomatic outreach also contributed significantly to that outcome. The ability of the NLF, and later PRG, to take on the trappings of at least a semi-legitimate government-in-waiting enabled state and non-state actors outside the communist bloc to champion its cause. Its mere existence as a nominally autonomous agent, moreover, allowed North Vietnam, the Soviet Union, and China to insist that the NLF must be represented in any peace talks with the United States and South Vietnam. That battering ram of a diplomatic strategy, alongside constant military pressure, forced the Nixon administration to accord it legitimacy as a political stake-holder in post-peace settlement South Vietnam.

Two other decolonization cases, those of Algeria and Angola, sketched briefly here for comparative purposes, underscore this broader pattern.[20] Nationalist claims-makers in each territory invariably adopted a strategy that blended armed resistance to the imperial power with urgent appeals to the international community. In each instance, those appeals were based on the righteousness of the pro-independence cause and the demonstrable fact that the claims-maker enjoyed widespread domestic support—and hence possessed legitimacy. Freedom fighters in Algeria, in the 1950s and early 1960s, and Angola, in the 1960s and 1970s, adroitly leveraged a mix of

[20] See also the chapter by Stephan Hensell in this volume on the long-standing decolonization struggle of the Polisario in Western Sahara.

armed opposition and diplomatic advocacy in their common quests for total independence, much as had their counterparts in Indonesia and Vietnam in the wake of World War II.

Outside of Vietnam, Algeria stands as the bloodiest of all the decolonization wars, a brutal conflict that claimed as many as half a million lives. There, an extraordinarily skilful diplomatic effort persevered, even in the face of a savage French counter-insurgency offensive. By the late-1950s and early 1960s, the world had almost completely transitioned into one in which anti-colonialism had assumed normative status in international politics. The new, anti-colonial norms represented a sea change in 'historical times', in the language of this volume, making the violent suppression tactics preferred by policy makers in Paris ever more problematic and ever more at odds with global opinion than anything faced by their predecessors vis-à-vis the Vietnamese insurgency just a decade earlier. When the Algerian independence-seekers formed the Provisional Government of the Algerian Republic, in 1958, it controlled hardly any territory. Yet within weeks it was recognized by thirteen countries, including China, North Vietnam, North Korea, Indonesia, and virtually all the Arab states. The FLN's leaders exulted when, in October 1960, Moscow accorded de facto recognition as well, promising to provide all possible Soviet aid. For its part, the United States, wary of being caught on the wrong side of history, was considerably less averse to pressuring its French ally than it had been in the early postwar years. The administrations of Dwight D. Eisenhower and John F. Kennedy were keenly aware that Algeria had become, in the words of historian Matthew Connelly, 'the pre-eminent anticolonial movement in a profoundly anti-colonial era' (Connelly 2002, 4). Consequently, they kept up the pressure on France to bow to the inevitability of an independent Algerian state. Algerian nationalists masterfully navigated the treacherous shoals of US–Soviet rivalry to gain a degree of legitimacy from each of the superpowers. Their diplomatic finesse and pragmatism, more than their military achievements, led directly to the Evian Accords of March 1962 which gave birth to a fully independent Algeria (Connelly 2002; Byrne 2016; Wall 2001).

In sharp contradistinction to the decolonization contests in Indonesia, Vietnam, and Algeria, each of which produced a single, hegemonic organization that spoke for the nationalist cause and attained strong domestic legitimacy among its internal constituents, Angolan freedom-fighters of the early and mid-1970s split into three factions. They clashed with each other—and with Soviet-supplied Cuban troops—rather than with a European colonial military force intent on preserving empire. Here, too, diplomacy played a significant role. Support from the so-called African 'front line' states, determined to help eradicate the last vestiges of empire on the continent, kept up steady pressure on the reactionary Portuguese government of Antonio Salazar. At the same time, early backing from a more assertive Soviet Union, intent on capitalizing on Africa's anti-imperialist surge for its own Cold War interests, fortified the Angolan independence-seekers. A regime change in Lisbon, in April 1974, settled the core issue of whether Angola would achieve unfettered sovereignty. It would. But a civil war among indigenous contestants, each drawing support from one of the superpowers, determined which one would rule. The particular shape and contours of the Angolan government that eventually claimed power and speedily won international recognition—the Moscow-backed Movimento Popular de Libertação de Angola (MPLA)—thus owed more to superior fighting prowess than to diplomatic manoeuvres (Westad 2007, 218–237).

The story of decolonization is in large measure the story of Third World nationalist movements' search for internal and external legitimacy and international recognition, the central concerns of this volume. The insurgencies that faced violent imperial resistance, such as those discussed above, endeavoured to achieve the bedrock goals of genuine independence and acceptance in the state-based international system by pursuing strategies that mixed armed resistance with diplomatic outreach. The relative importance of fighting and diplomacy differed in each case, of course, but the pragmatic mix of the two in the end proved efficacious to all four of these armed insurgencies. The United States and the Soviet Union constituted key audiences in each instance, since the superpowers possessed unparalleled influence within the wider international arena. The decolonization wars posed an unusually tricky conundrum for policy makers in Washington, caught as they were between the contradictory goals of supporting European allies and trying simultaneously to woo the emerging nations of the global South. Cold War priorities shaped the US response in each case, leading to support—however halting and reluctant—for Indonesia's bid for independence, open opposition to Vietnam's, and an embrace of the independence cause in both Algeria and Angola. The Soviet Union faced no such dilemma. Indeed, it eagerly supported national liberation movements across the Third World and wholeheartedly embraced the anti-colonial cause, especially in the post-Stalin era. Such a stance, Kremlin decision-makers calculated, aligned Moscow with the developing nations while weakening the West and tarring it with the brush of imperial aggrandizement—and racism. For their part, the independence movements of the decolonization era benefited from the East–West conflict, leveraging the prevailing bipolarity of world politics to enhance their prospects for achieving independence as quickly as possible. Structure and agency thus combined in manifold ways, creating unique historical opportunities for long-suppressed nationalist groups. They seized those opportunities with skill and foresight.

References

Asselin, Pierre. 2002. *A Bitter Peace: Washington, Hanoi, and the Making of the Paris Agreement.* Chapel Hill, NC: University of North Carolina Press.

Brigham, Robert K. 1999. *Guerrilla Diplomacy: The NLF's Foreign Relations and the Viet Nam War.* Ithaca, NY: Cornell University Press.

Bush, George H. W. 2004. 'Introduction'. In *At the Abyss: An Insider's History of the Cold War*, edited by Thomas C. Reed, pp. 1–3. New York: Ballantine Books.

Byrne, Jeffrey James. 2016. *Mecca of Revolution: Algeria, Decolonization, and the Third World Order.* New York: Oxford University Press.

Connelly, Matthew. 2002. *A Diplomatic Revolution: Algeria's Fight for Independence and the Origins of the Post-Cold War Era.* New York: Oxford University Press.

Duiker, William J. 1995. *Sacred War: Nationalism and Revolution in a Divided Vietnam.* New York: McGraw Hill.

Gaiduk, Ivya V. 1996. *The Soviet Union and the Vietnam War.* Chicago, IL: Ivan R. Dee.

Hess, Gary R. 1978. 'The First American Commitment in Indochina: The Acceptance of the "Bao Dai Solution", 1950'. *Diplomatic History* 2: pp. 331–350.

Hess, Gary R. 1987. *The United States' Emergence as a Southeast Asian Power, 1940–1950.* New York: Columbia University Press.

Hornbeck, Stanley K. 1948. 'The United States and the Netherlands East Indies'. *Annals of the American Academy of Political and Social Science* 255: pp. 132–133.

Logevall, Fredrik. 2013. *Embers of War: The Fall of an Empire and the Making of America's Vietnam.* New York: Random House.

McMahon, Robert J. 1981. *Colonialism and Cold War: The United States and the Struggle for Indonesian Independence, 1945–49.* Ithaca, NY: Cornell University Press.

McMahon, Robert J. 1999. *Limits of Empire: The United States and Southeast Asia since World War II.* New York: Columbia University Press.

Prados, John. 2009. *Vietnam: The History of an Unwinnable War, 1945–1975.* Lawrence, KS: University Press of Kansas.

Wall, Irwin M. 2001. *France, the United States, and the Algerian War.* Berkeley, CA: University of California.

Wiebes, Cees and Bert Zeeman. 1992. 'United States "Big Stick" Diplomacy: The Netherlands between Decolonization and Alignment, 1945–1949'. *International History Review* 14: pp. 45–70.

Westad, Odd Arne. 2007. *The Global Cold War: Third World Interventions and the Making of Our Times.* New York: Cambridge University Press.

White, Theodore H. 1947. 'So the Dutch Are at War'. *New Republic*, 4 August: p. 8.

Wolf, Charles. 1948. *The Indonesian Story: The Birth, Growth and Structure of the Indonesian Republic.* New York: John Day.

Zhai, Qiang. 2000. *China and the Vietnam Wars, 1950–1975.* Chapel Hill, NC: University of North Carolina Press.

3
Western Sahara, the Polisario, and the Struggle for International Legitimacy

Stephan Hensell

Introduction

The struggle over Western Sahara is one of the oldest ongoing conflicts.[1] It is not only a violent struggle that pits an armed group against a powerful state along the largest functioning military border in the world but also a diplomatic dispute that fuels continued antagonisms in the Maghreb involving a large refugee population, the United Nations (UN), the African Union (AU), and major western states such as the US, France, and Spain. Western Sahara is a largely uninhabited desert region, roughly the size of the UK, that stretches along the Atlantic coast bordering Morocco in the north and Mauritania in the southeast. As part of the Spanish colonial possessions in Africa, it also used to be known as the Spanish Sahara. Spain left the territory in 1976 but since then the former colony is disputed between Morocco and the Frente Popular para la Liberación de Saguia el-Hamra y Río de Oro (Frente Polisario). Having its origin in the anti-colonial nationalism of the Sahrawi people, the Polisario sees itself as a liberation movement that continues to fight for decolonization after the departure of the Spanish colonial power. The people of Western Sahara, according to the Polisario, have never attained independence because Morocco occupied the territory after Spain left in 1976. Western Sahara is, thus, 'Africa's last colony'. The Polisario seeks to liberate Western Sahara in order to allow for the self-determination of the Sahrawi people and to establish an independent state, the Saharawi Arab Democratic Republic (SADR). However, Morocco claims that Western Sahara is part of the Moroccan state. From the Moroccan perspective, there is no such thing as 'Western Sahara', there is only a Moroccan Sahara. Consequently, the SADR is a 'fictitious entity' (Institut Royal des Etudes Stratégiques 2021, 43). The contradiction between Moroccan irredentism and Saharawi nationalism is at the heart of the Western Sahara conflict, which is strongly influenced by French and US support for Morocco and Algerian support for the Polisario.

[1] Research for this chapter has been made possible by a grant from the German Research Foundation (Deutsche Forschungsgemeinschaft, DFG), project no. 424426045.

Stephan Hensell, *Western Sahara, the Polisario, and the Struggle for International Legitimacy*. In: *Armed Groups and the Politics of International Legitimation*. Edited by: Stephan Hensell and Klaus Schlichte, Oxford University Press.
 DOI: 10.1093/9780198974529.003.0003

The struggle of the Polisario is also a struggle for legitimacy. Throughout its long fight for independence, the Polisario has put forward various claims to legitimacy regarding their movement, their armed struggle, and the state they seek to establish.[2] In doing so, it has consistently pursued and attained international legitimacy, albeit with varying success. The Polisario has acquired legitimacy for itself as a liberation movement. It is widely recognized as a representative of the people of Western Sahara and as a party to the conflict. However, the group has been rather unsuccessful in also attaining legitimacy for the de facto state it has established and which is largely governed from exile in Algeria. While a number of governments have recognized the SADR as an independent and sovereign state, many others have not. The Polisario has obtained bilateral recognition for the SADR mostly in Africa and Latin America but a majority of states have not extended recognition to the Sahrawi Republic, among them major international powers such as the US, China, or Russia as well as all European and most Arab states. Moreover, several states that used to recognize the SADR have frozen or suspended their recognition in recent years. Being a recognized movement that governs a largely unrecognized state, the international legitimacy of the Polisario has remained limited.[3]

Why has the Polisario failed to achieve full international legitimacy among a majority of states? This chapter argues that the politics of legitimation in the Western Sahara conflict has at its heart a powerful claim to legitimacy trumped over time by realism and disinterest. International legitimacy is time bound. It emerges in a particular historical context but cannot necessarily be reproduced in another. The Polisario's claims to legitimacy were widely validated at a certain point in time but seem to have lost relevance later on. International legitimacy does not travel easily across historical times. The historical time of late decolonization was favourable for the liberation movement, who claims to have legitimacy according to the legally recognized right of self-determination and independence that has been granted to all colonial people.[4] The legitimating idea of the anti-colonial struggle based on principles of international law resonated within the UN and a constituency formed by a leftist European solidarity movement and states from the Third World that shared the historical experience of colonialism and fighting empires. Being part of an international anti-imperial movement in the late 1970s to mid-1980s, the group could count on attention, solidarity, and support from like-minded allies. It was also the time when the Polisario achieved its major military and diplomatic victories. However, the group has rather lost out as the conflict has dragged on after the end of the Cold War.

[2] The Polisario and the SADR are closely related and difficult to disentangle. The secretary general of the Polisario is also the president of the SADR. However, for analytical purposes both actors need to be differentiated. In the case of Western Sahara, one could analyse the quest for international legitimacy from the perspective of an armed group or from a de facto state. The question then is who is more important? Where is the agency located? In line with the assumption of this book that armed groups seek international legitimacy for themselves or the institutions they govern, this chapter sees the Polisario as the central actor. The movement not only dominates the SADR but also has a greater realm of political action.

[3] Morocco's sovereignty claims over Western Sahara, to be sure, are likewise widely unrecognized.

[4] On the role of decolonization for the legitimation of armed groups see the Introduction in this volume. See also the chapter by Robert McMahon on the anti-colonial rebellions in Indonesia and Vietnam.

The chapter makes two core observations in that regard. First, realist considerations have always been a powerful counterweight to legitimacy claims in the case of Western Sahara. Morocco has at all times rejected the Polisario's claim and in doing so, it has been tacitly supported by the US, France, and Spain in the name of stability and realpolitik since the early days of the conflict. The Polisario has failed to challenge the power alliance between Morocco and its major international patrons. In the post-colonial age, realism tends to prevail with regard to Western Sahara. The de-recognition of the SADR through various states since the 1990s, in large part due to Moroccan diplomatic efforts, reflects new political interests. Today, Morocco is more successful than ever in pursuing its irredentist politics and compelling others to accept its sovereignty over Western Sahara. Second, claims to legitimacy that have been convincing in the past do not necessarily capture the imagination of future audiences. Making people interested, drawing attention to a problem, and mobilizing support for the cause of decolonization is a major challenge these days for the Polisario. Publics in important Western capitals are not attentive when it comes to the fate of the Saharawis and the Western Sahara. In the third decade of the 21st century, the Polisario continues to frame the issue of Western Sahara as a matter of decolonization, but it struggles to break the wall of silence surrounding the conflict. The group's international quest for legitimacy has apparently become less powerful. As a long-term case, the conflict over Western Sahara thus exemplifies the impact of changing historical times on armed groups. The Polisario's claim to legitimacy is still validated but it is more and more at odds with the realpolitik pursued by major international powers in the region and it seems to be increasingly fallen out of time in the post-colonial age.

This chapter seeks to reconstruct the evolution and reproduction of the international legitimacy of the Polisario and the SADR. It does so by analysing the regional and time-bound factors that have been decisive for its long-term development. In line with the theoretical conception of this volume, this paper focuses on practices and publics in the politics of legitimation. I am analysing the making and validating of the Polisario's claim to legitimacy as well as the respective role of the group's international audience and constituency. In the long-term struggle of the Polisario, all practices of claim-making and claim validation can be observed. The movement relied on *fighting* just as *mimicking* and *voicing* while states and international organizations reacted with *acknowledging*, *supporting*, or *officializing*.[5] The first section gives a brief overview of the history of the conflict. The next section seeks to illustrate the argument by reconstructing the Polisario's struggle for international legitimacy since the 1970s through two major periods: the time of confrontation (1973–1991) and the time of accommodation (1991–present).[6] I am tracing the evolution of the group's claim to legitimacy in the first period roughly covering two decades while focusing more on micro-moves and different locations of the conflict in the most recent years of the second period. This section also outlines the Moroccan counterclaim on the

[5] On these practices of making and validating a claim to legitimacy see the Introduction to this volume.
[6] I am following Jacob Mundy (2017, 69–73) here.

Western Sahara which has gained international traction over the last three decades. The text finishes with a brief conclusion and points towards an explanation on the Polisario's limited success in pursuing international legitimacy. The case study is based on field research and 36 interviews in Morocco, Algeria, Western Sahara, Belgium, and Germany in 2022 and 2023 as well as archival material and secondary literature.

The conflict over Western Sahara

The conflict around Western Sahara dates back to the period of decolonization in the early 1970s in which the territory was still a colony of Spain.[7] Modernization and the expansion of education under Spanish rule had facilitated the emergence of a national consciousness among a largely tribal Saharawi population. While tribal elders were co-opted in the colonial institutions, students became politicized in universities in Morocco, Syria, and Egypt in the mid-1960s and early 1970s by the prevailing radical atmosphere at university campuses at that time. Influenced by the example of other Third World national liberation movements in Africa and the Arab world, they started to promote the national independence of the Western Sahara (San Martín 2010, 47–86; Hodges 1983, 149–159). In 1973, Sahrawi nationalists in Western Sahara founded the Polisario Front as an anti-colonial movement that took up arms against the Spanish colonial power. As a secular-leftist movement that pursued national self-determination, the Polisario featured elements of Marxist-inspired revolutionary rhetoric, nomadic traditions, and indigenous democracy while aligning itself with the programmes of other African and Arab liberation movements active in the 1960s and 1970s (Zunes 1987; San Martín 2010, 90–102, 120; Zunes and Mundy 2022, 115–116). After Spain decided to withdraw from its colony, it relinquished its administration in late 1975 with the Madrid Accords, which divided the former Spanish Sahara between Mauritania and Morocco. Both states subsequently started to annex the territory without considering the political demands of the Polisario or the Sahrawi people living in that territory. The Polisario, in turn, reacted to the final departure of Spain with the proclamation of the Saharan Arab Democratic Republic (SADR) in February 1976, coupled with the announcement of an armed struggle to achieve the self-determination and independence of the Sahrawi people. In the wake of the annexation, most of the local Sahrawi population fled to neighbouring Algeria, whereas the Polisario started to fight Mauritania (until 1979) and Morocco (until 1991) in what it perceived as a continuation of the liberation struggle. Since the early days of the conflict, the major conflict parties received consistent external backing. The Polisario got political and military support from Algeria, whereas Morocco was supported militarily and diplomatically by the US and France (Zunes and Mundy 2022, 40–43, 59–88).

[7] For a thorough historical account of the conflict, see Hodges (1983); San Martín (2010); Boukhars and Roussellier (2014); Zunes and Mundy (2022). More specifically on the Polisario, see Ruiz Miguel (2022).

While the Polisario was on the edge of military victory in the early 1980s with its hit and run tactics in Western Sahara, the military tide has since turned against them. The reason for this was the defensive stance of its adversary. Morocco established a 2,700 kilometre-long fortified military barrier (the 'Berm') in the 1980s which effectively separated Western Sahara and kept the Polisario from entering the Moroccan controlled territory. The war thus ended in a military stalemate. Since the mid-1980s, Morocco has exercised de facto authority over the majority of the territory of Western Sahara (approx. four-fifths) including the Atlantic coast and the major cities. The Polisario controls one-fifth of Western Sahara (the 'liberated zones'), but it has its rear base in southwest Algeria, where the government of the SADR as well as a refugee population of approximately 120,000 Sahrawis are exiled and located in refugee camps and tent cities near Tindouf (Fernández-Molina and Porges 2020, 381).

In 1991, the UN proposed a Settlement Plan along with a ceasefire agreement that were both accepted by Morocco and the Polisario. The Settlement Plan foresaw a referendum for Sahrawi self-determination with the option of independence or integration of Western Sahara into Morocco as well as the establishment of the United Nations Mission for the Referendum in Western Sahara (MINURSO) with the mandate to implement the settlement proposals. Since the agreement on the UN Settlement Plan, however, the conflict has stalled due to disagreement regarding the electorate entitled to vote in the envisaged referendum. Several international mediation efforts between 2001 and 2003 by James Baker, former US secretary of state and UN envoy for Western Sahara, to resolve the impasse (Baker Plan I and II) were rejected either by the Polisario or Morocco (Zunes and Mundy 2022, 169–253). The Saharawis and the Polisario continue to seek a referendum on self-determination which includes the option of independence, whereas Morocco proposed an autonomy plan in 2007 as an alternative which only concedes autonomy for Western Sahara within the Moroccan state. Since the establishment of the ceasefire in 1991, second-generation Sahrawi nationalists have staged non-violent civil protests in the occupied territory from time to time (in 1999, 2005, 2010), however, without being able to challenge the Moroccan authority in the territory (Boukhars 2012). After almost 30 years of the ceasefire, renewed armed clashes between the Polisario and Morocco at the southern border of the desert territory led to the breakdown of the ceasefire in November 2020. The war of attrition has resumed, but it is now highly asymmetric and receives almost no attention. Morocco controls the air space in the 'liberated zones' through armed drones with a strategy of Anti-Access Aerial Denial, and the armed actions of the Polisario are limited to symbolic hit and run tactics along the Berm that go largely unnoticed. It is a 'silent war' as a Polisario has put it.[8]

[8] Interview with a diplomat of the Polisario Representation to Europe and the EU, Brussels, 6 February 2023.

The struggle for international legitimacy

In its long fight for independence and liberation, the Polisario has been active throughout different historical times. Starting with violent means in the age of decolonization and the Cold War, the group shifted its activities to civil resistance after the end of the superpower rivalry. A major turning point in the Polisario's struggle was the agreement on a ceasefire with Morocco in 1991. It not only ended the fighting but also raised the hope among the Saharawis that the UN would broker a peace deal leading to a referendum about the future of Western Sahara, which the Polisario was sure to win. The group's politics of legitimation evolved throughout these years. While the Polisario has made a consistent claim to legitimacy since the 1970s, the transition from war to peace in the 1990s led to a shift in the group's strategy, which can be characterized as a move from confrontation to accommodation (Mundy 2017, 75). It also brought about change in the practices of legitimation. In the beginning, confronting Morocco on the battlefield and the Polisario's military performance mattered. Fighting was the main practice of claim-making. However, after the ceasefire, accommodating the struggle within the UN-led peace process and the architecture of North Atlantic interests and values became more important. The practice of voicing now became the core of the group's legitimation strategy. With the establishment of the SADR, the Polisario has also relied on mimicking state behaviour as a claim to legitimacy throughout its struggle. Since the beginning of the conflict, states and international organizations, notably Algeria, the AU, and the UN reacted to these claims at various points in time with practices of claim validation such as acknowledging the Polisario as a legitimate representative, supporting it with resources or officializing the SADR through formal interstate diplomacy. Morocco, in contrast, has constantly opposed the legitimation of the Polisario and the SADR ever since the first days of the conflict.

The time of confrontation (1973–1991)

The evolution of the Polisario's claim to legitimacy can be seen as a reaction of Morocco's irredentist policies in the early 1970s, when Western Sahara was still a Spanish colony. After various leftist plots and two military coups, King Hasan II of Morocco made the 'recovery' of the 'southern provinces' of Morocco a centrepiece of his politics of survival, which consisted mainly in gaining control of the political agenda and keeping the army busy (Zartman 2014, 60–61). Playing on the vision of a 'Greater Morocco' that was put forward by the Moroccan independence movement of the 1950s, Hasan II framed the Saharan issue as a question of national identity and integrity in order to generate support for his Saharan campaign and rally the various political factions around the king (Miller 2013, 180–184). As a result, the legitimacy of the monarchy became increasingly associated with possession of the Sahara, thus raising the stakes in a war the king could not 'afford to lose' (Damis 1983, 174). The Moroccan invasion of the territory in 1975, however, triggered not only the war but also a powerful counterclaim of the Polisario that started to become a broad movement in the Spanish Sahara.

The Polisario seeks self-determination for the Saharawi people and independence for Western Sahara. In doing so, the Polisario has made and continues to make a claim to legitimacy regarding its movement, its armed struggle, and the state it seeks to establish in Western Sahara. Fundamentally, the Polisario sees legitimacy as a very important element in its struggle, and it differentiates between 'légitimité populaire' and 'légitimité internationale'.[9] On the local level, the group claims to have the full support of the Saharawi people, which is its domestic constituency. 'Le Polisario tire sa légitimité du soutien du peuple Sahraoui lui-même', as a member of the group explained.[10] The Polisario declares itself to be the sole and legitimate representative of the Saharawis. Moreover, the group claims its armed struggle is justified, which it sees as legitimate self-defence. It also claims that the Saharawis have a legitimate right or aspiration to self-determination and to exist as a free and sovereign people. Finally, it sees the SADR a legitimate political entity that should be the independent and sovereign state of Western Sahara. The Polisario makes an ideation-based claim to legitimacy that refers to international law by framing the conflict as an issue of unfinished or derailed decolonization. Western Sahara, as they perceive it, was never truly decolonized by Spain, and its population needs to exercise its right to self-determination in order to settle the conflict. The group accordingly refers to principles and rules of international law such as the right to self-determination or the intangibility of borders inherited from colonial times that are laid down in the UN charter and UN resolutions, the AU charter or the Montevideo Convention. From a legal point of view, according to the liberation movement, Western Sahara is a straightforward case. Thus, the group sees all of its actions including the armed struggle not only as legitimate but also as legal and in line with UN regulations. The Polisario also makes a performance-based claim that refers to successes and practical achievements of the state it governs. The Polisario highlights that the SADR, built from scratch in the refugee camps, has established the basic political structures for a future free and independent Saharawi state. It not only has a constitution, courts, and a parliament. The Saharawi republic also cares for the basic needs of its citizens. According to the Polisario, it provides education, health, and welfare while promoting women's emancipation and inclusion (San Martín 2010, 113–119). And an independent Saharawi state, so the argument, would be a guarantee of stability and security as well as an example of tolerance in North Africa. Finally, the fact that the claims of the Polisario have been validated by others through the diplomatic recognition of the Polisario and the SADR feeds back into the claim-making. 'Nous avons toute la légitimité du monde', as one leading Polisario expressed.[11]

[9] Interview with member of the National Secretariat of the Polisario/Ambassador of the SADR to Algeria, 6 March 2023, Algiers.

[10] Interview with member of the Polisario/head of AFAPREDESA (Association of Families of Sahrawi Prisoners and Disappeared), 26 February 2023, Rabouni refugee camp.

[11] Interview with member of the National Secretariat of the Polisario/Ambassador of the SADR to Algeria, 6 March 2023, Algiers.

The Polisario's major aim from the beginning has been to address these claims to an international public and seek their validation. Doing so, the liberation movement relied first of all on fighting as a practice of claim-making. The Polisario waged armed struggle against Spanish, Moroccan and Mauritanian forces and in seeking to impose its will with violent means, the group also demonstrated its military might. The Polisario carried out significant military operations against Moroccan forces capturing large amounts of weapons, military equipment, and soldiers. In two of its boldest attacks, the liberation movement also raided twice the Mauritanian capital, Nouakchott, in 1976 and 1977. With its combat prowess and spectacular successes on the battlefield, the Polisario not only presented itself as a serious contender for power but also raised awareness of their cause and legitimacy claims. Voicing also became an important practice in the Polisario's legitimation effort, starting with the proclamation of the SADR on 27 February 1976 in a night-time ceremony in a small village in the northeast of Western Sahara before several thousand Saharawis and a group of foreign journalists (Hodges 1983, 238). Having proclaimed the SADR, the Polisario also began mimicking the state by presenting itself as a future governmental actor and adopting state-like formal behaviour. The group sought to build the state locally and secure its recognition internationally. While the liberation movement developed a set of state-like institutions in the Saharawi refugee camps, it also established a network of embassies and external missions in order to represent the SADR and to internationalize the conflict.

Since the mid-1970s, the SADR but also the Polisario itself would gain international recognition. The regional level proved to be of paramount importance here as the crucial role of Algeria makes clear. As the leading anti-imperial power of the age of decolonization, Algeria had supported various liberation movements and guerrilla armies since the 1960s (Byrne 2016) and extended its support to the Polisario, too. Algeria provided an important backing for the liberation movement from the first days of the conflict and has guaranteed its political survival ever since. It supported the Polisario with arms, aid, safe haven, and diplomatic lobbying. Moreover, Algeria constantly challenged the Moroccan claims to Western Sahara in international forums. The Polisario also received military equipment from Libya between 1979 and 1982. The Polisario's legitimacy was thus prefigured on the regional level. Within the North African region, both Algeria and Libya not only supported but also actively solicited the international recognition of the SADR and acted as a gatekeeper to international institutions (Willis 2012, 300).

The claims of the Polisario were also acknowledged on the global level by international institutions. Already in 1966, the UN General Assembly had affirmed in a resolution the right of the Spanish Sahara to self-determination. Similar resolutions including the option of Western Saharan independence followed from 1967 onwards. On the eve of the withdrawal of Spain in 1975, the International Court of Justice formally rejected Morocco's and Mauritania's claims to Western Sahara and declared the right of Sahrawi self-determination as paramount. While the UN already listed Western Sahara as a Non-Self-Governing Territory in 1963, in 1979 it reaffirmed the right of the people of Western Sahara to self-determination and independence and the 'legitimacy of their struggle'. Moreover, the UN recognized the Polisario as 'the representative of the people of Western Sahara' (United Nations General Assembly 1979). As the Polisario had signed a peace treaty with Mauritania in 1979, it also gained formal recognition as a party to the conflict—a notion which was then carried forward in subsequent UN regulations including the UN Settlement Plan (Ruiz Miguel 2022, 58–60). The Polisario furthermore successfully pursued the bilateral recognition of the SADR by other states. By 1983, more than 50 governments mostly from Africa and Latin America had already formally recognized the SADR as a sovereign state (Hodges 1983, 308). Most importantly in that regard was the admission of the SADR to the Organisation of African Unity (OAU) as a full member in 1984 with the crucial help of Algeria. It not only officialized the SADR as formal member of a regional organization but it also constituted a de facto acceptance by all those African member states that had not yet recognized the SADR.

In retrospect, the period between 1976 and 1986 can be seen as a crucial decade. It was the time when the Polisario was most successful on the military front and inflicted heavy losses on its enemies. The group was considered one of the most effective guerrilla movements in Africa at that time (Hodges 1983, 332). On the diplomatic front, the admission of the SADR to the OAU in 1984 marked the high point of the Polisario's diplomatic fortunes. One of the founding fathers of the Polisario recalls these years as 'les années d'or'.[12] However, it was also the decade of a turning point. Starting in 1980, Morocco began to build defensive walls throughout Western Sahara. As a result, the military fortunes of the Polisario steadily declined and the group could no longer enforce its diplomacy with violent action. Finally, it was the decade when the global constituency of the Polisario and the SADR took shape. Right from the beginning of the armed struggle in 1973, the Polisario had identified closely with other nationalist guerrilla movements in the Third World such as in Vietnam, Eritrea, Guinea-Bissau, Angola, Mozambique, and particularly in Palestine (Hodges 1983, 153, 164). Ten years later, in 1983, the Polisario published a decennial brochure titled 'DIX ANS de lutte contre le colonialisme et l'expansionnisme' (Front Polisario 1983). Looking back at a decade of armed action since 1973, the Polisario saw itself on a good path towards international legitimacy. In a chapter devoted to the Polisario's diplomacy, the group made clear that it saw its fight as part of a broader liberation struggle in the Third World. 'Le Front Polisario est un mouvement qui est une partie du mouvement de libération arabe et africain et son action s'inscrit parmi celles des autres organisations de libération de par le monde' (Front Polisario 1983, 100). As a long-term member explained, the Polisario saw itself as well-placed within a wider anti-colonial liberation struggle of revolutionary leaders and movements. 'Nous étions très bien placés à l'époque ... les dirigeants qui menaient cette lutte anticoloniale étaient présents, étaient chefs du mouvement, chefs d'état ... ça faisait que nous [étions] bien accompagnés.'[13] In the context of decolonization, the

[12] Interview with member of the National Secretariat of the Polisario, 26 February 2023, Rabouni refugee camp.

[13] Interview with member of the National Secretariat of the Polisario, 2 March 2023, Bojadour refugee camp.

Polisario addressed a constituency of like-minded allies in the Maghreb and in Africa and more generally also the UN. Friendly states in Latin America and the Caribbean also became important. The shared experience of colonialism and liberation as well as cultural and language ties of the Hispanosphere helped to forge transatlantic relations and to bring about recognition in various cases, although this was achieved with the crucial help of Algerian diplomatic lobbying (Klich 1982). Cuba also actively collaborated with the SADR by graduating thousands of Saharawi students in the health sector in Cuban schools and by deploying medical brigades to the Sahrawi refugee camps (San Martín 2010, 147–150; Ruiz Miguel 2022, 49–50).

Gaining full recognition for the SADR especially in Africa and Latin America, the Polisario directed its limited diplomatic resources and staff to friendly Third World governments or the UN, but it rather failed to address the US and European governments. Representations in the US and Europe were confined to small delegations. Importantly, the Polisario did not lobby the US Congress or the White House, nor did it tailor its political message to a larger audience in Washington.[14] As a result, the US media remained remarkable silent on Western Sahara. News coverage in the Arab world was dominated by the Israeli–Palestinian conflict, while the Ford, Carter, and Reagan administrations aligned more or less overtly with Morocco out of strategic Cold War considerations (Hodges 1983, 355–365; Pazzanita 1994, 273–276; Mundy 2017). In Europe it did not look much better. While a leftist solidarity movement had been active since the early days of the conflict in organizing support for the Saharawis, the collaboration with political parties from the left turned out to be rather ambiguous.[15] Opposition socialist parties which championed the Polisario took on a more pragmatic approach once in power. Morocco came to be seen as an important partner, and political and economic interests outweighed ideational concerns as evidenced by the pro-Moroccan politics of Felipe González in Spain and François Mitterand in France (Hodges 1983, 348–354; Naylor 1987).

Overall, the Polisario pursued the internationalization of the conflict in the first two decades of the war. In doing so, it also successfully sought a validation of its claims to legitimacy, indicating a growing belief in the international legitimacy of the liberation movement and the SADR. The Polisario has become widely recognized as a legitimate representative of the Saharawi people by governments and international organizations and it has received support from solidarity groups and committees in several European states. The SADR has become a full member of the OAU and a founding member of its successor, the AU, where today it participates at all levels of the organization, signs international treaties, and even contributes to AU peacekeeping forces (Zunes and Mundy 2022, 178–179). By 1990, the list of states that had formally recognised the SADR had reached a respectable 74 (Pazzanita 1994, 271).

[14] Successful lobbying of the US government in Washington made a crucial difference for the support and officialization of other armed groups such as in the case of Angolan UNITA. See the chapter by Reyko Huang in this volume.

[15] On the role of solidarity activists in the legitimation of armed groups see also the contributions of Eline van Ommen on the FSLN and Andrea Oñate-Madrazo on the FMLN in this volume.

However, the Polisario failed to achieve recognition for the SADR in the US and Europe or from a permanent member of the UN Security Council. The SADR has also remained excluded from other regional organizations such as the Arab League and the Arab Maghreb Union (Fernández-Molina and Porges 2020, 387).

Moreover, the Polisario suffered various setbacks in its struggle for legitimacy. Since the mid-1990s, a number of states in Africa and Latin America started to suspend or withdraw their recognition of the SADR, thus signalling a shrinking constituency of the Polisario and its SADR. Of the 84 or so UN member states that have recognized the SADR at one time or another, around 43 have practically de-recognized it.[16] While the constituency of the SADR is still sizeable, it has become smaller since the 1990s, reminding the Polisario that the international legitimacy of their state remains open to revision and contestation. The movement also suffered from several high-level defections of leading Polisario members to Morocco (Ruiz Miguel 2022, 74–82).

The time of accommodation (1991–present)

After agreeing to the ceasefire in 1991, the Saharawis began to wait for the UN-organized referendum that would give them the chance to vote for the independence of Western Sahara. The 1990s were experienced by the Saharawis as a time of anticipation and hope. However, with the inertia of the peace process and the stalled negotiations since 2003, the Saharawis have become more and more frustrated with the UN. In the decades since 1991, the Polisario has also shifted its political strategy, with accommodation becoming an important approach. The group began to collaborate with the UN within the peace process but also started to align its politics of legitimation to the dominating liberal discourse of the post-Cold War age by highlighting standards and values that resonated with western governments. The Polisario emphasized the role of democratic norms and implemented reforms to democratize the political structure of the SADR. As the peace process dragged on, the liberation movement also relied more on shaming and de-legitimating Morocco. Thus, the issue of human rights in the occupied part of Western Sahara increasingly mattered in the Polisario's international discourse. Finally, the Polisario started to judicialize the conflict by initiating proceedings and filing cases at international courts against Morocco's economic exploitation of the Western Sahara (Ruiz Miguel 2022, 117–135). Voicing as a practice of claim-making moved into the centre of the Polisario's justification efforts. As the politics of international legitimation is a plurilocal phenomenon, the following section traces these practices through different localities where they can be observed.[17]

[16] There are no exact figures available as the numbers have been inferred in part from erratic political declarations and instrumentalized in the conflict's diplomatic struggles (Fernández-Molina and Porges 2020, 386).

[17] Apart from Geneva and New York, I have done research in all these localities in the course of field trips.

In southwestern Algeria, near Tindouf, where the SADR and the Sahrawi refugees are exiled, the Polisario is busy with welcoming and organizing a constant stream of foreign visitors, delegations, diplomats, and NGOs who seek to inform themselves about the situation in the refugee camps, visit families, express solidarity, carry out development projects, and meet local activists or officials from the SADR government. Major events are the festivities on the anniversary of the declaration of independence but also international sport and cultural events that aim at demonstrating solidarity and raising awareness for the Saharawi cause such as the 'Sahara Marathon' or the 'FiSahara' film festival. Even though the camps are located in the middle of the desert, there are annually between 3,000 to 4,000 visitors who arrive at peak times with charter flights going from France or Spain directly to Tindouf. Receiving visitors is not only a welcome opportunity for the Polisario to get international attention but also to make a performance-based claim that the SADR is capable in managing the affairs of the Saharawis, that it is running schools and hospitals, and that it can cater to the needs of a refugee population even under harsh desert conditions.

Another location of the Sahrawi struggle is Geneva, where the United Nations Human Rights Council is based. The human rights discourse has become more relevant for the Polisario since the 1990s. As Sahrawi activists in the occupied part of Western Sahara have at times staged civil protests in the last two decades that have been violently repressed by Morocco, the Polisario has tried to capitalize on these protests by highlighting the role of non-violent resistance and human rights in the Sahrawi national struggle (Zunes and Mundy 2022, 140–166; Mundy 2017, 71–73). However, the group has been slow to adapt. The Polisario has established a presence in Geneva since the end of the 1990s but it only became more effective in 2017, when the Geneva Support Group for the Protection and Promotion of Human Rights in Western Sahara was created. An Italian coordinator of the group recalls that at the first Geneva meeting between European and Saharawi human right defenders, less than 10 people sat around the table. The working group then started to build a network of NGOs and to professionalize the Saharawi human rights activists, showing them how to issue proceedings, submit complaints, and present reports for committees at the UN Council. As the coordinator explains: 'we are educating them.'[18]

Yet another location is Brussels, where the Polisario has a diplomatic representation. A large part of the work in Brussels consists of addressing a European public. Raising awareness about Western Sahara, disseminating the claim and seeking its validation through public acknowledgment is at the heart of the Polisario's diplomacy. Mobilizing solidarity and support is another core aim. Finally, de-legitimizing Morocco makes up a large part of the Polsario's diplomatic efforts. The liberation movement denounces the Moroccan 'occupation' in Western Sahara and highlights the illegality of the trade agreements between the EU and Morocco. The ambassador

[18] Interview with a Coordinator of the Geneva Support Group for the Protection and Promotion of Human Rights in Western Sahara, 3 December 2022, Berlin.

in charge of Europe and the EU coordinates representations of the Polisario across the continent and touring its capitals while also liaising with EU institutions. He meets governmental officials, members of parliaments and political parties, activists, and media in order to brief them about the latest developments, to encourage parliamentary or humanitarian action and to counter the Moroccan narrative. He is also active on Twitter/X. The reactions to these diplomatic activities are mixed. Most states, including the EU itself, claim to be officially neutral in the conflict. In the usual diplomatic wording, Western Sahara is seen as a non-self-governing territory, the final status of which will be determined by the outcome of the ongoing UN process. 'We support the UN process' is thus a typical expression that signals undecidedness or rather 'the position of no position'.[19] In fact, however, the Polisario ambassador faces three different blocs: a sympathetic minority of Nordic countries that lean towards the position of the Polisario, a majority of countries that are more or less indifferent and uninterested in the conflict, and a hostile minority that tacitly or openly supports Morocco such as France and Spain. Especially the interests of the latter two have largely kept the EU from taking a more active role in the conflict while there is a general temptation in Brussels to assume the Moroccan position because of various trade agreements between the EU and Morocco (Benabdallah 2009).

The representatives of the Polisario also collaborate with the European Parliament (EP) and its Intergroup on Western Sahara, which was established in 1982. The Intergroup is a semi-official working group of members of the EP that work closely with the Polisario on issues such as human rights in the occupied territories, the EU–Moroccan commercial agreements and more recently the 'Qatargate' scandal involving allegations that members of the EP took bribes from the governments of Qatar and Morocco. The Intergroup aims to move the parliament into to a more favourable political position regarding the Saharawi cause but also seeks to give more visibility to the struggle of the Polisario, for example with thematic conferences in the EP. The collaboration with the Intergroup can be seen as an attempt to validate the Polisario's claim by officializing the Polisario within the EU institutions. The Intergroup, however, does not formally speak in the name of the EP.

Challenging the EU–Morocco trade relations, the Polisario has achieved a major judicial victory at the European Court of Justice that ruled in favour of the Polisario in 2016, 2021, and 2024. The court decided that the Association and Liberalisation Agreements concluded between the EU and Morocco as well as the EU–Moroccan Trade and Fisheries Agreements are not applicable to Western Sahara as the territory cannot be considered as part of Morocco. Moreover, the court also confirmed that the Polisario is recognized internationally as the representative of the people of Western Sahara and has legal personality under public international law, a status which allows it to bring proceedings before the EU courts (Ruiz Miguel 2022, 153–156).

Brussels also hosts various NGOs such as 'Western Sahara Resource Watch' or 'Independent Diplomat' that seek to expose the exploitation of Western Sahara's

[19] Interview with a diplomat of the Polisario Representation to Europe and the EU, Brussels, 14 February 2023.

natural resources by Morocco and help the Polisario to professionalize its foreign policy. Brussels is also the headquarters of an NGO entitled European Conference of Solidarity and Support for the Saharawi People (EUCOCO). Led by an over eighty-year-old former Belgian senator and Third World activist, EUCOCO is the heart of an aging European solidarity movement. Since the 1970s, it has organized annual conferences, mostly in Spain but also in other European states. EUCOCO usually gathers long-term activists, most of whom have been politically engaged since the 1970s, members of the Polisario, diplomats from states who recognize the SADR or veterans of other liberation movements. Its aim is to discuss political strategies, mobilize support, and ultimately to reiterate and confirm the Polisario's claims to legitimacy. 'Les Sahraouis ont toute la légitimité et toute la légalité' as the former Belgian senator does not tire of saying.[20]

Another theatre of the conflict is the UN headquarters in New York. When the UN became actively involved in the Western Sahara issue in the late 1980s, it was a time of growing optimism. The thawing of the Cold War opened up new opportunities to settle long-standing conflicts, and the UN massively expanded its worldwide peacekeeping missions. In the memoires of Marrack Goulding, former head of the UN peacekeeping operations, it was the heyday of peacekeeping, a 'dizzy decade' (Goulding 2002, 18). The UN attempts to solve the Western Sahara conflict gained momentum shortly after the successful UN mission in Namibia spurred hopes that the UN would be also successful elsewhere. Building on prior OAU and UN peace initiatives, in 1991 the UN Security Council thus established MINURSO to organize a referendum on the future of the disputed territory. Since the beginning of the UN-led peace process, the Polisario counted on the UN and refused any political initiatives outside the UN, which it perceived as the organization that most clearly represented or embodied international legitimacy. As they expressed it: 'le Polisario a refusé toutes les propositions qui viennent du dehors des Nations Unies parce qu' aucune force politique ne peut garantir la légitimité internationale qui est ... encadrée par [les] Nations Unies.'[21]

However, the UN got caught up in a conflicting dual track approach. The norms of decolonization, self-determination, and independent statehood as they are affirmed through the UN General Assembly contrast with the limited crisis management and realpolitik of the UN Security Council, in which key members are concerned for the stabilization of Morocco (Roussellier 2014). The more than 30 years of UN involvement in the conflict is thus a story of dashed hopes. After a decade of stalled preparations for the referendum and the final failure of the Baker Plan in 2003, the UN limited itself to occasionally facilitating talks between the conflict parties, which, however, never produced any results. Other major conflicts started to hold the UN's attention, and with no open violence in Western Sahara, the issue became increasingly less important. The UN began to skirt the issue. As the former head of the

[20] Interview with leader of the EUCOCO, March 16, 2023, Brussels.

[21] Interview with member of the Polisario/Minister Adviser to the SADR Presidency, 1 March 2023, Smara refugee camp.

MINURSO was told: 'we send you there to avoid trouble, not to make trouble'.[22] As a result, the Polisario became deeply disappointed with the inaction of the UN. The Saharawis came to see themselves as victims of their goodwill and confidence in the UN that they by now simply perceive as the 'United Nothing'.[23] The Polisario ambassador to the UN is busy in New York with shaming Morocco, meeting with UN officials, or writing letters to the members of the Security Council. However, Morocco has a good standing. Apart from paying membership fees and contributing to peacekeeping missions, it also has powerful friends in the Security Council. France and the US support Morocco while Russia and China are watching from a distance and the UK has no stake in the outcome (Zunes and Mundy 2022, 219–253). The UN still lists Western Sahara as a non-self-governing territory but avoids any reference to the SADR (Fernández-Molina and Porges 2020, 379). And the Security Council is unlikely to promote a political solution that is against Moroccan interests. In New York and other diplomatic capitals, the Polisario is struggling against indifference and a wall of silence. In the eyes of diplomats, political attention for Western Sahara has always been limited and will remain so. The case is neither important nor virulent enough while there is no chance of finding a political solution in the medium or long term. The SADR and the Polisario are tolerated, but Western Sahara is generally seen as a 'cold case' that is 'on the top of no one's agenda'.[24]

The Moroccan counterclaim

As the main opponent of the Polisario, the Kingdom of Morocco has a diametrically opposed view on the Western Sahara. Today, the Moroccan government, but also parties across the political spectrum, perceive Western Sahara as an integral part of Morocco. An overwhelming majority of the population sees the integration of the territory into the kingdom as a fait accompli. In Rabat, where the Moroccan Ministry of Foreign Affairs is located, the Sahara has become a sort of foreign policy benchmark. In the words of King Mohammed VI: 'the Sahara issue is the lens through which Morocco looks at the world'. Seeing through this lens, Morocco distinguishes in a binary fashion between international partners and opponents. States with neutral or ambiguous positions concerning the Moroccanness of the Sahara are asked to clarify their positions 'in a manner that leaves no room for doubt'.[25] In line with this extraordinary commitment, Morocco's foreign policy is almost exclusively geared towards securing support for the Western Sahara issue. The kingdom pursues

[22] Interview with the former head of the MINURSO mission, 29 March 2023, Hamburg.

[23] Interview with a diplomat of the Polisario Representation to Europe and the EU, Brussels, 6 February 2023.

[24] Interview with an official from the German Embassy, 5 March 2023, Algiers; Interview with the former head of the MINURSO mission, 29 March 2023, Hamburg.

[25] HM King Mohammed VI delivers a speech to the Nation on the occasion of the 69th anniversary of the Revolution of the King and the People, 20 August 2022, https://www.maroc.ma/en/royal-activities/hm-king-mohammed-vi-delivers-speech-nation-occasion-69th-anniversary-revolution, accessed 14 April 2023.

a crucial lobbying effort to convince or force other states to accept the Moroccan perspective. Moroccan diplomats are active around the world in order to achieve an official or tacit recognition of its sovereignty over Western Sahara or at least a de-recognition of the SADR (Pazzanita 1994, 266–270; Fernández-Molina 2016, 63–64). Making a counterclaim while denying the Polisario's aspirations is at the heart of Moroccan foreign politics.

In doing so, Morocco can count on long-standing allies but also on new partners. Since its independence in 1956, Morocco has been an important strategic ally especially for the US and France. In the Cold War, the kingdom was supported for its pro-North Atlantic position as the US sought reliable allies in the broader Middle East region. After 9/11, the monarchy was seen as a major partner in the War on Terror (Mundy 2017). More recently, Morocco is also seen as indispensable in controlling migration towards the EU. In the last two decades, Morocco has also become a partner of the EU in the field of economic cooperation, trade as well as technical and development cooperation. In recent years, a significant convergence in policies towards Morocco has occurred on the part of the US, France, Spain, and the EU. Driven mostly by concerns over terrorism and migration, western governments have taken a much more conciliatory approach towards the monarchy which, in turn, has become more autonomous in its foreign policy choices (Willis 2012, 330). For various political, economic, and security reasons, major Western powers have always been interested in a stable, reliable, and pro-Western Morocco and this implied the tacit support of Morocco's claim on the Western Sahara that has become integral to the political survival of the monarchy (Zunes and Mundy 2022, 59–88). Apart from maintaining good relations to major Western powers, Morocco has also actively developed its stance as a 'puissance africaine' (Abourabi 2021) or middle power, and it has sought new partners on the African continent through all kinds of development cooperation.

In lobbying foreign governments, Morocco achieved a major diplomatic victory in 2020 when the US under the Trump administration abandoned its long-standing, officially neutral position and formally recognized Morocco's sovereignty over Western Sahara. Morocco has also convinced various African states to open diplomatic representations in the disputed territory. There are now around 25 consulates of African states in Western Sahara that are meant to validate the Moroccan claim over this territory. The tacit recognition of Morocco's authority in the Western Sahara is also reflected in the area of trade and resource exploitation. Various commercial deals with international actors and multinational corporations, notably the EU's fishery and trade agreement, include the territory of Western Sahara as part of Morocco (White 2015, 349–353).

Moreover, the monarchy has been successful in bringing various European states to see the problem of the Western Sahara from Rabat's perspective. A noteworthy case is Spain, which is constantly pressured by the kingdom, as Morocco can allow thousands of migrants to enter the Spanish enclaves of Ceuta and Melilla, which are located on the northern shores of Morocco's Mediterranean coast. Morocco and Spain have recently strengthened their diplomatic ties, but the new partner is seen as an 'uncomfortable neighbour' in Spain.[26] A result of Morocco's proactive foreign policy has been that Spain, but also other European states, have endorsed the Moroccan autonomy plan as a solution to the conflict. The autonomy plan, co-created with the US and proposed in 2007 by the Moroccan government, favours an autonomy status for the Sahara within Morocco. The kingdom is promoting the plan as the 'most credible, just and pragmatic' solution to the conflict in official diplomatic statements, and it judges other governments according to their position regarding the autonomy plan. Even though the plan has been rejected by the Polisario as it no longer foresees a referendum over the future of the territory, it tends to be favoured as a possible conflict resolution scenario by important states such as Spain or France. It has not yet become part of the official political positions of the majority of states, 'but they are talking about it', as a UN official put it.[27] Finally, Morocco constantly delegitimizes the Polisario with accusations of terrorism and crime (Ruiz Miguel 2022, 157–173). De-recognizing the Polisario, the kingdom also tries to bilateralize the conflict as a dispute between Morocco and Algeria. According to that view, the conflict over Western Sahara has nothing to do with a people that seeks self-determination. The Polisario is neither an indigenous nor independent movement but rather the front desk of the Algerian Military, a 'construction de l'Algérie' or more simply: 'il n'y a pas de Polisario'.[28] The kingdom argues that the conflict is in reality a dispute between Morocco and Algeria and could be resolved by direct talks between the two.

Another final venue of the conflict is Western Sahara itself or more precisely the part that is controlled by Morocco. In the major settlements such as El-Ayoun or Dakhla, Morocco maintains a heavy security apparatus that closely monitors the population including foreign visitors and that represses any civilian protest. Moreover, Morocco is pushing for the 'Morocconization' or integration of the territory into the Moroccan kingdom. It does so through a policy of normalization by modernizing the infrastructure, promoting businesses including tourism, attracting foreign investments, and providing advantages for settlers such as state jobs or welfare programmes. The monarchy wants to win the hearts and minds of the Saharawi resident population. At the same time, Morocco is seeking to de-internationalize and silence the conflict. Apart from the UN mission MINURSO, there is virtually no international presence in Western Sahara and foreign engagement with the actors on the ground has been very limited, apart from a few official UN trips to the territory. Journalists or NGOs are not welcome. As a result, the conflict and ongoing developments such as the significant political protests since 1999 in Western Sahara have received scant or no media coverage and, thus, no international attention. In the Moroccan press, the Sahara issue is usually trivialized as 'l'affaire' or 'le dossier'.

[26] Interview with an Efe News correspondent, 20 September 2022, Rabat.
[27] Interview with UN official from MINURSO, 28 February 2023, Tindouf /Rabouni.
[28] Interview with a former official of the Moroccan Directorate General of National Security, 23 May 2022, Rabat.

Conclusion

The Polisario is a liberation movement which has successfully acquired international legitimacy for itself as a group but failed to gain full recognition for the state it governs. The SADR has attained international legitimacy in the eyes of a significant quantity of states, but it did not acquire system-wide recognition. The history of the politics of legitimation in the Western Sahara conflict shows that the Saharawi liberation movement has upheld a consistent claim to international legitimacy over half a century. However, the constituency validating that claim has tended to become smaller over the decades. The period of armed confrontation in the first one-and-a-half decades of the conflict was the time when the group was most successful in military and diplomatic terms. Yet, after the agreement of a ceasefire in 1991, the Polisario fortunes started to decline as the conflict became less virulent and publicly visible. In the post-Cold War phase of accommodation, the emphasis on democracy, human rights, and non-violent civil resistance won the Polisario some sympathy here and there, but this did not translate into a fully recognized and independent state of Western Sahara.

Historical times matter for the evolution of armed groups' international legitimacy and this chapter has made two core observations in that regard. The first one is that realpolitik has constantly been a powerful counterweight to the Polisario's claim. While this is true since the early days of the conflict, one can also discern a shift in historical perspective. Since the end of the Cold War, realist considerations tend to prevail. The Polisario was initially well positioned with its quest for recognition. Self-determination as a powerful legitimating idea resonated in the historical time of decolonization. This context was favourable for the Polisario's ideation-based claim for international legitimacy as it allowed the group to gain legitimacy among like-minded Third World allies and within a European solidarity movement that used to be part of the Global Left in the Cold War. Fighting for self-determination and independence was not only a just cause but also a transboundary project, and the Polisario believed that internationalizing the conflict would pay out. However, Morocco has always made a powerful counterclaim for the recognition of its sovereignty over Western Sahara. And in doing so, it was and still is supported by the US, France, Spain, and for that matter, also the EU in the name of realpolitik and strategic interests. From Cold War narratives to the post-9/11 War on Terror and today's EU migration policies, Morocco is seen by major international powers as a reliable and valuable partner whose destabilization by an independent Western Sahara needs to be avoided. Morocco's stability has always been more important than Western Sahara's rights (Mundy 2017, 54). Today, realist considerations seem to be more dominant and justifiable than ever in the case of Western Sahara. The struggle for Western Sahara can be characterized following Max Weber's classical dictum as a conflict between ideas versus interests in which the latter have over time trumped the former.

Another observation of this chapter is that the claim of the Polisario bears a historical imprint that imposes limits on the group's politics of legitimation. The age of late decolonization not only favoured the Polisario's claim but also left a deep historical imprint that makes for strong continuities regarding claims, discourse, actors, political positions, and power constellations.[29] Today, the reference to decolonization continues to be important for the Polisario and its constituency. The Polisario and its supporters frame the conflict as a case of unfinished or stalled decolonization. They see Spain, the former colonial power, as still responsible for the Western Sahara or 'Africa's last colony' that deserves independence and self-determination. The UN continues to view the conflict as a question of decolonization and lists Western Sahara as a non-self-governing territory. Former colonized states and colonial powers continue to play a major role in the conflict. States in Africa and Latin America support the SADR, and the governments of France and Spain back Morocco whereas the grassroots in both countries champion the Polisario. The reference to decolonization is carried through until today by the Polisario. The historical forces of decolonization have left a deep imprint on the Polisario's claim to legitimacy that cannot easily be adapted for instrumental purposes according to changing circumstances. As the long struggle under harsh local conditions shows, the Saharawis in the Algerian refugee camps and the fighters of the Polisario are true believers in what they perceive as a just cause. However, when seeking legitimacy, armed groups need to capture the imagination of their public, and this imagination is time bound. Their claims to legitimacy, in order to get validated, need to resonate with the dominating legitimating ideas of the time, be it the liberation from imperial rule or the protection of human rights. Yet, times have changed and the struggle of the Polisario belongs in a way to a different world historical context, the era of decolonization that was so important for armed groups in gaining international legitimacy.[30] The Polisario has been strong in 'their' historical time. However, today it is fighting a highly asymmetric war with outdated Cold War weaponry against an adversary that relies on modern drone warfare. With an aging leadership and solidarity movement, the Polisario is also struggling to keep up momentum while closely following the recognition scoreboard in an international context that has become less responsive to the cause of liberation movements.[31] Today, the struggle of the Polisario seems to have fallen out of time and is unable to break the wall of silence and political disinterest surrounding the conflict. The claim of the Polisario is still reproduced and validated. It allows the group to keep the question of Western Sahara open, but it increasingly falls on deaf ears while the pragmatic decisions of governments have come to dominate the Western Sahara issue.

[29] On the presence of the past in international politics see also Schlichte and Stetter (2023).

[30] On the age of decolonization and its role for the legitimation of armed groups see the Introduction to this volume.

[31] The recent international solidarity with Palestine in the wake of the Israel–Gaza war since 2023, however, could be seen as a counterexample. On the armed struggle for liberation in Palestine see also the chapters by Christian Olsson on Fatah and Leila Seurat on Hamas in this volume.

References

Abourabi, Yousra. 2021. *La politique africaine du Maroc: Identité de rôle et projection de puissance*. Leiden: Brill.

Benabdallah, Karima. 2009. 'The Position of the European Union on the Western Sahara Conflict'. *Journal of Contemporary European Studies* 17(3): pp. 417–435.

Boukhars, Anouar. 2012. *Simmering Discontent in the Western Sahara*. Washington DC: Carnegie Endowment for International Peace.

Boukhars, Anouar and Jacques Roussellier, eds. 2014. *Perspectives on Western Sahara: Myths, Nationalism, and Geopolitics*. Lanham, MD: Rowman & Littlefield.

Byrne, Jeffrey James. 2016. *Mecca of Revolution: Algeria, Decolonization and the Third World Order*. Oxford: Oxford University Press.

Damis, John. 1983. 'The Western Sahara Conflict: Myths and Realities'. *Middle East Journal* 37(2): pp. 169–179.

Fernández-Molina, Irene, 2016. *Moroccan Foreign Policy under Mohammed VI, 1999–2014*. London: Routledge.

Fernández-Molina, Irene and Matthew Porges. 2020. 'Western Sahara'. In *Routledge Handbook of State Recognition*, edited by Gëzim Visoka, John Doyle, and Edward Newman, pp. 376–390. London: Routledge.

Front Polisario. 1983. *DIX ANS de lutte contre le colonialisme et l'expansionnisme (20 mai 1973—20 mai 1983)*. Ministère de l'Information et de la Culture de la R.A.S.D.

Goulding, Marrack. 2002. *Peacemonger*. Baltimore, MD: John Hopkins University Press.

Hodges, Tom. 1983. *Western Sahara: The Roots of a Desert War*. Westport, CT: Lawrence Hill.

Institut Royal des Etudes Stratégiques. 2021. *Glossary of the Moroccan Sahara Issue/Glossaire de la question du Sahara marocain*. Rabat.

Klich, Ignacio F. 1982. 'Les efforts du Front Polisario en Amérique latine et dans les Caraïbes'. *Le Monde Diplomatique*, July, p. 7.

Miller, Susan Gilson. 2013. *A History of Modern Morocco*. Cambridge: Cambridge University Press.

Mundy, Jacob. 2017. 'The Geopolitical Functions of the Western Sahara Conflict: US Hegemony, Moroccan Stability and Sahrawi Strategies of Resistance'. In *Global, Regional and Local Dimensions of Western Sahara's Protracted Decolonisation: When a Conflict Gets Old*, edited by Raquel Ojeda-Garcia, Irene Fernández-Molina, and Victoria Veguilla, pp. 53–78. New York: Palgrave Macmillian.

Naylor, Phillip C. 1987. 'Spain and France and the Decolonization of Western Sahara: Parity and Paradox, 1975–87'. *Africa Today* 34(3): pp. 7–16.

Pazzanita, Anthony G. 1994. 'Morocco versus Polisario: A Political Interpretation'. *The Journal of Modern African Studies* 32(2): pp. 265–278.

Roussellier, Jacques. 2014. 'The Evolving Role of the United Nations: The Impossible Dual Track?' In *Perspectives on Western Sahara: Myths, Nationalism, and Geopolitics*, edited by Anouar Boukhar and Jacques Roussellier, pp. 119–140. Lanham, MD: Rowman & Littlefield.

Ruiz Miguel, Carlos. 2022. *El Frente Polisario: Desde sus orígenes hasta la actualidad.* Córdoba: Almuzara.

San Martín, Pablo. 2010. *Western Sahara: The Refugee Nation.* Cardiff: University of Wales Press.

Schlichte, Klaus and Stephan Stetter, eds. 2023. *The Historicity of International Politics: Imperialism and the Presence of the Past.* Cambridge: Cambridge University Press.

United Nations General Assembly. 1979. Resolution 34/37 (Document A/RES/34/37), 34th Session. New York.

White, Natasha. 2015. 'Conflict Stalemate in Morocco and Western Sahara: Natural Resources, Legitimacy and Political Recognition'. *British Journal of Middle Eastern Studies* 42(3): pp. 339–357.

Willis, Michael J. 2012. *Politics and Power in the Maghreb: Algeria, Tunisia and Morocco from Independence to the Arab Spring.* London: Hurst.

Zartman, I. W. 2014. 'Morocco's Saharan Policy'. In *Perspectives on Western Sahara: Myths, Nationalism, and Geopolitics*, edited by Anouar Boukhar and Jacques Roussellier, pp. 55–70. Lanham, MD: Rowman & Littlefield.

Zunes, Stephen. 1987. 'Nationalism and Non-Alignment: The Non-Ideology of the Polisario'. *Africa Today* 34(3): pp. 33–46.

Zunes, Stephen and Jacob Mundy. 2022. *Western Sahara: War, Nationalism, and Conflict Irresolution.* 2nd edn. Syracuse, NY: Syracuse University Press.

4
The FSLN and the Politics of Legitimation in the Cold War

Sandinista Revolutionaries and Transnational Activism

Eline van Ommen

In October 1977, Nicaraguan opposition newspaper *La Prensa* published a communiqué signed by some of the country's most prominent men.[1] It denounced the anticommunist dictatorship of Anastasio Somoza Debayle, which ruled the small Central American country through a combination of repression, corruption, and pragmatic alliances. Responding to the high levels of political violence used by Somoza's notorious Guardia Nacional (National Guard), the group of wealthy businessmen, priests, writers, and intellectuals called for an end to the 'violence deployed from above to silence all signs of discontent' in Nicaragua. At the same time as rejecting Somoza's right to govern, the Group of Twelve (as it came to be known) voiced its support for the leftwing guerrillas of the Frente Sandinista de Liberación Nacional (FSLN), who had unsuccessfully tried to overthrow the Somoza dictatorship since the early 1960s. The search for 'permanent and effective peace' in Nicaragua, the Twelve wrote, could only succeed with the participation of the FSLN, which for more than a decade had 'struggled nobly to achieve change in Nicaragua' and showed 'a greater degree of political maturity every day'.[2]

This manifesto by members of Nicaragua's elite was a watershed moment in the FSLN's search for legitimacy, contributing to its transformation from a small and isolated armed group into a popular mass movement with significant international backing. On 19 July 1979, this movement finally toppled the Somoza regime, ushering in a decade of radical revolutionary change, political upheaval, and international intervention. Beyond Nicaragua, the Sandinista victory served as a source of inspiration for the Left. For the first time since the 1959 Cuban Revolution, an anti-imperialist national liberation movement had come to power in Latin America, a region traditionally seen as part of the United States' so-called 'sphere of

[1] Portions of this chapter were adapted from Eline van Ommen, Nicaragua Must Survive: Sandinista Revolutionary Diplomacy in the Global Cold War. Copyright 2024 by the University of California Press. Used by permission of the publisher.

[2] The National Archives, Kew, United Kingdom (hereafter, TNA), FCO 99/44, Translation of Statement by 'the Twelve' as published in *La Prensa*, on 21 October 1977.

Eline van Ommen, *The FSLN and the Politics of Legitimation in the Cold War*. In: *Armed Groups and the Politics of International Legitimation*. Edited by: Stephan Hensell and Klaus Schlichte, Oxford University Press. © Eline van Ommen (2025). DOI: 10.1093/9780198974529.003.0004

influence'. Many believed—falsely, it would later turn out—that it would not be long before the leftist guerrillas of the Frente Farabundo Martí para la Liberación Nacional (Farabundo Martí National Liberation Front, FMLN) would triumph over the US-backed regime in El Salvador.[3] On the other side of the political spectrum, anticommunist states and politicians perceived the Sandinista victory as evidence of the dangerous spread of Soviet and Cuban communism in Central America. In the decade that followed the revolution's triumph, these actors—which included the United States, Central American militaries, and the Argentine and Chilean dictatorships—continued to use extreme levels of violence, political pressure, and economic sanctions to 'roll back' communism in the region. The human costs of the Central American civil wars of the 1980s were enormous; around 75,000 were killed in El Salvador, 200,000 in Guatemala (most of them of indigenous descent), and 50,000 died during the Contra War in Nicaragua. Many more were displaced, wounded, and traumatized. In this context, the Sandinistas could not hold onto power. In February 1990, the FSLN lost the election against a broad opposition coalition and gave up power, bringing an end to the war, the revolution, and the US-imposed economic sanctions.

The Nicaraguan Revolution was a national project, in which revolutionaries set out to radically transform their society, but it was also an international and transnational event, shaping individual lives, social movements, and foreign policies in Central America and beyond. In the years leading up to the insurgents' victory, the project of building revolutionary momentum took place in the mountains, the countryside, and the cities, where guerrillas and sympathizers risked their lives to denounce Somoza and mobilize people for the Sandinista cause. Crucially, the fight also took place in the international arena, as Sandinistas roamed the globe in search of material support and attempted to convince foreign audiences of the validity and legitimacy of their armed struggle. In this period, the FSLN built international alliances and formed relationships with a wide range of state and non-state actors, varying from government leaders and social democrats to human rights campaigners, solidarity activists, and feminist organizations. These actors made concrete contributions to the FSLN's campaign by providing financial aid, political support, and, in some cases, military training and weapons. As an armed group, the FSLN was a key actor in the Nicaraguan civil war, but it was also an international actor, with profound consequences for international politics in the Cold War.

In the late 1970s, the FSLN thus succeeded in obtaining the international legitimacy needed to isolate and overthrow the Somoza regime. The Socialist International (SI), a prominent international organization of socialist, labour, and social democratic politicians, invited Sandinista representatives, such as the priest Ernesto Cardenal, to its well-attended meetings. Latin American governments, such as Mexico, Costa Rica, Cuba, Panama, and Venezuela, called for Somoza's departure and backed the Sandinista struggle by providing the FSLN with visibility, refuge, and

[3] See the chapter by Andrea Oñate-Madrazo on the FMLN in this volume.

financial backing.[4] In June 1979, the Panamanian government offered the Sandinistas an international platform by allowing Miguel d'Escoto, a member of the Group of Twelve, to join its delegation to the Organization of American States (OAS). At the OAS, the Sandinistas' unofficial 'ambassador to all nations' called upon the international community to break off diplomatic relations with the Somoza regime and recognize instead the provisional new government in exile, which was led by the FSLN.[5] By openly siding with the Sandinista guerrillas over the Somoza dictatorship, politicians and governments in the Americas and Europe validated the FSLN's claim to legitimacy on the international stage, thereby contributing significantly to the movement's eventual victory.

But how did it get this far? What caused these states and parties to become part of the Sandinistas' international constituency? To answer these questions, this chapter examines the FSLN's campaign for international legitimacy in the late 1970s, highlighting practices of claim-making that the revolutionaries used to further the FSLN's political legitimacy. It argues that the FSLN and its activists focused on making claims to legitimacy through *voicing*. Addressing a broad international audience and presenting the Sandinista movement as a moderate group with clever arguments regarding the Cold War, human rights, democracy, and social justice was at the core of this claim-making. Political leaders and parties in Latin America and Western Europe validated these claims by *acknowledging* the FSLN as a legitimate political actor and by *supporting* the group, thus broadening the FSLN's international constituency and paving the way for its official recognition.[6] Of course, governments in Latin America and Europe also had their own pragmatic reasons for backing the FSLN, as they responded to domestic pressure, worried about Cold War instability, and sensed an opportunity to challenge US power in the region.

Because the FSLN ultimately fought for power over the Nicaraguan state, it was crucial for the guerrilla group to be recognized as a legitimate political actor by other states. But to get to this point, the insurgents first had to cultivate relationships with non-state actors, who acted as intermediaries between the FSLN and the international community. Solidarity activists in the Americas and Europe, who were not officially part of the FSLN, were at the heart of this strategy. They influenced public opinion, introduced Sandinistas to influential journalists, politicians, and civil society groups, and actively lobbied governments in their own countries to recognize the FSLN as the legitimate representative of the Nicaraguan people.[7] In doing so, these early members of the FSLN's transnational constituency paved the road towards the Sandinistas' transformation into a legitimate political actor on the global stage.

[4] For more on the support of Latin American states for the Sandinistas' revolutionary struggle, see Sánchez Nateras (2022) and Jarquín (2024).

[5] Goshko, John M. 1979. 'OAS Weighs Nicaragua Move'. *The Washington Post*, 22 June; Hovey, Graham. 1979. 'US Proposals on Nicaragua Crisis Meet Sharp Criticism from O.A.S.' *The New York Times*, 23 June.

[6] On these practices of claim-making and claim validation as well as the role of audiences and constituencies see the Introduction to this volume.

[7] On the role of solidarity networks in the legitimation of armed groups see also the contributions of Andrea Oñate-Madrazo on the FMLN and Stephan Hensell on the Polisario in this volume.

By zooming in on the role of non-state actors—drawing particular attention to solidarity activists—in the FSLN's international diplomatic efforts, this chapter contributes to our understanding of the politics of legitimation of armed guerrilla movements in the Cold War context in two critical ways. First, it demonstrates that the FSLN consciously, pragmatically, and successfully developed a global campaign to increase its international legitimacy, with significant consequences for revolutionary trajectories in Nicaragua, Central America, and the global Cold War. In making this point, the chapter engages with new scholarship that highlights the influence of Global South actors in shaping international politics in the Cold War era. Second, the chapter highlights the importance of analysing the domestic and grassroots dynamics of the international politics of legitimation. While the FSLN developed a truly global campaign, it was up to local activists and Sandinista diplomats to ensure these diplomatic efforts fell onto fertile ground. This required intimate knowledge of, amongst other things, a country's history, position in the Cold War, and political culture. In other words, by tailoring the FSLN's message in such a way that it appealed to various domestic audiences, these grassroots organizers made the Sandinistas' revolutionary diplomacy visible, tangible, and impactful. Moreover, the Sandinistas' international campaign was intimately connected to what was happening in Nicaragua itself, as the FSLN attempted to use—relatively small—military victories to further its international support base and, vice versa, used transnational alliances to obtain weapons, money, and a safe haven for its guerrilla fighters. As we shall see below, the road to the Sandinista Revolution was a 'transboundary' process in which the international, transnational, and domestic interacted with one another (Lawson 2019, 8).

A new international strategy

Starting out as a guerrilla movement with a small support base, the FSLN only developed its successful international campaign after more than a decade of military defeats and isolation. The FSLN was founded in the early 1960s by a group of students, peasant leaders, and working-class youth who were inspired by the Cuban Revolution, as well as determined to do something about the high levels of inequality and poverty they encountered within Nicaragua itself. Driven by nationalism as well as anti-imperialism, the guerrillas named their organization after Augusto César Sandino, a rebel leader who fought against the US Marines occupying Nicaragua in the late 1920s and early 1930s.[8] The young Sandinistas dreamed of themselves as the armed vanguard of a popular movement. In reality, the FSLN lacked support and, as a result, went from one military defeat to the next (Borge 1992, 138). In particular, in 1963 and 1967, two failed Sandinista operations, resulting in the death of many guerrillas, alerted the surviving FSLN leadership of the necessity of a new strategy.

Agreeing on an effective strategy, however, turned out to be impossible and the movement split up in three factions. The Tendencia Guerra Prolongada Popular

[8] For more on Sandino, see Bendaña (2016), Gobat (2005), and McPherson (2014).

(Prolonged People's War Tendency) argued that the FSLN, before the conditions of a popular insurrection could be met, should first focus on building a support base in rural areas. The Tendencia Proletaria (Proletarian Tendency) disagreed, believing in collaboration with the working class, the so-called proletariat. The third faction—known as the terceristas—was the Insurrectional Tendency. This ultimately more powerful faction primarily targeted the urban areas, where they organized highly visible military actions meant to provoke a popular uprising. More than the other groups, the terceristas understood the importance of political legitimacy, spectacle, and a broad anti-Somoza opposition movement, led by the FSLN. Tercerista leader Daniel Ortega elaborated on this latter point in 1979, arguing that, by collaborating with Nicaragua's business and economic elites, 'we seek to conserve the political hegemony of the FSLN and [...] avoid the possibility of the bourgeoisie becoming the political leader of an anti-Somoza front' (Ortega quoted after Miranda 1979, 117).

As became clear in the second half of the 1970s, the tercerista plans were the most effective, leading to new recruits, popularity, and legitimacy. The occupation of the National Palace in August 1978 was a particularly successful example of the tercerista strategy. After a small group of Sandinistas took 3,000 members of the National Congress hostage, Somoza agreed to release political prisoners, aired a pro-Sandinista message on the national radio, and paid 50,000 dollars. More than a military action, commander Dora María Téllez later explained, the raid was a successful 'political manoeuvre' to create an insurrectional culture within Nicaragua, convincing many to align themselves with the Sandinista cause (Téllez cited in Arias 1980, 163). Rather than fighting for territorial gains alone then, the Sandinistas used violence as a means of communication, strengthening the movement's claim to legitimacy.[9]

By 1977, the Sandinista movement had expanded significantly. Yet, it was by no means certain that the FSLN would achieve its objective of taking over the Nicaraguan state. Somoza refused to give up power, preferring a continued war with the guerrillas to a negotiated solution with non-Sandinista opposition figures such as *La Prensa* editor Pedro Joaquín Chamorro. Somoza opted for further polarization, a US State Department official commented in October 1978, as he believed that 'we will have no choice but to support him' if it comes to a 'Somoza vs. Marxists' situation.[10] Somoza's calculations about the Cold War's influence on US foreign policymaking were not entirely wrong. Jimmy Carter's administration in the United States did worry about a Soviet-aligned and Cuban-backed government in Central America, even considering a military intervention to keep the leftist Sandinistas out of power. Within the anti-Somoza movement, too, there were challenges to the FSLN's claim to political legitimacy. Fearing a Sandinista takeover, other opposition groups were willing to preserve the National Guard, as well as other institutions closely connected to the dictatorship if it meant that Somoza would step down. The Sandinistas

[9] On fighting as a practice of making a claim to legitimacy see the Introduction to this volume.

[10] FRUS, Volume XV, Central America, 1977–1980, Document 124, Memorandum from the Executive Secretary of the Department of State (Tarnoff) to the President's Assistant for National Security Affairs (Brzezinski), undated.

fiercely rejected this so-called 'Somocismo without Somoza' (Ramírez 2012, 81). They continued to aim for the dictator's military overthrow and, by extension, FSLN leadership over the anti-Somoza struggle and future transformation of Nicaragua.

To that end, the Sandinistas complemented their domestic strategy with an international component. The idea was simple: using the recognition of the international community—particularly Latin American governments—as a tool to force Somoza out and transform the FSLN into a legitimate political force, capable of governing Nicaragua when the opportunity presented itself. In 1977, the tercerista leaders tried to put this strategy into action for the first time. The plan was that, on 13 October, the Sandinistas would call for a general insurrection, prompting Nicaraguans to take to the streets and risk their lives to confront the National Guard. The FSLN would then launch an attack from neighbouring Costa Rica—where many Sandinistas lived in exile—and capture a small strip of land, benefiting from the chaos and distracted National Guard. Using this fact as evidence of the FSLN's legitimacy and military strength, a revolutionary junta, led by entrepreneur Felipe Mántica, would reach out to foreign government leaders and seek diplomatic recognition from the international community.[11]

The plan failed spectacularly, as the guerrillas could not maintain control of the captured territory, making painfully clear that the FSLN did not yet have the strength and manpower needed to defeat the National Guard. The international component of the tercerista strategy, however, proved more durable. The members of the Costa Rican-based revolutionary junta, continued to support the FSLN, but now as part of the abovementioned Group of Twelve, using their wealth and international connections to support the military struggle and significantly expand the FSLN's international constituency. Beyond furthering the Sandinistas' legitimacy, this had tangible consequences for the FSLN's financial and military standing. The writer Sergio Ramírez, for example, managed to get in touch with Venezuelan president Carlos Andrés Pérez by asking Gabriel García Márquez, the famous Latin American author, to deliver a message on the Sandinistas' behalf. After meeting Mántica in Caracas, the president promised to give the FSLN $100,000 per month. Every four weeks, corporate lawyer Joaquín Cuadra Chamorro, the Sandinista 'shadow finance minister', who was recruited into the FSLN by his son Joaquín Cuadra Lacayo, travelled incognito to Venezuela to pick up a suitcase of bills for the young revolutionaries, who could use the money to buy weapons, medicine, and clothing (Ramírez 2012, 86). These practices of supporting can be seen as a validation of the FSLN's claim to legitimacy.

As part of their legitimation strategy, the FSLN appointed ambassadors, installed a revolutionary government, and developed an attractive political programme, thus mimicking the state.[12] The organizational structure employed by the Sandinistas in the late 1970s thus served a practical function, ensuring that the FSLN could implement its international strategy more effectively. Sandinista diplomats were key

[11] Several former Sandinistas, including Ernesto Cardenal, Gioconda Belli, and Sergio Ramírez, speak about this failed attempt to topple Somoza in their memoirs.

[12] On mimicking as a practice of claim-making see the Introduction to this volume.

in getting international audiences to acknowledge and validate the FSLN's claim to legitimacy, as they created alliances and spread positive information about the guerrillas' objectives. The priest and poet Ernesto Cardenal, to give one example, took on the role of travelling ambassador, harnessing his prestige for the Sandinista cause. Seeking good relations and contacts with governments, politicians, intellectuals, and activists around the world, he was constantly on the move, causing his comrades to joke that he would probably 'die in a plane crash' (Cardenal 2005, 43).

The revolutionary diplomats that travelled across the globe employed several tactics and arguments to mobilize artists, journalists, feminists, and various religious groups for the Sandinista cause. Indeed, one of the key characteristics of the FSLN's revolutionary diplomacy was its malleability, as representatives tailored their message to their audiences' preferred interests and preferences. At the heart of this creative strategy, however, was a clear attempt to move beyond the FSLN's previous image of an armed group of Cuban-inspired communists. In private conversations, press interviews, and propaganda materials, Sandinista diplomats stressed the FSLN's moderate tendencies and played down the guerrillas' connections to the Soviet Union, the Eastern bloc, and Cuba. The FSLN, they argued repeatedly, was guided by the three pillars of political pluralism, nonalignment, and a mixed economy. It was 'quite untrue', a Sandinista representative told British officials during a private meeting on 28 June 1978, that the 'extreme Left' was the 'only alternative to Somoza'. On the contrary, they promised, the Nicaraguan revolutionaries were planning to implement 'moderate policies'.[13] Similar promises were made in the weeks leading up to the revolution's triumph on 19 July 1979, when the FSLN shared a written communique with the OAS, which guaranteed democratic freedoms and elections after Somoza's departure (Schmidli 2022, 83).

There were many benefits to presenting the revolutionary struggle as one that transcended the Cold War. In previous decades, the discourse of anticommunism—powerfully propagated by the United States and its regional allies—had successfully delegitimized elected governments, political parties, popular movements, and armed groups. This resulted, amongst other events, in the 1954 Guatemalan coup d'état, the 1973 overthrow of Salvador Allende's government in Chile, and the imprisonment, torture, and disappearance of thousands of Latin Americans that were accused of holding communist sympathies. The 1959 Cuban Revolution quickly resulted in the country's regional isolation and economic dependency on the Soviet Union. Cuban advocacy and support for leftwing armed movements in Latin America drastically declined in the late 1960s, as Fidel Castro was forced to realize his adventurous policies did not result in more successful armed revolutions in the Americas. On the contrary, in the mid-1970s, Latin America's most powerful states, including Brazil, Chile, and Argentina, were ruled by anticommunist dictatorships that did not hesitate to hunt down their critics across borders, even targeting opponents living in exile in Europe and the United States. In other words, for the FSLN, being part of the

[13] TNA, FCO 99/350, Note of a meeting held by Miss M. L. Croll with spokesmen for the provisional government of reconstruction of Nicaragua on 29 June 1979, 2 July 1979.

polarized Cold War landscape was not only extremely dangerous, but Cuba's inability to provide guerrilla movements with the weapons and funding they needed also meant that there was little chance for a successful armed uprising without a more expansive support network.

Moving beyond Cold War rhetoric, then, allowed the FSLN to reach out to new audiences and broaden its transnational constituency to include mainstream politicians and human rights groups. The Socialist International, for example, explicitly rejected the bipolar Cold War system, presenting social democracy as a reasonable alternative to both Soviet communism and US capitalism. Many social democratic Latin American leaders, recognizing that Cold War thinking—and the foreign interventions and violence this has inspired—had been disastrous for the region, were SI members and supportive of the FSLN (see Van Ommen 2024). In the international arena, and specifically in the Americas and Western Europe, the mood was also different from earlier decades. Tired and disillusioned by the bloodshed of previous decades, most notably during the Vietnam War, activists and politicians in the 1970s increasingly embraced ideas and ideologies that transcended Cold War binaries. The rise of human rights language—and its claim to universality and political neutrality—was crucial in this regard.[14] As historians such as Samuel Moyn, William Michael Schmidli, and Debbie Sharnak demonstrated, leftwing activists in the 1970s effectively used human rights to denounce their political opponents, accusing anticommunist dictators—with good reason—of grossly violating their citizens' rights. On the other side of the political spectrum, conservative groups also starting using human rights language, if only to denounce curbs on political freedoms in the Soviet Union more effectively. By presenting the FSLN as an organization that fought for human rights and democracy against an oppressive dictatorship, Sandinistas could thus more effectively reach out to audiences that otherwise would have dismissed the guerrillas as communists.

The FSLN further strengthened the idea that it was more moderate than other Latin American guerrilla groups by integrating religion into their revolutionary diplomacy. Different from the Cuban revolutionaries, who had a notoriously difficult relation with the Catholic Church, the Sandinista guerrillas argued that religion and revolution were not mutually exclusive. As alluded to above, the Group of Twelve included several priests, including the brothers Ernesto and Fernando Cardenal, and Miguel d'Escoto, a Maryknoll priest who studied in the United States. In part, the fact that many Catholic priests supported the Sandinista cause can be explained by the rise of liberation theology in Latin America, which promoted solidarity with the poor and the idea that the Church should bring about positive social change. It is also reflective of the Sandinista strategy to build a broad support base and seek political legitimacy from a wide range of international actors and organizations. As such, on 13 February 1978, Ernesto Cardenal sent a letter to 'priests and pastors in Germany', asking them to 'help the cause of liberation' in Nicaragua. Explaining that the Church

[14] For a similar role of human rights for armed groups' legitimation see the chapters by Andrea Oñate-Madrazo on the FMLN and Klaus Schlichte on the NRA in this volume.

in Nicaragua had already 'taken the side of justice', Cardenal asked his colleagues to do the same thing. Building on the idea of a transnational religious community, Cardenal wrote that the Nicaraguan priests that support the Sandinistas 'feel connected to Father Delp and Pastor Bonhoeffer, who stood up against injustice in their country and did not fear death because they wanted life'.[15] By mentioning these two dissidents, who were murdered by the Nazi regime in 1945, the priest drew a parallel between the German and the Sandinista resistance, and the crucial role that Christians played in both these movements. By aligning with the Catholic Church, then, Sandinistas not only transcended Cold War boundaries but also created new opportunities to expand their transnational constituency.

As we have seen above, the FSLN predominantly targeted audiences in the Americas and Europe. This made sense, as Latin American countries were strategically placed, which was useful for smuggling money and weapons, and their leaders were important diplomatic actors with the ability to further the FSLN's political legitimacy in the Americas and beyond. Western Europe was important, too, not just as a source of money but also because, in the Cold War framework, these countries were seen as the—perhaps more reasonable and moderate—allies of the United States. If Western European peoples and governments backed the Sandinistas' claim to legitimacy, the FSLN reasoned, this would assuage fears in the rest of the capitalist world—and specifically the United States—that the guerrillas were Soviet-inspired communists. Through its outreach to Western Europe, then, the FSLN provided evidence for its claim that it valued political pluralism and a nonaligned foreign policy.

Nevertheless, the United States, which had backed the anticommunist Nicaraguan dictatorship throughout the entire Cold War area, was the most powerful country in the region. Realizing that a shift in US foreign policy could make Somoza's fall much more likely, Sandinista diplomats thus established—albeit limited—contact with US officials. In conversations with US diplomats, Sandinista representatives tried to convince them that it would not be detrimental to the United States' national interests if Somoza was replaced by a government that included the FSLN. Once again seeking to move beyond the Cold War framework, the revolutionaries promoted the idea of a 'moderate' Sandinista movement. On 29 August 1978, for example, Sergio Ramírez, Emilio Baltodano, and Miguel d'Escoto (all members of the Group of Twelve), had a secret meeting with Richard Feinberg at the US embassy in the Nicaraguan capital, Managua. During the meeting, the Sandinista diplomats explained that the FSLN was not a radical communist movement. According to Ramírez, the document that the guerrillas released after the abovementioned raid of the National Palace was, 'moderate if poorly written by the combatants themselves'. Furthermore, they added that the Sandinistas 'were not anti-US'. In fact, Ramírez told Feinberg, one of the references to the United States in the document was even 'favorable'. However, the Nicaraguans pointed out, this could change rapidly if the US did not 'remove' Somoza from office

[15] International Institute of Social History, Amsterdam, the Netherlands, Informationsbüro Nicaragua Wuppertal Collection, not catalogued, Ernesto Cardenal to the priests and pastors of Germany, 13 September 1978.

and 'bloodshed continued'. Obviously, they warned, 'an FSLN that came to power after militarily defeating the GN would be radical'. While US officials were never fully convinced by the Sandinistas' claim to be moderate, the fact that this meeting took place at all should be seen as evidence of the FSLN's expanding international legitimacy.[16] By conversing with the Group of Twelve, even if the meeting took place in secret, US officials acknowledged that the FSLN was a political actor, a relevant party to a conflict that the US hoped to resolve. Ramírez strategically underlined this point in the meeting itself, arguing that 'there were only three forces in Nicaragua—the National Guard (GN), the private sector, and the Sandinistas—and all had to be included in any solution'.[17]

The FSLN's transformation into a legitimate political actor on the international stage impacted its military capabilities and popularity within Nicaragua. As other opposition groups were pushed to the sidelines and a negotiated solution to the country's crises seemed increasingly unlikely, Nicaraguans associated opposition to the dictatorship with allegiance to the guerrillas and support for armed struggle. Somoza was clearly aware of this trend, as he complained about the 'growing respectability' and 'legitimization' of the Sandinista movement to the US ambassador in Managua in February 1978.[18] In private, US officials agreed with Somoza that the FSLN's revolutionary diplomacy—and specifically its ability to bring the guerrillas international legitimacy—ultimately could make the difference in the battle over the Nicaraguan state. 'The real source of strength for those who want to replace Somoza is international public opinion and legitimacy', Robert Pastor of the National Security Council wrote to his colleagues. Somoza is 'so worried about the OAS', Pastor explained, because he knows that 'the real power is in international legitimacy'.[19] And by the late 1970s, it was increasingly clear that the FSLN was succeeding in obtaining that power, which would ultimately put them in charge of the Nicaraguan state.

Exile, solidarity, and revolutionary diplomacy

The Sandinistas' international strategy could not have been successfully implemented without the support of thousands of solidarity activists, acting as grassroots ambassadors for the revolutionary cause. In the late 1970s, these activists were part of a loosely organized transnational network that assisted the Sandinistas' search for international legitimacy and material aid. As we have seen above, the FSLN had a

[16] The Carter administration's foreign policy towards Nicaragua in the late 1970s is a perfect example of its inability to successfully combine human rights and Cold War concerns. See, for more, Schmidli (2012).

[17] FRUS, Volume XV, Central America, 1977–1980, Memorandum of Conversation, Managua, 29 August 1978.

[18] FRUS, Volume XV, Central America, 1977–1980, Document 67, Telegram from the Embassy in Nicaragua to the Department of State, 8 February 1978.

[19] FRUS, Volume XV, Central America, 1977–1980, Document 94, Memorandum from Robert Pastor of the National Security Council Staff to the President's Deputy Assistant for National Security Affairs (Aaron) and the President's Assistant for National Security Affairs (Brzezinski), undated.

clear vision of how it could best sell the revolution abroad. Yet, as this section demonstrates, local activists were key in transforming this broader vision into posters, events, and narratives that appealed to various domestic audiences. Benefitting from in-depth knowledge of their own countries' histories and political landscape, these grassroots ambassadors greatly contributed to the FSLN's campaign for international legitimacy.

One of the first pro-Sandinista solidarity committees was founded in Mexico. Already in 1974, the Nicaraguan poet and FSLN militant Francisco de Asís Fernández asked Thelma Nava, a Mexican poet, to create a solidarity committee that could provide 'international legitimacy' to the Sandinistas' military actions in Nicaragua. The group consisted of prominent Mexican intellectuals, Nicaraguan exiles, and artists (Sánchez Nateras 2020). In the years that followed, Nicaragua solidarity groups were created in many other countries in Europe and the Americas, including Costa Rica, Argentina, Spain, France, the Netherlands, the United Kingdom, France, and West Germany. In 1978, there existed around seventy solidarity committees in Western Europe (Christiaens 2014). While some committees were founded spontaneously by leftwing activists and students, the real driving forces behind the pro-Sandinista networks were Nicaraguans living abroad, such as exchange students, refugees, and expats, as well as travelling Sandinista diplomats. Tercerista commander Humberto Ortega, for example, asked the exiled Nicaraguan poet Gioconda Belli, who lived in the Costa Rican capital of San José, to coordinate the Costa Rican solidarity committee and mobilize regional support for the FSLN's cause. 'Solid support networks' were important for Sandinista militants, Belli remembers, 'especially since Costa Rica was the rear guard for the struggle in Nicaragua' (Belli 2003, 161). Similarly, in Western Europe, Sandinista diplomats such as the Spanish priest Angel Barrajón (who previously lived in Nicaragua), Enrique Schmidt Cuadra, a Nicaraguan exile who lived in West Germany, and the Nicaraguan engineering student Tomás Arguello Chamorro, who lived in the United Kingdom, encouraged local activists to create pro-FSLN solidarity groups and united these committees into a functioning transnational network, which coordinated their campaigns on a Western European scale.

Sandinista ambassadors, who were based in various European and Latin American countries, received requests and information, so-called comunicados, from the FSLN's international committee, the Comisión Exterior in San José, which was led by Gioconda Belli and Melania Agüero. They then passed this information on to local solidarity committees in newsletters and periodicals, who used it to raise awareness about the Sandinistas' military struggle, calling on audiences in the Americas and Europe to provide the young guerrillas with money, medicine, and political support (Helm 2014, 602). Like the revolutionaries, solidarity activists were aware of the importance of international legitimacy for the Sandinista struggle. On 22 February 1978, for example, a representative of the Mexican solidarity committee travelled to Western Europe, bringing with him a list of objectives, which included the international isolation of Somoza and fundraising for weapons and medicine. At the top of the list, however, was the 'recognition for the FSLN as the legitimate representative of the Nicaraguan people'.[20] Even though these activists were not officially part of the FSLN, they consciously worked for the Sandinista cause, considering themselves to be part of the wider revolutionary project.

As such, solidarity activists often adhered to a more radical political ideology than the broader FSLN constituency, which included, as we have seen above, social democrats and human rights activists. What separated many solidarity activists from other pro-Sandinista organizations was their explicit support for armed struggle. The Mexican solidarity committee, for example, distributed flyers with the text: 'A bullet for Somoza. Give moral support. Give financial support. One bullet costs the FSLN 25$. Maybe your bullet will hit Somoza' (quoted in Sánchez Nateras 2020). Sandinista representative Ángel Barrajón, too, spoke about the importance of donating money to the FSLN at a well-attended solidarity event in London, reminding the two hundred people in the audience that the battle for the Nicaraguan state would be won by 'those who had the most and best arms'.[21] The fact that the FSLN was an armed movement meant that—despite the moderate discourse that many Sandinistas employed—it attracted supporters that identified with the radical Left. Solidarity groups in the United Kingdom and the Netherlands, for instance, often consisted of Latin American exiles who had previously been part of militant groups, such as the Chilean Movimiento Izquierda Revolucionaria (Revolutionary Left Movement, MIR). Latin American studies professor Klaas Wellinga, one of the founders of the Dutch solidarity committee, had previously acted as representative for the MIR in Western Europe. Enrique Schmidt Cuadra, too, had worked for the Chile solidarity movement in the mid-1970s. This was reflective of a wider trend. At a time when a revolutionary triumph in Nicaragua was more likely than the overthrow of the anti-communist dictatorship in Chile, many activists who had previously been on the radical flank of the Chile movement decided to join the Nicaraguan cause instead.

Sandinistas and solidarity activists, however, understood that radical language would alienate significant sections of the audiences they hoped to mobilize. In many of their campaigns and events, they thus adopted a moderate and humanitarian tone; claiming that they were raising funds for Nicaraguan refugees in Costa Rica or denouncing the human rights violations of the Somoza regime. Specifically, FSLN representatives stressed to local activists, solidarity committees should point out that the Sandinista movement was 'democratic, popular, and nationalist', and explicitly avoid discussions about its 'socialist character'. In meetings, solidarity activists strategically focused on commonalities and shared interests with their target audience, refusing to engage in 'discussions of principle'. This pragmatic approach also meant that activists did not always ask for international recognition or outright political support for the FSLN. For example, if it was unlikely an organization or politician would

[20] Ángel Barrajón Private Archive, Managua, Nicaragua, letter from Thelma Nava to European solidarity committees, 22 September 1978.

[21] TNA, FCO 99/187, Minter to Perceval, 1 November 1978.

openly side with the Sandinistas, activists focused on the isolation of Somoza instead, detailing the violent repression and corruption that were rampant in Nicaragua.[22]

Solidarity committees complemented individual meetings with an extensive media campaign, raising money for adverts and other promotional materials. In the months leading up to the revolution, the FSLN called upon solidarity groups to buy space for a big advert in the 'most important national newspapers' of their countries, such as *Le Monde*, the *Harald Tribune*, *La Stampa*, *de Standaard*, and the *Frankfurter Allgemeine*. In the advert, the committees explained that the FSLN was the 'vanguard' of the Nicaraguan people, which fought for justice, democracy, and human rights.[23] In their materials, solidarity activists linked the Sandinista struggle to themes that resonated in their own countries. This contributed to heightened levels of interest in Nicaragua, as audiences grew aware that this was not some distant event, but rather something that they could make a tangible contribution to. In some countries, this connection was relatively easy to make. In neighbouring Costa Rica, for instance, solidarity activists promoted a 'jornada de solidaridad con el Pueblo de Nicaragua' in December 1978 by pointing out that their country's 'national sovereignty' was being violated by the National Guard, which frequently crossed the border to hunt down Sandinista guerrillas. According to the flyer, which explains in detail why Somoza was trying to 'provoke a conflict' with his neighbour, Costa Ricans could defend their 'sovereignty by supporting the Sandinista struggle'.[24] Similarly, in the United States, solidarity activists focused on the US government's continued support for 'Somoza's bloody regime'. Undoubtedly, the activists noted in a pamphlet printed on 10 April 1978, 'President Carter's support to Somoza is what maintains him in power'. Creating a powerful sense of urgency and responsibility, the pamphlet explained that 'for every weapon and soldier provided and trained by the US government, there will be at least one Nicaraguan killed who is fighting for his liberty'. As such, readers were urged to write to US Secretary of State Cyrus Vance, demanding that 'the US respect Nicaragua's right to self-determination and that it not continue to side with Somoza against the oppressed people of Nicaragua'.[25] Building a connection with Western European audiences was less straightforward. Next to calling on governments to break off diplomatic relations with Somoza, solidarity committees decided to focus on the fact that Europeans and Nicaraguans had lived through periods of 'oppression and dictatorship', as well as experience with 'fascism'. By implicitly referring to the Second World War, then, solidarity committees attempted to invoke the idea of a shared history. As such, the Western European secretariat of the solidarity network announced in a newspaper advert in May 1979, it was the 'duty' of European peoples' to give 'active support to the struggle of the Nicaraguan people'.[26]

[22] IISG, INW, Bericht und Ergebnisse des Europäischen Treffens der Nicaragua-Solidaritätskomitees in Utrecht/Holland Vom 30. März—1. April 1979.

[23] IISG, INW, Klaas Wellinga to Western European solidarity committees, 8 May 1979.

[24] Senate House Library, Bloomsbury, United Kingdom (hereafter SHL), ER320/PAM/13/4, Flyer, Costa Rica Solidaria: Organo del Comite costarricense de Solidaridad con el pueblo de Nicaragua, date unknown.

[25] SHL, ER320/PAM/1, Pamphlet, The Nicaraguan Struggle Continues, 10 April 1978.

[26] *De Volkskrant*, 18 May 1979.

In addition to voicing as a powerful practice of claim-making, solidarity committees used cultural production as a means to mobilize support for the Sandinista cause. In part, the success of these cultural campaigns traces its roots to the work of Latin American exiles done in the years before. Seeking to escape the violent Southern Cone dictatorships, which hunted down suspected communists across borders, many Latin American leftists went into exile to Western Europe in the 1960s and early 1970s. There, their campaigns against the anticommunist regimes raised public interest in the Americas, as audiences were shocked to hear about the dictatorships' human rights abuses, most notably torture, murder, and forced disappearances.[27] Latin American exiles used political music, art, and literature to grow public awareness about political struggles in Argentina, Chile, Uruguay, Paraguay, and Brazil. The Chilean cause was particularly popular because, after the military coup against the democratically elected government of Salvador Allende on 11 September 1973, thousands of Chileans fled to Europe, where they created solidarity committees to work against the Pinochet regime. The popular Chilean band Inti-Illimani, for example, was forced into exile after the 1973 coup and, in the mid-1970s, they performed hundreds of times in various European cities, raising money for Chileans that were persecuted by the dictatorship (McSherry 2017). Joan Jara, the British wife of Víctor Jara, a Chilean artist who was tortured and murdered by the dictatorship in the days following the 1973 coup, similarly ensured that his music was played in hundreds of concert halls and on the radio throughout Europe and the Americas. In the Netherlands, the Kultuur Kollectief Latijns Amerika (Latin American Cultural Collective, KKLA) translated and distributed works by Latin American authors such as Julio Cortázar and Gabriel García Márquez. They also sold records and organized art exhibitions, concerts, and poetry nights. The musicians who performed at these events delivered a clear political message, addressing social injustices, romanticizing guerrilla warfare, and detailing the suffering of those who were targeted by the dictatorships (Jiménez Botta 2023).

Pro-Sandinista solidarity groups were thus able to build on the work of Southern Cone exiles, which had raised public interest in Latin American affairs, when they started translating and selling Sandinista novels, poetry, art, and music in the late 1970s. Indeed, many in Western Europe already held sympathetic views of Latin American leftists, while they were deeply suspicious of US-backed anticommunist states. Progressive publishing houses, such as the abovementioned Latin American Cultural Collective, were key in propagating the Sandinista cause through the distribution of literature. Hermann Schultz, for example, was the director of *Peter Hammer Verlag*, a publishing house based in the city of Wuppertal, which published translated literature from Africa and Latin America. Next to his work as an editor, Schultz was also a central figure in the transnational solidarity movement for Nicaragua, coordinating campaigns on a West German and Western European scale. Notably, the headquarters of the West German solidarity movement, the Informationsbüro Nicaragua, was based in Wuppertal, and Schultz corresponded frequently

[27] For more on this, see Kelly (2018) and Perry Fauré (2020).

with Sandinista ambassadors Ernesto Cardenal and Sergio Ramírez, whose literary work he greatly admired. By translating and distributing the work of pro-Sandinista artists, people like Schultz contributed greatly to the Sandinistas' international standing. Gioconda Belli's poetry, for example, was increasingly translated, sold, and read by audiences outside Latin America after she won the prestigious Casa de las Américas Prize for her powerful poetry collection *Línea de fuego* (Firing Line), which glorified motherhood, revolution, and resistance. In her book, she makes no secret of her FSLN membership. Indeed, one of the book's three sections, called *Patria o muerte* (homeland or dead), was dedicated to her FSLN comrades (Arenal 1981, 24). Similarly, Sergio Ramírez's novel, *¿Te dio miedo la sangre?* (published in English under the title *To Bury Our Fathers*), which traces Nicaragua's long struggle against imperialism, also had clear political connotations. It appeared in multiple editions and languages, further propelling Nicaragua—and the Sandinista cause—into the international limelight.

Perhaps more important than literature for the Sandinistas' international legitimacy was music. The Nicaraguan Revolution, as Sophie Esch writes, did not have a song—as many other revolutions did—but rather 'an entire soundtrack', with seemingly an individual song for each moment in the revolution's history. The music of the Sandinista Revolution was performed by the singer-songwriters Carlos and Enrique Mejía Godoy, who composed powerful songs about firearms, strikes, love, religion, repression, and revolution. In doing so, the brothers Mejía Godoy romanticized and legitimated the FSLN's armed struggle, which impacted audiences in Nicaragua and around the globe. In 1973, they released 'Cristo ya nació en Palacagüina', which tells the story of Jesus Christ from a Nicaraguan perspective. Born to parents who lived in abject poverty, the song narrates, this little boy did not want to grow up to be a carpenter. Instead, he dreamed of becoming a guerrilla fighter, a 'guerrillero'. In retelling this classic story, the Mejía Godoy brothers created a powerful connection between religion and armed struggle, which resonated in the deeply Catholic country of Nicaragua and across the world. Several years later, in 1979, their music became more explicit in its support for armed struggle, highlighting the growing legitimation of the Sandinista movement. The 1979 album *Guitarra Armada*, for example, included clear instructions on battle tactics, advice on how to handle firearms, and information on homemade explosives (Esch 2018, 130–133).

Next to advising Nicaraguans on the practicalities of armed struggle, pro-Sandinista music helped to legitimate the FSLN's used of violence abroad. Indeed, due to the revolutionary content of their music, the Mejía Godoy brothers performed and recorded most of their music outside Nicaragua. In the late 1970s, for example, Carlos Mejía Godoy gave multiple concerts in Mexico, while Luis Enrique often performed in Costa Rica, where he lived in exile since 1974. In 1977, Carlos and his band fled Nicaragua and settled in Spain, where they were hosted by the progressive priest Victoriano Aritzi (Ágreda Portero 2021). From that moment onwards, the musicians gave many concerts in Western Europe, mobilizing new audiences and raising money for the Sandinista cause. Solidarity activists played key roles in coordinating these shows. In October 1978, for example, solidarity activists organized an extensive international tour for Carlos Mejía Godoy and his band *Los de Palagüina*, performing in multiple cities, such as Berlin, Madrid, Zaragoza, Groningen, and Rotterdam.[28] In the Netherlands, solidarity activists cooperated with a Chilean cultural organization to make the tour a success, allowing Godoy and his band to speak about the 'suffering but struggling people of Nicaragua'. The performances clearly contributed to the FSLN's international legitimacy, as Godoy gave multiple speeches on 'behalf of the Sandinista National Liberation Front'.[29]

More often than not, pro-Sandinista events combined political speeches with discussions about literature, musical performances, and film showings. By combining politics with culture, solidarity activists created a powerful and emotionally charged atmosphere, which allowed audiences to feel connected to the Nicaraguan struggle. For example, the documentary *Nicaragua: patria libre o morir* (*Free Homeland or Death*), filmed in Costa Rica in 1978, and produced in collaboration with solidarity activists from Colombia, the United States, and Mexico, was often shown at solidarity events. The movie premiered on 21 February 1979, on the forty-fourth anniversary of the death of Sandino, and gave audiences an intimate view of the Sandinista guerrillas who risked their lives in the fight against Somoza. The FSLN was closely involved in the production and distribution of this movie, calculating that it would help to raise 'money to purchase arms'. Next to raising funds, the documentary was a powerful tool to make audiences aware of the bravery, popularity, and skills of the young Sandinista fighters, as well as their high levels of popularity amongst the suffering Nicaraguan population.[30] Indeed, as Emilio Rodriguez Vazquez of the Nicaraguan film institute INCINE commented later in an interview, the film had been incredibly effective 'in generating international solidarity' for the Sandinista cause (Burton et al. 1980, 28–31). Using powerful audiovisual materials, then, solidarity activists created an appealing narrative of a popular movement that fought against a violent dictatorship, thereby furthering the Sandinistas' international legitimacy.

The multifaceted nature of the FSLN's revolutionary diplomacy allowed the guerrillas to build connections with a range of progressive organizations, including women's groups and feminist networks. Indeed, the FSLN made a conscious effort to present itself as a progressive organization that promoted gender equality and women's rights. The FSLN, as historian Friederike Apelt writes, actively constructed 'emancipatory gender images' and disseminated them 'all over the world' to further the movement's international legitimacy and attract women to the Sandinista cause (Apelt 2015, 175). The FSLN's historic programme, first published in 1969, had already promised that the revolution would end 'the odious discrimination that women have been subjected to compared to men'.[31] Indeed, more than in any other Latin American guerrilla movement, Sandinista women were active guerrilla fighters,

[28] Instituto de Historia de Nicaragua y Centroamérica, Managua, Nicaragua, ASD-064, Informaciones sobre Nicaragua, Edición 6/78 Octubre, Publicado por el Comité de Solidaridad con Nicaragua en la República Federal de Alemania.

[29] *De Volkskrant*, 27 October 1978.

[30] The full documentary is on YouTube: https://www.youtube.com/watch?v=XKBAOfvrglQ

[31] 'The Historic Programme of the FSLN', as published in Bruce (1985).

making significant contributions to all aspects of the FSLN's military struggle. Most famous was the guerrilla commander Dora Maria Téllez, who led the occupation of the National Palace in 1978, demonstrating that Nicaraguan women rejected oppression—not only by Somoza, but also by traditional gendered expectations. These emancipatory ideas were powerfully reproduced in pro-Sandinista solidarity posters, photographs, music, and literature. The song 'Venancia', by Luis Enrique Godoy, for example, was about a young Nicaraguan woman who became a guerrilla after her brother was murdered by the National Guard. Similarly, Gioconda Belli's literary worlds were populated by 'rebellious, independent, and sexually self-assured women', who are willing to make sacrifices and fight for justice, equality, and freedom (Barbas-Rhoden 2003, 48). Combined, these artistic expressions helped to convince audiences in the Americas and Europe that the FSLN was a legitimate organization that fought for national liberation and against social injustices in Nicaragua, and beyond.

By mobilizing audiences for the Sandinista cause, these grassroots ambassadors thus paved the way for official recognition of the FSLN by governments and political parties in the Americas and Europe. As we have seen above, the Socialist International and progressive Latin American leaders were on the front lines of legitimating the FSLN through practices of acknowledging and supporting. In June 1979, for example, the Andean Pact nations of Venezuela, Colombia, Peru, Bolivia, and Ecuador announced that the Sandinistas were 'legitimate combatants', seeking to build 'a true representative democracy, freedom and justice in Nicaragua'.[32] Western European social democrats backed the FSLN, too. The ruling Sozialdemokratische Partei Deutschlands (SPD), for instance, supported the FSLN by providing political training and financial aid, while the Dutch labour party openly called on people to support the Frente. Overall, though, Western European governments were more hesitant than the Latin Americans when it came to recognizing the FSLN. While critical of Somoza, they generally waited with providing official recognition until after the revolutionary triumph on 19 July 1979, highlighting once again how important military gains and domestic realities are in the politics of legitimation (see Mujal-Leon 1987).

Conclusion

The revolutionary diplomacy of the Sandinistas, creatively implemented by Nicaraguans and solidarity activists, transformed the FSLN into a legitimate actor on the global stage, backed by powerful Latin American governments, social democratic politicians, and progressive organizations around the world. This international legitimacy had tangible consequences for the insurgents' military capabilities, as it resulted in more weapons, money, and the international isolation of the Somoza dictatorship.

[32] Riding, Alan. 1979. 'Rebels in Nicaragua Name Five to Form Provisional Junta'. *The New York Times*, 18 June.

On 19 July 1979, the Sandinista revolutionaries took charge of the Nicaraguan state, and the new revolutionary junta was immediately recognized by governments in the Americas, Europe, and beyond. These governments also provided the junta with aid to rebuild and transform Nicaraguan society. The transnational constituency that the FSLN built in the late 1970s also shaped the revolution's later trajectory. Various groups had different—sometimes opposing and utopian—visions of what the revolution should look like now that the FSLN was in power. These expectations influenced the way in which activists and governments engaged with the revolutionary project, and the policies that the Sandinistas implemented. In many ways, the Sandinistas struggle for international legitimacy in the late 1970s set the stage for the chaotic and violent decade of the 1980s, ultimately resulting in the FSLN losing power after elections in 1990.

This chapter analysed the quest of the FSLN for international legitimacy in the late 1970s from various perspectives. It highlighted the creativity and adaptability of the Sandinistas' foreign policy, which resulted in a diverse transnational constituency. While recognizing that state actors were key players for the Sandinistas' international legitimacy, the chapter argued for the importance of non-state actors in the foreign policy of insurgent movements. In particular, it showed that solidarity activists—acting as grassroots ambassadors—were key in translating and tailoring the Sandinista message to appeal to audiences around the world. While solidarity activists were political actors, their success in popularizing the Sandinistas abroad should be seen as a successful example of voicing legitimacy claims, using a powerful mixture of cultural appeal, appealing narratives, and pragmatic arguments. As audiences became enamoured with the young guerrillas, politicians in the Americas and Europe increasingly realized that they could count on the public's support if they backed the FSLN. As such, solidarity activism not only put the Sandinista cause on the political agenda, but it also encouraged state actors to recognize the guerrillas as legitimate actors on the international stage.

References

Ágreda Portero, José Manuel. 2021. 'Internacionalistas, Activistas y Brigadistas: La Red Transnacional de Solidaridad con Nicaragua desde el Estado Español, 1978–1991'. PhD diss., Universidad de Santiago de Compostela.

Apelt, Friederike. 2015. 'Between Solidarity and Emancipation?: Female Solidarity and Nicaraguan Revolutionary Feminism'. In *Making Sense of the Americas: How Protest Related to America in the 1980s and Beyond*, edited by Jan Hansen, Christian Helm, and Frank Reichherzer, pp. 175–196. Chicago, IL: Chicago University Press.

Arenal, Electa. 1981. 'Two Poets of the Sandinista Struggle'. *Feminist Studies* 7(1): pp. 19–27.

Arias, Pilar. 1980. *Nicaragua: Revolución: Relatos de combatientes del frente sandinista*. Mexico City: Siglo Veintiuno Editores.

Barbas-Rhoden, Laura. 2003. *Writing Women in Central America: Gender and the Fictionalization of History*. Athens, OH: Ohio University Press.

Belli, Gioconda. 2003. *The Country under My Skin: A Memoir of Love and War*. New York: Anchor Books.

Bendaña, Alejandro. 2016. *Sandino: Patrica y Libertad*. Managua: Anamá Ediciones.

Borge, Tomás. 1992. *The Patient Impatience*. Willimantic, CT: Curbstone Press.

Bruce, Marcus, ed. 1985. *Sandinistas Speak: Speeches, Writings, and Interviews with Leaders of Nicaragua's Revolution*. New York: Pathfinders Press.

Burton, Jullianne, Emilio Rodriguez Vazquez, and Carlos Vicente Ibarra. 1980. 'Filmmaking in Nicaragua from Insurrection to INCINE: An Interview with Emilio Rodriguez Vazque and Carlos Vicente Ibarro'. *Cinéaste* 10(2): pp. 28–31.

Cardenal, Ernesto. 2005. *La revolución perdida. Memorias III*. Buenos Aires: Fondo de Cultura Económica de Argentina.

Christiaens, Kim. 2014. 'Between Diplomacy and Solidarity: Western European Support Networks for Sandinista Nicaragua'. *European Review of History* 21(4): pp. 617–634.

Esch, Sophie. 2018. *Modernity at Gunpoint: Firearms, Politics, and Culture in Mexico and Central America*. Pittsburgh, PA: University of Pittsburgh Press.

Gobat, Michel. 2005. *Confronting the American Dream: Nicaragua under US Imperial Rule*. Durham, NC: Duke University Press.

Helm, Christian. 2014. 'Booming Solidarity: Sandinista Nicaragua and the West German Solidarity Movement in the 1980s'. *European Review of History* 21(4): pp. 597–615.

Jarquín, Mateo. 2024. *The Sandinista Revolution: A Global Latin American History*. Chapel Hill, NC: UNC Press.

Jiménez Botta, Felix A. 2023. 'Popular Culture and Celebrity Activism in the British Solidarity Campaign with Sandinista Nicaragua, 1979–1990'. *Journal of Transatlantic Studies* 21: pp. 30–53.

Kelly, Patrick William. 2018. *Sovereign Emergencies: Latin America and the Making of Global Human Rights Politics*. Cambridge: Cambridge University Press.

Lawson, George. 2019. *Anatomies of Revolution*. Cambridge: Cambridge University Press.

McPherson, Alan. 2014. *The Invaded: How Latin Americans and Their Allies Fought and Ended US Occupations*. New York: Oxford University Press.

McSherry, J. Patrice. 2017. 'The Political Impact of Chilean New Song in Exile'. *Latin American Perspectives* 44(5): pp. 13–29.

Miranda, Pedro. 1979. 'Sandinista Perspectives: Three Differing Views: Interview with Daniel Ortega 1979'. *Latin American Perspectives* 6(1): pp. 114–118.

Mujal-Leon, Eusebio. 1987. 'The West German Social Democratic Party and the Politics of Internationalism in Central America'. *Journal of Interamerican Studies and World Affairs* 29(4): pp. 89–123.

Perry Fauré, Mariana. 2020. *Exilio y Renovación: Transferencia política del socialismo chileno en Europa Occidental, 1973–1988*. Santiago de Chile: Ariadna Ediciones.

Ramírez, Sergio. 2012. *Adiós Muchachos: A Memoir of the Sandinista Revolution*. Durham, NC: Duke University Press.

Sánchez Nateras, Gerardo. 2020. '"¡Nicas y Mexicanos Solidarios como Hermanos!": El Movimiento Mexicano de Solidaridad con Nicaragua (1974–1979)'. *Secuencia* 108: pp. 6–33.

Sánchez Nateras, Gerardo. 2022. *La última revolución: La insurrección sandinista y la Guerra Fría interamericana*. Mexico City: Secretaría de Relaciones Exteriores.

Schmidli, William Michael. 2012. '"The Most Sophisticated Intervention We Have Seen": The Carter Administration and the Nicaraguan Crisis, 1978–1979'. *Diplomacy & Statecraft* 23(1): pp. 66–86.

Schmidli, William Michael. 2022. *Freedom on the Offensive: Human Rights, Democracy Promotion, and US Interventionism in the Late Cold War*. Ithaca, NY: Cornell University Press.

Van Ommen, Eline. 2024. *Nicaragua Must Survive: Sandinista Revolutionary Diplomacy in the Global Cold War*. Oakland, CA: University of California Press.

5

Tequila Diplomacy

Mexican–FMLN Relations in the Early Salvadoran Civil War

Andrea Oñate-Madrazo

On 7 April 1993, former Salvadoran guerrilla leader Joaquin Villalobos met with the then Mexican president, Carlos Salinas de Gortari. An iconic photograph of this gathering shows a jubilant baby-faced Villalobos handing a pensive Salinas an AK-47 Kalashnikov automatic rifle. The Soviet-made weapon had been given to Villalobos by Cuban leader Fidel Castro in the early 1980s to support the Salvadoran revolutionary leader in his fight against the US-trained-and-armed Salvadoran army. Now, Villalobos was gifting it to Carlos Salinas as a signal of his outmost gratitude and appreciation 'for Mexico's role in helping to bring peace and a democratic transition in El Salvador'.[1] The fact that the Soviet weapon made its way to support the Salvadoran guerrillas through Cuba, and ended up in the hands of Mexico's head of state after the civil war in El Salvador, speaks volumes about how the Salvadoran armed group, Frente Farabundo Martí Para la Liberación Nacional (FMLN), sought out and acquired international legitimacy through an audacious transnational diplomacy that targeted diverse audiences. By 1993, when Villalobos gifted Salinas his rifle, the FMLN had abandoned its armed struggle and was recognized, both at home and abroad, as a legitimate political party of opposition in El Salvador. As this chapter will argue, Mexico had played a consequential role in this process.

The Salvadoran Civil War between the state and army on one side, and the FMLN on the other, unfolded from January 1981 through January 1992 in the smallest of the Central American nations. The military forces, which alleged to embody 'the state' and uphold 'law and order', were backed politically and diplomatically by Washington. They fought to maintain the status quo and protect the interests of the country's economic and political elite. Their national support base consisted primarily of business owners and landowners, the mainstream political parties, and the military apparatus. The second group, the FMLN, was an umbrella organization of leftist armed groups. It consisted of the country's different leftist elements—four guerrilla movements, the Communist Party, and mass-based organizations of workers, peasants, and university students. The FMLN also had a civilian counterpart, the Democratic Revolutionary Front (FDR), made up of politicians disenchanted with

[1] Villalobos, Joaquín. Interview with author. Mexico City, Mexico, 30 July 2011.

Andrea Oñate-Madrazo, *Tequila Diplomacy*. In: *Armed Groups and the Politics of International Legitimation*. Edited by: Stephan Hensell and Klaus Schlichte, Oxford University Press.
DOI: 10.1093/9780198974529.003.0005

the elite–military alliance and government-sponsored violence, social democrats, the Jesuit Central American University, the University of El Salvador, and the coordinating structures of the social movement controlled by the guerrillas (Martínez Álvarez 2010, 16). Together, the FMLN-FDR claimed to champion the interests of the oppressed, repressed, and impoverished *pueblo salvadoreño*—Salvadoran people—by advancing social and economic justice.

The conflict in El Salvador left more than 75,000 dead and over a million displaced. While it was a national conflict in which the belligerents fought over control of the state, the civil war was also, from its outset, an international conflict at the heart of the global Cold War. During this time, El Salvador became the theatre where international actors fought out ideological battles and attempted—by influencing El Salvador—to shape broader regional and international dynamics. In turn, the Salvadoran FMLN capitalized on foreign involvement to further its own bids for political power. Since its birth, the FMLN sought out international legitimacy. Just as the Salvadoran insurgents depended on M-16 and FAL rifles to fight the state in the countryside at home, they simultaneously relied on press releases, international institutions, and transnational organizations as they jockeyed state and non-state actors for recognition on the international stage. Between the FMLN's formation in 1980 and the end of the civil war in 1992, the guerrillas built a formidable diplomatic apparatus that secured them acknowledgment, support, and officialization from myriad state and non-state actors of differing ideological persuasions. These included not only leftist revolutionary actors and those allied with the Eastern camp, but also European Western democracies and social democrats; the Socialist International; United States activists, intellectuals, and progressive Congress members; Mexican diplomats; and the Non-Aligned Movement, to name only a few. Spearheaded by the FMLN's Political Diplomatic Commission (CPD), which was based in Mexico City, the Salvadoran revolutionaries *cum* diplomats transcended the Cold War divide and appealed to a diverse body of international audiences. These international efforts alongside their military accomplishments on the domestic front proved crucial to the Salvadoran insurgents' quest to become legitimate political actors, both at home and abroad, and rightful representatives of the Salvadoran people.

When the Salvadoran war finally came to an end in 1992, it was through a peace treaty in which the FMLN and the Salvadoran state sat across from each other at the negotiation table as equals. The end of the conflict was mediated by a high-profile international delegation and by the United Nations (UN), marking the first instance in which the United Nations—arguably the most respected international institution at the time—brokered a peace agreement between warring parties to a civil war. Facilitators to the negotiations included UN Secretary General Javier Perez de Cuellar, Latin American and European Ambassadors to the UN, and high-ranking members of the United States government. While over the course of the war the FMLN secured notable military victories, acquired territorial control in important areas of the country, and successfully mimicked the state in the areas under its control by providing services and administering public institutions, the insurgents were unable to secure an all-out military victory over the US-backed Salvadoran state and did not seize political control at the national level. The success of their efforts to secure recognition and legitimacy on the international stage, however, were indisputable. Over the course of the decade the FMLN had effectively become a parallel state in the international arena. Both military and political endeavours were essential to the FMLN's ultimate ability to secure a position at the negotiating table—on an equal footing with Salvadoran elites—and to foster a transition to a more inclusive political order in their country.

This chapter examines the decisive role that Mexican state actors played in legitimating the FMLN on the international stage in the first years of the Salvadoran Civil War. It draws particular attention to the importance of Mexico supporting the insurgents by providing a haven for their CPD and to the Mexican government's efforts to secure international recognition for the FMLN as a legitimate force of opposition through the Franco-Mexican Declaration of 1981. Based in Mexico's capital, the FMLN's Diplomatic Commission made a claim to international legitimacy by *mimicking* a state's foreign ministry and by *voicing* its political messages to an international audience. The Mexican government, in turn, validated these claims through *supporting* the Salvadoran guerrillas with refuge and through *acknowledging* and *officializing* the FMLN via the Franco-Mexican Declaration.[2] Mexico, thus, played a critical role in the FMLN's bid to become a legitimate political actor in international politics. Mexico's agency in this process highlights the significance of legitimation by regional actors and evidences how regional legitimacy can then expand into broader international recognition for armed groups. Drawing on newly available documents from a private FMLN archive, and interviews with former leaders of the Salvadoran revolutionary movement and Mexican government, this chapter furthers our understanding of the politics of legitimation of armed movements, specifically within the context of the late Cold War in Latin America.

First, it elucidates the importance of the FMLN's international diplomatic efforts in furthering their legitimacy and clout. Second, it adds to the growing body of scholarship on the importance of regional actors, beyond the United States and Cuba, in shaping the history of Latin America's Cold War. Third, in uncovering how the Mexico–FMLN connection shaped the FMLN's trajectory, and also increased Mexico's regional clout and helped to offset some domestic opposition to the regime, this chapter sheds light on the ways in which the politics of legitimation of armed groups can be transformative not just for the armed groups seeking legitimacy, but also for established states that are validating their claims. Fourth, 'Tequila Diplomacy' speaks to the potential costs of securing international legitimacy for the armed groups being legitimated. While the officialization obtained through the Franco-Mexican Declaration was a crucial win for the FMLN, this public recognition limited the military options available to the insurgents and worked to curtail their political and economic goals. Finally, while the FMLN's dealings with Mexico and the diplomatic efforts of their CPD were seemingly of a non-military nature, the FMLN's claim-making through fighting and military strength at home was integral to their

[2] On these practices of claim-making and claim validation, see the Introduction to this volume.

securing claim-validation on the international stage. The FMLN was simultaneously an armed movement contesting the power of the state at home through violence and a transnational political and diplomatic actor shaping and being shaped by the broader geopolitical order of the late Cold War.

Mexico as haven and the CPD

When different Salvadoran revolutionary factions met in Cuba in 1979 and 1980 to discuss the terms of their union, they set out to build a political diplomatic body to work in tandem with their insurgent military organization. In November 1980, the incipient FMLN created its CPD. It began with a mere seven members and was presided over by social democratic politicians and FDR members Guillermo Ungo and Rubén Zamora. It had one representative from each of the five guerrilla organizations of the FMLN and was subdivided into areas of operation and countries of interest. Its original purpose was to contact foreign governments and prepare to form a provisional government to take power when the FMLN's impending insurrection succeeded. The CPD was entrusted with lobbying foreign governments, building relations with social actors abroad, and promoting the FMLN's legitimacy internationally by presenting it as a genuine representative of Salvadoran interests. In this sense, the CPD was intended to mimic a nation's foreign ministry and it became the most important avenue through which the FMLN voiced its political claims to diverse international audiences.

The Commission was subdivided into areas of operation and countries of interest. As early as 1981 it began to send delegates abroad as foreign attachés and Salvadoran exiles were designated as representatives of the FMLN abroad.[3] CPD attachés were entrusted with identifying potential supporters in governments, progressive political parties, and sympathetic members of civil society, and they kept the CPD leadership informed about developments in the countries they were posted. They voiced the FMLN's claims to a diverse array of international audiences by drawing attention to the exclusionary political and economic orders in El Salvador that had long been protected and perpetuated by an elite–military alliance, and by exposing the horrific violence propagated by the military regime against anyone considered to be sympathetic to the guerrillas. CPD members did this by sustaining private meetings with potential sympathizers, holding fundraisers, sponsoring cultural events, and hosting film screenings. In effect, the role played by FMLN foreign attachés closely paralleled the duties of official diplomats thereby mimicking the de facto state on the international arena, save for the notable difference in budgets and—depending on where they were posted—the need for FMLN diplomats to keep a low profile. Although the CPD's central office was stationed in Mexico, and its leadership spent most of its time travelling abroad, the Commission was in constant communication with FMLN military leaders on the ground in El Salvador. ERP leader Joaquín Villalobos recalls

[3] Villalobos, Joaquín. Interview with author. San Salvador, El Salvador, 22 July 2012.

that throughout his years stationed in the department of Morazán he could and did communicate with CPD leaders frequently and that this was also the case for the other military factions of the FMLN.[4] In effect, the CPD and the five military groups were two pillars of a united organization and both military campaigns and political diplomatic strategies were formulated in concert with one another.

The decision by Salvadoran insurgents to form a Diplomatic Commission was deeply influenced by the trajectory of the Nicaraguan revolutionary organization, Frente Sandinista de Liberación Nacional, FSLN.[5] In the months leading up to the Nicaraguan Revolution's triumph in July 1979, the FSLN allied with a group of prominent intellectuals, businessmen, and progressive politicians known as *El Grupo de los Doce* (the Group of Twelve). This group, which formed in 1977, helped the FSLN secure recognition abroad by fostering an international audience that would condemn the Somoza regime and consider the FSLN a genuine representative of the Nicaraguan people. The political support the FSLN received from the *Grupo de los Doce* was crucial. Not only did this group acknowledge the FSLN as a political actor, but by including economic elites and prominent politicians it also helped to assuage fears abroad that the FSLN was a radical Marxist group and a Soviet proxy. In the aftermath of Nicaragua's revolutionary triumph, the *Grupo de los Doce* played a key role in the legitimation of the new government both domestically and internationally. The role that the *Grupo de los Doce* played in the Sandinista Revolution served as a referent for the Salvadoran insurgents.[6] While not the subject of this paper, the impact of the Nicaraguan Revolution on the Salvadoran FMLN was extensive. The Nicaraguan Revolution was an ideational referent and source of inspiration for Salvadoran insurgents and, more importantly, the revolutionary regime was a critical military ally of the FMLN. The FSLN's contributions to the FMLN's military prowess over the course of the 1980s was surpassed in importance only by Cuba.

The FMLN's first large-scale military operation in January 1981, optimistically called the *final offensive,* failed to take control of the country and did not overthrow the Salvadoran regime. This severely shook the Salvadoran insurgent leadership. They had expected a swift military victory through insurrection, as had been the case for their Nicaraguan allies less than two years prior. When this did not materialize, the FMLN was forced to complement its military operations with a parallel civilian track and the CPD's functions changed accordingly. While still responsible for conducting international relations, the objective of the CPD now also became to galvanize international acknowledgement and political and diplomatic support for the Salvadoran armed movement with the goal of officializing and legitimating it abroad (Martinez 1997). With these objectives in mind, CPD members reached out to foreign governments, international institutions, and non-state actors, and the

[4] Villalobos, Joaquín. Interview with author. San Salvador, El Salvador, 22 July 2012.

[5] See the chapter by Eline van Ommen on the FSLN in this volume.

[6] Joaquín Villalobos and Ana Guadalupe Martínez both mentioned the Nicaraguan *Grupo de los Doce* as a crucial referent for the FMLN in its decision to create a political diplomatic commission. Villalobos, Joaquín. Interview with author. Mexico City, Mexico, 30 July 2011; Martínez, Ana Guadalupe. Interview with author. San Salvador, El Salvador, 16 June 2012.

CPD began sending representatives abroad as foreign attachés and using Salvadoran exiles as diplomatic representatives of the FMLN.[7]

While the military factions of the FMLN were based in El Salvador, the CPD required an extraterritorial base from which its members could travel freely and communicate with foreign actors. By 1980, when the FMLN and its CPD came into existence, the levels of repression in El Salvador were such that anybody suspected of sympathizing with the guerrillas could be rounded-up, imprisoned, tortured, 'disappeared', or killed by the Salvadoran army, police, or right-wing paramilitary death squads, as journalist Mark Danner graphically described (Danner 1994, 25–26).

Travel privileges for Salvadorans were restricted and the government closely monitored citizens' private communications. Thus, the FMLN needed to find a politically stable country, sympathetic to the Salvadoran opposition, from which the CPD could coordinate its affairs. They found a willing and suitable host in Mexico, a country with a revolutionary history of its own and an established reputation for welcoming foreign leftist dissidents.

During the Cold War, Mexico pursued a multi-faceted foreign policy. While the country shared a 2,000-mile border and strong economic ties with the United States, it maintained open diplomatic relations with the Soviet Union and its allied countries. When the Cuban Revolution of 1959 shook the continental axiom and the United States took great strides to isolate the island economically and politically, Mexico publicly stood by the Castro administration, never ceasing commercial and diplomatic relations with the country. In fact, Mexican presidents tried to capitalize on the revolutionary momentum in the aftermath of the Cuban Revolution by publicly standing alongside the new government.[8] The same was true when Chilean president Salvador Allende launched his social democratic experiment in the southern cone. After Allende came to power in 1970, Mexican president Luis Echeverría travelled to Santiago and embraced the Chilean experiment as if it were a natural successor to the Mexican Revolution of 1910 (Riding 1989, 346).

Maintaining diplomatic relations with the United States and with Cuba in the midst of the Cold War required audacity, skill, and often double-dealing on the part of Mexican diplomats, but it served the Mexican administration well.[9] Beyond sending the message to domestic audiences that Mexico stood by its fellow revolutionary nations, an unwritten pact was established between Cuban leader Fidel Castro and his Mexican counterparts: so long as Mexico stood by Cuba, Castro would not offer any assistance or support to subversive and insurgent movements in Mexico. Until the end of the twentieth century, both kept their end of the bargain. Beyond using radical rhetoric and diplomatically embracing leftist regimes in the region, Mexican leaders exhibited a marked receptivity towards foreign liberal and leftist dissidents

[7] Sancho, Eduardo. Interview with author. San Salvador, El Salvador, 22 July 2012.

[8] For a recent and particularly insightful work discussing Mexico's relationship with the Cuban Revolution and involvement in the regional Cold War, see Keller (2015).

[9] The Cuban Embassy in Mexico and the homes of Cuban diplomats were tapped by the CIA with Mexico's consent, and the passenger list of all flights going to Havana from Mexico City were overseen by the CIA. See Iber (2014), Morley (2008), and Castañeda (1994).

and political refugees fleeing the right-wing dictatorships that took hold in the southern cone in the 1970s and 80s. When their time came, the Salvadoran guerrillas benefitted from Mexico's open doors and used the country as a safe base to rest and hold meetings, and as a springboard to travel to other countries. An anecdote by Joaquín Villalobos proves illustrative regarding Mexico's lax attitude towards foreign dissidents. On a trip sometime in the 1970s Villalobos was stopped at Mexico City's Benito Juárez Airport and his luggage, which was full of revolutionary propaganda and pamphlets, searched. 'I was scared shitless', he recalls, 'in El Salvador that would secure my imprisonment, torture and disappearance'. To his relief (and disbelief), the Mexican officer doing the searching seemed enthralled and asked sheepishly if Villalobos could give him some pamphlets as personal souvenirs. After Villalobos more that willingly acceded, he was sent on his way.[10]

Eduardo Sancho, an FMLN commander and head of the *Resistencia Nacional* (RN), one of the five factions making up the FMLN, recounts that the time he and his comrades spent in Mexico in the 1970s was formative and consequential. It was here that the FMLN established its first contacts with the international press thereby gaining a platform to voice its claims abroad. The Mexican publications *Uno Más Uno* and *Jornada* took an interest in the young revolutionaries, interviewed them, and helped put them in touch with news outlets and reporters from other parts of the world. In Sancho's words, 'Mexican reporters gave us, for the first time ever, an international voice'.[11] The FMLN's relationship with the foreign press became increasingly important in later years as the Salvadoran war became the subject of scrupulous monitoring outside of El Salvador and the battle for international legitimacy took centre stage. The FMLN would draw upon these contacts with the foreign press in the effort to broadcast and disseminate their actions and ideals in a positive way and contrast them with the material actions of the Salvadoran military. Furthermore, during their early visits to Mexico, Salvadoran guerrillas built relations with sympathetic activists and non-governmental organizations. These solidarity networks would in later years support the FMLN's efforts in various ways.[12] Throughout the 1980s, they organized charity events to help collect funds for the Salvadoran insurgents, offered guerrillas places to stay when they travelled to Mexico, helped put the Salvadorans in touch with similarly sympathetic groups in other parts of the region, and some even went to El Salvador where they joined the ranks of the FMLN and fought the Salvadoran military alongside their FMLN comrades.[13]

Mexico's willingness to receive the international left served not only the foreign dissidents being welcomed but also the Mexican regime. In part, this openness

[10] Villalobos, Joaquín. Interview with author. Mexico City, Mexico, 30 July 2011.

[11] Sancho, Eduardo. Interview with author. San Salvador, El Salvador, 12 July 2010.

[12] On the role of solidarity networks in the legitimation of armed groups, see also the contributions of Eline van Ommen on the FSLN and Stephan Hensell on the Polisario in this volume.

[13] Amongst the Mexican nationals who joined the FMLN was Bernardo Gonzalez, a Mexican doctor who travelled to El Salvador and served as doctor and surgeon for the ERP leadership that was based in the region of Morazán throughout the entire 1980s. His wife at the time, Guadalupe Gonzalez, also a Mexican national, served as one of the FMLN's chief representatives in Washington D.C. through the FMLN's CPD office. Gonzalez, Bernardo. Interview with author. Mexico City, Mexico, 28 July 2011 and Gonzalez, Guadalupe. Interview with author. Mexico City, Mexico, 4 August 2011.

was intended to serve as proof of Mexico's alleged revolutionary nature, but it also gained Mexican leaders the sympathy of the international left. Dissidents that came to Mexico felt grateful to the regime that welcomed and protected them. Comparing conditions in Mexico to the lack of political freedom and levels of repression in their own countries, Mexico's one-party state seemed a very favourable alternative to authoritarian right-wing military dictatorships elsewhere in the region. Some of the leftist exiles that came to Mexico after fleeing the terror and repression of right-wing dictatorships elsewhere in the region joined the university ranks and shared their positive views of conditions in Mexico with young and impressionable students. So too did activists and foreign guerrillas who met with Mexicans eager to be in close proximity to anything smelling of revolutionary gunpowder. Maria Socorro Álvarez, an activist Mexican student who eventually went to fight with the FMLN's ERP faction in El Salvador, remembers how the foreign radicals she and her comrades admired and sought out spoke positively of conditions in Mexico and encouraged activists like herself to direct their revolutionary fervor and militancy abroad rather than at home.[14] Whether the attenuation of domestic opposition was something that Mexican leaders intended a priori, or rather a fortunate by-product of a policy guided by other considerations, the result was that Mexico's embrace of foreign leftists helped it to co-opt a sector of its own population.

The Mexican administration's first contacts with the Salvadoran opposition were in the late 1970s.[15] Gustavo Carvajal, who served from 1979–1981 as leader of Mexico's ruling party, the Institutional Revolutionary Party (PRI), claims that the decisive moment in the Mexican–FMLN nexus came in July 1980, when during a state visit to Cuba Mexican president José López Portillo was directly asked by Fidel Castro to consider supporting the bourgeoning revolutionary movement in El Salvador.[16] Fidel Castro's call to López Portillo was met with predictable acceptance. Gustavo Carvajal was then instructed to meet with the leader of the Salvadoran Communist Party, Shafik Handal. Subsequently, Mexican foreign minister Jorge Castañeda de la Rosa and Carvajal held a series of gatherings with Salvadoran guerrilla leaders to discuss the terms of their support.[17] Mexico explicitly conditioned its backing on the FMLN agreeing to three preconditions: first, the FMLN's activities in Mexico would be solely political and Mexico would not offer military support; second, the FMLN would not establish relations with dissident movements in Mexico and would refrain from supporting or encouraging them in any way; third, if the FMLN broke

[14] Álvarez, Maria Socorro. Interview with author. Mexico City, Mexico, 23 July 2011.

[15] On 12 October 1979, Mexico's PRI created COPPPAL, a transnational organization of Latin American parties considered to be 'of the left.' According to COPPPAL's first president, Gustavo Carvajal, the organization was intended to strengthen the Latin American left and, by extension, Mexico's clout in the hemisphere. Scholar Fernando Pedrosa contends that the creation of COPPPAL cannot but be seen as a strategy by Mexico's PRI—who was not a member of the Socialist International—to ensure that it did not lag behind European social democracies who were making strides in the Americas. Carvajal, Gustavo. Interview with author. Mexico City, Mexico, 25 July 2011; Pedrosa (2015).

[16] Carvajal interview with author. Mexico City, 25 July 2011.

[17] Carvajal, Gustavo. Interview with author. Mexico City, Mexico, 25 July 2011; Sancho, Eduardo. Interview with author. San Salvador, El Salvador, 22 July 2012.

the first two rules they would be swiftly expelled from the country.[18] The Salvadoran revolutionary leaders agreed and promptly began setting up their diplomatic office in Mexico City's Colonia Roma. From this cosmopolitan Latin American capital and with the support of the Mexican government, the FMLN would build and coordinate its nexus of revolutionary diplomats.

Mexico's preconditions offer fascinating insight into the dynamics underlying Mexican foreign policy and how the Salvadoran conflict played into the domestic politics of a regional power. Granting the FMLN a haven on Mexican soil offered the host a series of advantages. It augmented Mexico's clout in the region by making it one of the few powers that had open relations with all the key actors in the Salvadoran conflict: the Salvadoran state, the US, Cuba, and Nicaragua under the FSLN. Mexico was thus in a privileged position to serve as intermediary between the different warring factions and, as will be shown, to influence the FMLN and coax it towards a more moderate platform of a liberal rather than a socialist order. Conditioning the FMLN to limit its activities to the strictly diplomatic realm served to avoid confrontations between Mexico and Washington. It also helped to ensure that Mexico did not become a point of transit for weapons meant for El Salvador that could fall into the hands of domestic opposition groups seeking to challenge the Mexican regime. Southern states in Mexico such as Chiapas, Oaxaca, and Guerrero, could have been convenient points of transit for weapons destined for El Salvador and these were geographic spaces where the Mexican State had limited physical presence and its legitimacy was weak. The second precondition—not establishing relations with dissident movements in Mexico or supporting or encouraging them in any way—was a means to control the impact that the Salvadoran revolutionary effort could have on Mexican politics and society. Mexico's acknowledgement and conditional support of the FMLN, therefore, helped boast the legitimacy of both actors.

The Franco-Mexican Declaration

Less than a year after the CPD set up its diplomatic headquarters in Mexico City, the FMLN and Mexican administration crafted an audacious diplomatic manoeuvre of great significance for the Salvadoran rebels. On 27 August 1981, the governments of both Mexico and France issued a joint pronouncement on the FMLN that came to be known as the Franco-Mexican Declaration. The document, elaborated by the Mexican and French foreign ministers, Jorge Castañeda de la Rosa and Claude Cheysson, respectively, and signed by French president Francois Mitterrand and

[18] Mexico's preconditions for the FMLN were conveyed to the author, both by FMLN members and by Mexican public servants, in a series of separate interviews. Amongst the FMLN members were Martínez, Ana Guadalupe. Interview with author. San Salvador, El Salvador, 21 August 2010; Sancho, Eduardo. Interview with author. San Salvador, El Salvador, 22 July 2010 and Jovel, Francisco. Interview with author. San Salvador, El Salvador, 16 July 2010. From Mexico, Montaño, Jorge. Interview with author. Mexico City, Mexico, 27 July 2011; Sepúlveda, Bernardo. Interview with author. Mexico City, Mexico, 3 August 2011; Bartlett, Manuel. Interview with author. Mexico City, Mexico. 27 July 2011.

Mexican president José López Portillo, recognized the FMLN as a legitimate representative political force in El Salvador, subject to the rights and responsibilities that this entailed in accordance with the UN Charter. The document stated:

> Taking into account the extreme gravity of the situation in El Salvador and the country's need for fundamental changes in its social, economic and political spheres, both governments recognize the alliance of the FMLN and FDR as a representative political force willing to assume the resulting rights and obligations. Consequently, it is legitimate for the alliance to participate in the unfolding of negotiations necessary for reaching a political solution to the Salvadoran crisis.[19]

This formal and public recognition by France and Mexico of the FMLN's legitimacy as a representative political force of opposition constituted the Salvadoran revolutionaries' first decisive political–diplomatic breakthrough. It marked the first instances in which they were officialized by two well-established and internationally recognized nation-states. Furthermore, by stressing the primacy of a political negotiated solution to the conflict and acknowledging that the Salvadoran insurgency would need to participate in negotiations, the Franco-Mexican Declaration countered the United States and Salvadoran governments' attempts to present the crisis as a purely military conflict between a legitimate government and an illegitimate terrorist organization.

Given its far-reaching implications, the Franco-Mexican Declaration expectedly provoked a strong international reaction by both supporters and opponents. News of the proclamation made headlines in major newspapers across the globe and gave the FMLN a heretofore unknown international platform. The governments of Norway, Sweden, Holland, Ireland, and the Democratic Republic of Germany were quick to back the Mexican and French presidents, as were the 43 members of the Socialist International. In turn, El Salvador's president, Napoleon Duarte, embarked on a diplomatic tour to denounce the actions of France and Mexico as a foreign intervention into El Salvador's domestic affairs. In an effort to counter the validation the FMLN had received, Duarte travelled to the United States, gave several press conferences, and spoke against the Franco-Mexican Declaration at the Organization of American States. The United States Department of State published a communiqué rejecting the idea of a political solution in which the FMLN and its civilian counterpart, the FDR, would participate, stressing instead that Washington would back an electoral solution led by the Salvadoran military junta, and that they guaranteed President Duarte that US military advisors in El Salvador would stay put until provocations by foreign actors subsided (Covarrubias 2013, 53). Furthermore, the Christian Democratic president of Venezuela, Luis Herrera Campins, a close ally of

[19] 'Texto de la Declaración Franco-Mexicana sobre El Salvador.' 1989. In *México En Centroamérica: Expediente de Documentos Fundamentales, 1979–1986*, edited by Raúl Benítez Manaut and Ricardo Córdova Macías, 1st edn, Serie Antologías/Centro de Investigaciones Interdisciplinarias En Humanidades. México, D.F: Universidad Nacional Autónoma de México [Centro de Investigaciones Interdisciplinarias en Humanidades]: pp. 45–46.

Duarte and the Reagan administration, led efforts to consolidate wider international condemnation in the region by mobilizing governments allied with Salvadoran president Napoleon Duarte or those who feared that the FMLN's newfound international stature could galvanize other insurgent movements in the region. President Herrera Campins' labours gave way to the Declaration of Caracas in September 1981. This document, signed by the governments of Venezuela, Colombia, Bolivia, Argentina, Chile, Guatemala, Honduras, Paraguay, and the Dominican Republic, stated that third parties were under no obligation to recognize the Salvadoran guerrillas as legitimate belligerents and could continue dealing exclusively with the Salvadoran government (Ceberio 1981).

While contentious, the Franco-Mexican Declaration helped shape the subsequent history of the FMLN and of Mexican involvement in Central America for the remainder of the decade. Mexico sent a clear message of where it stood on the Salvadoran crisis—acknowledging the claims of the FMLN and favouring a political negotiated solution. It demonstrated that it was willing to take risks—most notably antagonizing the United States and other regional governments—to promote such a platform. This sent a compelling message to the Salvadoran insurgents that Mexico was a serious political ally. Furthermore, up to this point Mexican administrations had been staunch defenders of the principle of non-intervention in the international arena. In this case, however, regardless of what may have been argued at the time to stave off criticism by opponents of the Declaration, Mexico was prioritizing other ethical, political, and diplomatic principles. Allusions to the grave human rights crisis in El Salvador in the document and explicit reference to this by Mexican president López Portillo in subsequent interviews on the Declaration, suggested that peace in Central America, social change in El Salvador, and respect for human rights in the region were being prioritized over abstract principles of national sovereignty.[20] Mexico's shifting attitude towards sovereignty was reflective of broader global trends and changing perceptions regarding the primacy of international principles such as human rights and peace in the international arena.[21]

For the Salvadoran insurgents, the Declaration gave their organization unprecedented international visibility and allowed them to develop a more active political diplomacy. Recognition by a liberal non-aligned country like Mexico and a social democracy like France, and the subsequent backing of this recognition by Europe's western democracies, helped to offset the image of the Salvadoran insurgents that the Duarte and Reagan administrations wished to propagate—that of a radical Marxist terrorist organization. While this benefitted the FMLN by bolstering its international legitimacy, it also compelled the organization to tone down its Marxist rhetoric and moderate some of the more radical elements of its economic platform, lest they risk

[20] Extracts from press conference given by López Portillo to Alan Riding and Flora Lewis of the *New York Times* on 23 September 1981 can be found in Benítez Manaut and Córdova Macías, *México En Centroamérica*. In his memoires, Jorge Castañeda Gutman asserts that this is what the Declaration meant in terms of Mexico's shifting stance on non-intervention (Castañeda 2014).

[21] For these broader shifts and their effects on the legitimation of armed groups, see the section on historical times in the Introduction to this volume.

scaring off their new international allies. Furthermore, recognition as a legitimate insurgent force implied that the FMLN was now bound by the Geneva Convention and if it failed to abide by the laws of war in its military proceedings it would be held to account, and violations could carry important political and diplomatic costs. The international legitimacy that the Franco-Mexican Declaration bestowed on the FMLN was neither unconditional nor permanent. To keep it and deepen it, the Salvadoran armed movement would need to convince international audiences, particularly Western and non-aligned nations, that its use of violence was a last resort. Appearing open to a political solution was thus crucial.

Gaining the acknowledgement and diplomatic support of Western European democracies also made it critical for the FMLN to demonstrate that, unlike its opponent who routinely tortured, disappeared, and killed suspected dissidents and alleged guerrilla sympathizers, the Salvadoran insurgents respected the human rights of civilians and combatants and conducted their military operations in accordance with the Geneva Convention. As several scholars have persuasively argued, the 1970s was a critical period in the history of human rights. During this decade, the discourse of human rights gained unprecedented traction largely due to the activities of non-state actors. US president James Carter, who came to office in 1977, rode the human rights wave and made human-rights promotion a cornerstone of his presidency and foreign policy rhetoric. While the policies he advanced and supported in Latin America often did not prioritize human rights concerns, the impact of his discourse was consequential. Documenting and denouncing the human rights abuses of specific state and non-state actors became a major strategy to delegitimize the violators, particularly on the international stage (see, for example, Moyn 2012; Schmidli 2013; Iriye et al. 2012). When the FMLN began its quest for international legitimacy, human rights had an undeniable normative value on the global stage and Mexico understood this well. To gain international recognition and support, particularly from the non-Eastern camp, the Salvadoran insurgents needed to present themselves as committed to upholding human rights and contrast this with the actions of the de facto state they were fighting.

This was also critical to curtail the vast amounts of resources the United States was giving to the Salvadoran army. Denouncing the human rights violations of the Salvadoran army and allied death squads was at the heart of efforts to mobilize the United States' Congress to cut off aid to El Salvador (see Sikkink 2007; Smith 1996).[22] For the FMLN, the discourse of upholding human rights mattered to the extent that it had traction on the international stage. The language used in their press releases, interviews, and communiqués shows that this became more important as the war went on. Failing to secure an absolute military victory that would bring them to power, the FMLN relied increasingly on international acknowledgement as a legitimate political actor. Beyond international legitimacy, at home the FMLN saw the praxis of defending human rights in pragmatic terms. How they treated the civilians

[22] For similar cases of denouncing the opponents' human rights violations, see the contributions of Klaus Schlichte on the Ugandan NRA and Eline van Ommen on the FSLN in this volume.

under their control would determine the loyalty and support they received from the local population; thus, human rights allegiance was critical to their domestic legitimacy. As part of the Salvadoran Peace Accords of 1992, a commission comprised of three international commissioners appointed by the United Nations secretary general was charged with investigating serious acts of violence occurring in the country since 1980, the nature and effects of the violence, and to recommend strategies for promoting national reconciliation. Based on collected testimonies acquired between July 1992 and March 1993, the Truth Commission for El Salvador attributed 85% of the acts of violence to state agents. Approximately 5% of the acts of violence were attributed to the FMLN (see Truth Commission: El Salvador 2016).

It is pertinent to note that the Franco-Mexican Declaration was not a foreign or exogenous imposition on the FMLN, but rather a joint diplomatic strategy devised in collaboration with Mexico and France. Eduardo Sancho tries to credit the idea behind the Franco-Mexican Declaration to the FMLN. He recounts that Salvadoran insurgent leaders wanted international recognition as a representative political force and they reached out to Gustavo Iruegas, a Mexican diplomat and FMLN friend, for this purpose. In Sancho's recollection, the Mexican foreign ministry looked favourably upon this, but also understood that the motion would be highly contentious, particularly give the Reagan administration's commitment to defeating the FMLN militarily and delegitimizing it on the international stage. Feeling that the support of a great Western power outside of the region would be key, Mexico reached out to France.[23] Given the social democratic nature of President Mitterrand's administration, his political advisor Régis Debray's sympathy with insurgencies in Latin America, and the amicable relationship between Mitterrand and Mexican foreign minister Castañeda, France seemed like an opportune partner. The fact that the FDR was led by prominent social democrats Guillermo Ungo and Rubén Zamora doubtless carried favour with the Mitterrand administration. It provided an ideological common background between the French administration and the civilian wing of the armed movement. More importantly, the fact that Mexican foreign minister Castañeda, a close friend of Mitterrand's, vouched for the FMLN as a legitimate political force doubtless carried favour. Furthermore, in recognizing the FMLN, the Franco-Mexican Declaration placed a negotiated political solution to the Salvadoran war front and centre. France, along with other European democracies at the time, had a vested interest in pushing back against the re-escalation of Cold War conflicts and rearmament. Ronald Reagan's ascent to the US presidency had caused fears in Western Europe that détente and the hard-won accomplishments on arms control it entailed would be abandoned in favour of renewed military confrontations that Europe would get dragged into. Acknowledging the FMLN in the framework of endorsing a political solution to what was quickly becoming a Cold War battle made practical sense for France.

CPD leader Salvador Samayoa provides a different account in his book *La Reforma Pactada*. According to Samayoa, Jorge Castañeda Gutman, the son of the Mexican

[23] Sancho, Eduardo. Interview with author. San Salvador, El Salvador, 12 July 2010.

foreign minister and an advisor to his father on the Central American crisis, was the intellectual author of the declaration. Castañeda Gutman first consulted the matter with Régis Debray, a renowned Marxist intellectual sympathetic with Latin American revolutionary movements and a close advisor to Mitterrand. Then Castañeda Gutman conferred with representatives of the FDR and FMLN (Samayoa 2007). In his memoires, Castañeda Gutman recalls that Colombian writer Gabriel García Márquez—a world-renowned novelist who was then living in Mexico—helped convince the newly elected Mitterrand government to support and consummate the joint Declaration on El Salvador (Castañeda 2014). Precisely who deserves credit for concocting this strategy seems less relevant than the fact that the Franco-Mexican Declaration was the product of negotiations and diplomatic manoeuvering involving many actors, and that the Salvadorans were protagonists helping craft this groundbreaking document.

Mexican diplomats knew full well that the polemic declaration could help the FMLN's legitimacy through recognition and officialization. In a private letter between Gustavo Iruegas and Castañeda de la Rosa, Iruegas told Castañeda that Mexico had a historic responsibility to give the Salvadoran revolutionaries 'that which their ideological partners—practically their comrades in arms—Cuba and Nicaragua could not give: anoint them with legitimacy'.[24] By granting its support, Mexico also hoped to moderate the economic and political goals of Salvadoran insurgents. At its root, the Franco-Mexican Declaration was a call for a middle-of-the-road solution to the crisis in El Salvador; a negotiated compromise between the military dictatorship and the guerrillas. This would not occur if the FMLN were perceived as a dogmatic Marxist organization reluctant to make compromises with the sitting Salvadoran regime and unwilling to accept anything save an absolute revolutionary victory. Historian William M. LeoGrande eloquently describes this in his exhaustive work on Central American–US relations during the last decade of the Cold War as follows:

> The Mexicans certainly did not favor the establishment of pro-Soviet Marxist-Leninist regimes in Central America, although, based on their experience of peaceful coexistence with Cuba, they were confident they could live with whatever revolutionary government might emerge. They believed the best strategy for blunting Cuban and Soviet influence was to support the social democrats among the region's revolutionaries rather than side with the forces of conservatism. By trying to hold back the process of social change in Central America, the United States would only make the inevitable break with the past more cataclysmic and radical. By introducing large amounts of military aid, Washington itself was turning Central America into an arena of East–West conflict, drawing Cuba and the Soviet Union in. 'The United States will not make the Vietnam syndrome disappear by repeating it', warned one Mexican official.
>
> (LeoGrande 2000, 98)

[24] Iruegas, Gustavo to Jorge Castañeda, San Salvador, 10 June 1981. Gustavo Iruegas Personal Archive, p. 5. Cited in Covarrubias (2013).

Ana Guadalupe Martínez, an FMLN commander and member of the CPD, notes that Mexico took effort to pull on the FMLN's more moderate strings. Martínez stresses that this was also the case with Western European democracies. In her words:

> Mexico knew that we were revolutionaries ... but they also knew that this phenomenon (revolution) would not be repeated elsewhere in the continent. They understood that Mexico needed to help us tone down our rhetoric and mentality, and so they would take us aside and talk to us about democracy and institutions and political parties ... in essence how democratic institutional systems functioned. The Europeans would do the exact same thing with us.[25]

Although the Mexican government wished to moderate the FMLN, it did not want to compromise the organization's integrity. Ultimately, Mexico—and the myriad countries that to varying degrees sympathized with and supported the FMLN—was cognizant of the appalling crisis in El Salvador under the military junta and they wished to see genuine transformations in the Central American country.[26] A series of internal documents from an FMLN private archive shed light on how Mexico juggled moderating the FMLN and bringing it to the negotiating table while simultaneously trying to bolster its strength. In February 1981 Foreign Minister Castañeda de la Rosa held a private meeting with two leaders of the CPD, Salvador Samayoa and Guillermo Ungo, and with Eduardo Sancho who was a member of the revolutionary organization's governing body (*Comandancia General*, CG). The purpose of this gathering was to discuss a potential meeting between FMLN leaders and representatives of the US government. While not explicitly stated, the documents suggest that members of the US administration reached out to Castañeda and asked him to orchestrate a gathering between them and the Salvadoran armed movement.[27] Beyond leaving little doubt about the deep level of trust the FMLN placed in Castañeda—they praise his sympathy with their cause, his genuine interest in supporting the FMLN, and stress that the FMLN trusts that Castañeda would not pressure them to negotiate prematurely, or under terms that were unfavourable to their interests—the documents convey that Castañeda told FMLN members that their best card at that moment was the organization's military strength. Their claim-making through fighting would

[25] Martínez. Ana Guadalupe. Interview with author. 21 August 2010. A similar argument was related to the author by Mexican public servant Jorge Montaño, who served as Mexican ambassador to the United Nations between 1989–1993, during the Salvadoran Peace Process. Montaño, Jorge. Interview with author. Mexico City, Mexico. 25 July 2011.

[26] When Mexico and France recognized the legitimacy of the FMLN, the human rights crisis in El Salvador was appalling. In just the past seven months, between January and July 1981, around 10,000 people had been murdered and over 300,000 Salvadorans had fled the country. These numbers are even more staggering when one considers that the population of this small country was approximately 4.5 million. The number of dead and displaced are cited in Covarrubias (2013, 47).

[27] In a letter for Minister Castañeda signed by CDP members Salvador Samayoa, Eduardo Sancho, and Guillermo Ungo, the Salvadorans thank Castañeda for his willingness to serve as intermediary between the FMLN and the US government and his efforts to orchestrate some type of dialogue between the two parties. The letter stresses that the FMLN has on repeated occasions voiced its openness to initiate a process of dialogue and inquires what the subject matter of the meeting with the US would be, and the rank of US government officials expected to be present at the meeting. 'Carta para el Ministro Castañeda de Salvador Samayoa, Eduardo Sancho y Guillermo Ungo'. FMLN private archive. 9 February 1981. D1004-01-08.

determine the FMLN's ability to be taken seriously and have leverage at a possible future negotiation. Consequently, Castañeda advised the Salvadorans not to suspend their military activities and went even further by mentioning that it would serve them well to have their first meeting with US officials with a resounding military victory under their belts.[28] In his memoir, Castañeda Gutman, the son of Castañeda de la Rosa who was himself deeply involved behind-the-scenes in Mexican–FMLN relations at this time, recalls that the same advice was given a few months later before the Franco-Mexican Declaration as the Salvadorans were told that a striking military victory in the days or weeks around the Declaration would be advisable (Castañeda 2014).

The documents mentioned above shed light on a number of important dimensions of Mexico's role in the conflict, US–FMLN relations, and the evolution of the Salvadoran guerrillas' quest for international legitimacy through diplomacy. First, the exchanges between high-ranking Mexican officials and FMLN leaders show that from the very early stages of the Salvadoran Civil War, Mexican diplomats were actively seeking to help mediate between warring parties and that they did so hoping to facilitate a genuine—if not entirely revolutionary—transition in El Salvador. The Mexican government was in good standing with both the United States and the Salvadoran insurgents, and this bestowed Mexico with a degree of clout in the conflict that was not open to other FMLN allies in the region, like Cuba and Nicaragua. This reality proves illustrative regarding the degree to which non-aligned third parties could quietly and sometimes not so quietly exert their influence in regional and global affairs during the Cold War. Second, Castañeda's advice that the FMLN prioritize and step up its military strength evidences Mexico's understanding that at this early stage of the civil war, the Salvadoran insurgent's quest for international legitimacy rested largely, if not entirely, on their military might. To what extent the FMLN was genuinely open to negotiating in the months after its 1981 'final offensive' is debatable. The organization was in fact deeply divided on this matter in the early years of the war, and divisions amongst the FMLN leadership over this issue almost fractured the organization. The role that international actors played in navigating these divisions and keeping the FMLN united would prove vital. What is pertinent to stress here is that, as Mexican diplomats accurately noted to FMLN leaders, in 1981 they did not yet have political and social capital in the international arena. Fighting was the primary way to make their claims heard. This would change markedly in the following years as the international acknowledgement, support, and officialization for the FMLN mounted and the armed movement threw increasing weight behind its rebel diplomacy seeking international recognition as a representative political force worthy of being treated accordingly. Third, the Reagan administration's interest in meeting with Salvadoran combatants and the FMLN's openness to this possibility in the early months of 1981 reflects that the relationship

[28] 'Informe de Entrevista con Canciller Jorge Castañeda.' De Misión Especial Comandancia General/Comisión Politico Diplomática para Comandancia General. FMLN private archive. 9 February 1981. D1005-01-09-01.

between the Reagan administration and the FMLN was more fluid and complex than the antagonistic rhetoric used by both would have suggested. This deepened significantly in later years when the FMLN directly lobbied US Congress members through a covert diplomatic office in Washington D.C.

Conclusion

By the early 1990s, when negotiations to end the brutal civil war in El Salvador took on a much more serious character, the FMLN had established itself as an internationally recognized legitimate force of opposition. Over the course of the previous decade the insurgent organization stretched its diplomatic wings across the globe and effectively became a parallel state in the international arena. This was in no small measure due to the FMLN's diplomacy and the efforts of its CPD to legitimate the rebels on the international stage. As this chapter has demonstrated, Mexico's support in these efforts was vital, especially at the beginning of the war. Attesting to Mexico's continued involvement in the conflict and eventual peace process, the final signing of the Salvadoran Peace Accords in January 1992 took place in Mexico City's Chapultepec Castle and the accords are known as the *Acuerdos de Chapultepec* or Chapultepec Peace Accords. When former FMLN guerrilla leader and one of the signers to the accords, Joaquín Villalobos, gave Mexican president Salinas his prized AK-47 a year after the ceasefire, the former insurgent was paying homage to the fact that, in his view, Mexico's support had been integral to the FMLN securing the legitimacy necessary to transform El Salvador.

'Tequila Diplomacy' has examined Mexico's role in the FMLN's quest for international legitimacy in the early years of the armed movement's history. It has stressed the importance of Mexico, a consequential regional power, providing a haven for the revolutionaries' CPD and its effort to secure international recognition for the FMLN as a legitimate force of opposition through the Franco-Mexican Declaration of 1981. This history evidences some of the ways in which Mexico advanced the legitimacy of the Salvadoran insurgency and the implications this had not just for the FMLN but also for Mexico. Recognizing and supporting the Salvadoran insurgents increased Mexico's own regional clout and helped to offset some potential domestic opposition to its one-party state. For the FMLN, Mexico's support carried indisputable political and diplomatic benefits, particularly in the early years of the organization's history. And yet, allowing the guerrillas to conduct their diplomatic operations from Mexico City and spearheading the acknowledgement and officialization of the armed movement amongst Western powers worked to moderate some of the more revolutionary goals of the insurgency. The peace that was signed by the FMLN and Salvadoran state in 1992 ended not only a gruesome civil war, but also an exclusionary and authoritarian political system dominated by the military and economic elites, which had been in place for almost sixty years. This momentous accomplishment would not have been possible without the efforts of the FMLN. As part of the peace accords, the FMLN became an established political party that could compete in elections and hold public office. The political reforms that paved the way for contested democratic elections, however, were not accompanied by a meaningful restructuring of the economic order that had long fostered staggering socio-economic disparities and poverty in the country. One cannot divorce realities in El Salvador today from the neoliberal peace that was signed in 1992. By the time the negotiations began in earnest, the FMLN's revolutionary ambitions had, as Mexico hoped, become more centrist. While their quest for international legitimacy and their proximity to Mexico played an important part in this story, it would be myopic and ahistorical to attribute this shift in the FMLN's platform exclusively or even primarily to Mexico's clout. Multiple domestic and international realities which are not the subject of this chapter played a role in the changing nature of the FMLN's ideological, political, and economic platforms, not the least of which was the collapse of the Soviet Union in 1989, the ousting of the Nicaraguan FSLN from power in 1990, and the alleged triumph of capitalism and electoral democracy as the pillars of the post-Cold War order. Simply put, the world in the early 1990s, when the FMLN agreed to lay down its arms, looked very different to what it had a decade earlier and what may have seemed possible and imminent in 1981 was no longer tenable.

References

Castañeda, Jorge G. 1994. *Utopia Unarmed: The Latin American Left after the Cold War.* New York: Vintage Books.

Castañeda, Jorge G. 2014. *Amarres Perros: Una Autobiografía.* Kindle edn. Alfaguara.

Ceberio, Jesús. 1981. 'Caracas critica la declaración franco-mexicana sobre El Salvador'. *EL PAÍS*, 4 September.

Covarrubias, Ana. 2013. 'La Declaración Franco-Mexicana Sobre El Salvador'. *Revista Mexicana de Política Exterior*, Special issue, pp. 39–62. https://revistadigital.sre.gob.mx/index.php/rmpe/issue/view/27.

Danner, Mark. 1994. *The Massacre at El Mozote.* 1st edn. New York: Vintage.

Iber, Patrick. 2014. 'Paraíso de Espías. La Ciudad de México Y La Guerra Fría'. Nexos, April. http://www.nexos.com.mx/?p=20004

Iriye, Akira, Petra Goedde, and William I. Hitchcock, eds. 2012. *The Human Rights Revolution: An International History.* Oxford; New York: Oxford University Press.

Keller, Renata. 2015. *Mexico's Cold War: Cuba, the United States, and the Legacy of the Mexican Revolution.* New York, NY: Cambridge University Press.

LeoGrande, William M. 2000. *Our Own Backyard: The United States in Central America, 1977–1992.* Chapel Hill, NC: University of North Carolina Press.

Martínez Álvarez, Alberto. 2010. *From Revolutionary War to Democratic Revolution: The Farabundo Martí National Liberation Front (FMLN) in El Salvador.* Berlin: Berghof Conflict Research.

Martinez, Ana Guadalupe. 1997. 'Central American Peace Process'. Yale–UN Oral History Project, 21 June.

Morley, Jefferson. 2008. *Our Man in Mexico: Winston Scott and the Hidden History of the CIA*. Lawrence, KS: University Press of Kansas.

Moyn, Samuel. 2012. *The Last Utopia: Human Rights in History*. Cambridge, MA: Belknap Press of Harvard University Press.

Pedrosa, Fernando. 2015. 'Partidos Políticos Y Acciones Trasnacionales: El Comité Para América Latina Y El Caribe de La Internacional Socialista (1976–1983)'. *Revista Izquierdas* 22(Enero): pp. 48–77.

Riding, Alan. 1989. *Distant Neighbors: A Portrait of the Mexicans*. New York: Vintage.

Samayoa, Salvador. 2007. *El Salvador: La Reforma Pactada*. 750 copies edn. San Salvador: UCA Editores.

Schmidli, William Michael. 2013. *The Fate of Freedom Elsewhere: Human Rights and U.S. Cold War Policy toward Argentina*. Ithaca, NY: Cornell University Press.

Sikkink, Kathryn. 2007. *Mixed Signals: U.S. Human Rights Policy and Latin America*. Ithaca, NY; Bristol: Century Foundation.

Smith, Christian. 1996. *Resisting Reagan: The U.S. Central America Peace Movement*. Chicago, IL: University of Chicago Press.

'Truth Commission: El Salvador'. United States Institute of Peace. Accessed 11 August 2016. http://www.usip.org/publications/truth-commission-el-salvador.

6

'When we became stronger, they listened more'

The Politics of Legitimation of the NRA in Uganda (1981–1986)

Klaus Schlichte

Introduction

In hindsight, the relatively short war of the National Resistance Army (NRA) between 1981–1986 looks like an easy case.[1] It followed a textbook version of Maoist guerrilla warfare, starting in rural areas and ending with overtaking the capital Kampala. The war also did not become an issue in the enduring Cold War, and it ended after five years with a clear-cut victory for the NRA, the military wing of the National Resistance Movement (NRM), bringing to power Yoweri K. Museveni, who has remained head of state ever since.

As will be argued in this chapter, the NRA's pathway[2] was not quite so easy, and it was not just military might that brought the NRA to power. It was also its reputation as being a much more disciplined force than the slowly degrading forces of the old regime, led by Milton Obote. The human rights violations by government forces became internationally known around 1984, and in summer 1985, Obote was unseated by a military coup d'état. Other than this erosion of legitimate rule, favourable global historical timing and strategic engagement with neighbouring governments allowed for the NRA's success. The NRA operated a twofold threshold of global historical timing: the immediate post-colonial role of Great Britain waned in the course of the war, and with the changes in the Soviet Union, humanitarian issues became more prominent. Both changes worked to the advantage of the NRA.

[1] The quote in the title is from Matthew Rukikaire, head of the NRA's External Committee (1981–1986), Interview with the author, Kampala 17 February 2023. I would like to thank Jude Kagoro for his valuable comments on earlier versions of this paper. The other authors of this volume delivered a host of valuable comments as well. Research for this chapter was possible through funding from the German Research Foundation (DFG), project no. 424426045.

[2] For a good and succinct description of the NRA's trajectory, Kasfir (2005) is still the best account. The NRA leader and later president of Uganda, Yoweri Kaguta Museveni, has produced his own version that is a great primary source for Uganda's recent political history (Museveni 1997).

Klaus Schlichte, *'When we became stronger, they listened more'*. In: *Armed Groups and the Politics of International Legitimation*. Edited by: Stephan Hensell and Klaus Schlichte, Oxford University Press. © Klaus Schlichte (2025).
DOI: 10.1093/9780198974529.003.0006

The case of the NRA displays all the practices of claim-making and claim validation outlined in the Introduction to this book. The NRA practised *fighting*, *voicing*, and, following Maoist ideas, *mimicking* of the state. External actors reacted quite differently to the NRA's war. With regard to claim-validation, *acknowledging* and *supporting* was practised only by the government of Libya. The government to neighbouring Kenya at least tolerated the presence of NRA-members on its territory as long as they did not act politically in the open. *Officializing*, finally, occurred only during the last few months of the war, when the attacked regime and its armed forces disintegrated.[3] Neighbouring African governments were the first to accept the leadership of the NRA as official interlocutors. Western governments either took a very long time to establish contacts, as in the case of Great Britain or West Germany, or they stood back for a long time, as with the US administration under Ronald Reagan.[4] Generally, international recognition was driven much more strongly by the regional dynamic than by 'great power' decisions. All regional governments and great powers recognized the NRA as the legitimate government within weeks of it taking Kampala in January 1986.

For reasons of simplicity, in the development of the NRA and its warfare three phases can be distinguished, each of which show different patterns of how the rebels made their claims and how and when these were validated by others. Its claim to political power was already voiced during the elections of December 1980, which the NRA considered 'rigged', so that, still under the name 'Ugandan Patriotic Movement', they retreated into the bush and began their attacks on military bases. In this initial period, from the first attacks in February 1981 until mid-1982, other political groups joined and the re-named NRA slowly gained strength and established its structures not only militarily but also as an organization, including building up its external relations. Apart from modest external support, the NRA mainly acted under harsh conditions in rural areas in central Uganda. It operated along a classic hit-and-run strategy and provisioned itself with arms mainly acquired from attacks on police stations and army barracks. In this early period, external relations were not considered particularly important.

In the second period from 1982 to 1984, the NRA was put under pressure as the sitting regime in Kampala gained support from other governments, primarily in the form of military aid by a Commonwealth group led by the British. Militarily, the NRA entered a critical period. During this time the government forces became ever more ill-disciplined, while human rights abuses became frequent and also internationally publicized. The regional arena mattered most for claim-validation, with London as a side-arena. To put it in the words of a former leading cadre: 'On the international scene, little was known about the movement and its leaders. ... The exception was Tanzania' (Kategeya 2020, 102).

[3] Cf. the Introduction of this volume for the definition of practices of claim-making and claim-validation used in this chapter.

[4] For another case in which the crisis of a sitting government during a civil war pushed international actors to recognize the leader of an armed group instead, see the chapter by Wolfram Lacher on Khalifa Haftar and the LAAF in this volume.

In a third period, the practice of voicing became more important. The NRA invited journalists to the war arena, encouraging them to report on the government's massive human rights violations, and in both the British and the US political arenas the situation in Uganda became an issue. Faced with a weakened government, the NRA launched bigger attacks and made a swift move across Western Uganda, bringing larger parts of territory under its sway. With the military coup against President Obote in summer 1985, the final countdown of the old regime began. Officializing by East African governments took place in early summer in preparation of mediated peace negotiations. Officialization by Western governments, however, took much longer, and happened often only after the NRA had conquered the capital Kampala in January 1986.

The military might that the NRA accrued as the regime crumbled during the first half of 1985 was the precondition for this officialization of relations, first carried out by East African governments. There was the 'normative power of the factual', as the state theorist Georg Jellinek (1905) put it, that preceded the officialization of relations. In the case of Britain and Germany, however, it is remarkable to see how long they stuck to the practice of not recognizing armed opposition groups and continued to consider the old regime the legitimate government of Uganda. This even stretched into the period of negotiations in late 1985. Recognition by Western governments was only granted once the NRA had conquered the capital in January 1986, when Museveni was inaugurated in a ceremony as head of state.

The following analysis is based on diplomatic archives in Berlin and London, on interviews with former NRA cadres, on case literature, and on informal conversations with numerous office holders, diplomats, and 'ordinary' people in Uganda in a series of research stays between 2014 and 2023. I will first delineate the practices of claim-making applied by the NRA itself, including some interaction with three Western governments, namely the British, the Swedish, and the Americans. In a second section, I will look closer at the internal debates in the British, West German, and East German foreign offices in order to identify tipping points in the governments' policy towards the NRA. The diplomatic archives of the two German states also allowed for insights into the Cold War dynamics—or the absence of such dynamics—surrounding the NRA's rebellion.

The NRA's diplomacy

The NRA practiced all the forms of claim-making distinguished in the Introduction to this volume. Military attacks can also be considered a basic form of communication, not just towards the enemy but targeting a wider audience as well. It was only over time, though, that the NRA became more active in its verbal communication to the outside world. In this practice of voicing, the 'External Committee' played an important role, as did the direct action of two leading representatives: the future president Yoweri Museveni and his long-time confidant Eriya Kategeya (cf. Kategeya 2020; Museveni 1997, 164–172).

The External Committee

Two historical backgrounds prepared a broad range of regional connections for the NRA when it took up arms in February 1981. First, Uganda, like Kenya and Tanzania, had been part of a colonial British East Africa, so that via civic administration, higher educational institutions, businesses and churches, as well as through migration, numerous connections between the political arenas of the successor states of these British colonies existed. The East African region was thus the first external political platform for Ugandan politics.

Second, during the years of the rule of Idi Amin (1971–1979), a large proportion of the political class of Uganda fled the country, many of them spending years in Tanzania and Kenya. In addition to the standard diaspora network that any country exhibits, there were also multiple connections to the political milieus of Kenya and Tanzania. Because of the close relationship between Milton Obote and Julius Nyerere, Tanzania became the hub of Ugandan opposition against Idi Amin's military dictatorship. Museveni claims to have formed a political group in Tanzanian exile, the 'Front for National Salvation' (Fronasa), as early as 1971 (Museveni 1997, 54). A number of later core members of the NRA were already active in this group during the 1970s and gained practical experience in military training in camps of one of the 'front states' in the fight against South Africa, namely in Mozambique, which was already ruled by the Frente de Libertação de Moçambique (FRELIMO), a socialist anti-colonial liberation movement (Golooba-Mutebi 2008, 11). Fronasa then fought alongside the Tanzanian army when it deposed Idi Amin in 1979. The NRA's genealogy is thus part of a much longer history of warfare and liberation movements in the region of Eastern and Southern Africa, a network that has left its traces up to current times (cf. Roessler and Verhoeven 2016).

The NRA's early diplomatic efforts remained on a low level and two personalities were key in these endeavours. Eriya Kategeya, a close friend of Museveni since their childhood days, led an additional diplomatic wing of the NRA. Both Kategeya and Museveni travelled back and forth between the bush war arena, Nairobi, London, and their exiled families in Sweden. Formally though, Matthew Rukikaire, about a decade older than the core group of the NRA, but also a South-Western Ugandan, led the 'External Committee' formed in February 1981 with its base in Nairobi, where he acted as a businessman with international connections.

The first and immediate aim of the NRA's external outreach was the acquisition of arms and financial support (Kategeya 2020, 91; Rukikaire 2019, 206). The mobilization of diaspora communities in East Africa, Europe, and later the US had this purpose in mind. Only gradually did the External Committee engage in classical diplomacy, using diaspora connections to gain access to foreign governments (Ngoga 1999, 234). In the meantime, it could draw on the established practice that the neighbouring governments of Kenya and Tanzania would tolerate the presence of Ugandan political oppositionists. Even though this did not amount to the open provision of resources, this toleration can still be considered part of the practice of supporting or claim validation as outlined in the Introduction of this book. This diaspora support was crucial for making valid claims in the eyes of foreign governments. Pro-active engagement was thus needed in order to create support for the NRA and to streamline the image among diaspora communities. The External Committee prepared the ground and maintained a network here.

The NRA had formalized its 'foreign policy' early on, in the same month during which the armed struggle began. Many Ugandan exiles working in different positions in Kenya contributed to the struggle financially and with logistical resources. From Nairobi, the External Committee coordinated the activities of NRA sympathizers in Canada, Scandinavia, the US, Southern Africa, and even Papua New Guinea (Rukikaire 2019, 187).

In the internal organization of the NRA, political backing for the armed struggle was considered to be a key task in addition to those tasks covered by other committees including a 'Civilian Committee' in the war area, the 'Political and Diplomatic Committee', headed by Eriya Kategeya, a 'Committee on Information and Propaganda', and finally an overarching 'Executive Committee' as the NRA's supreme governing body (Rukikaire 2019, 188–189). With these practices, the NRA was already mimicking a state, as many successful armed groups did (cf. Schlichte 2009, ch. 6).

Apart from this toleration of the NRA in Kenya, there was very little direct support from other governments. Only the Libyan government indicated its readiness to support the rebels via the Libyan Embassy in Nairobi. Early on in 1981, arrangements were made that finally led to the delivery of a consignment of arms via Burundi and Rwanda in empty petroleum lorries (Rukikaire 2019, 191).[5] In 1982, Libya also trained about one hundred NRA cadres who had no prior military training.[6] However, the 'External Committee' was not without internal cleavages. Within the NRA, a leftist tradition prevailed as can be seen both in its initial discourse but also in the adoption of Maoist conceptions of 'protracted warfare' (Ngoga 1999). Beyond such general military conceptions, socialist leanings remained contested within the NRA to the extent that it did not seek socialist countries' direct support.[7]

Despite a general orientation along the lines of African socialism as it was promulgated and practiced in Tanzania and Zambia, the NRA thus did not connect with socialist great powers. Its internal political divisions did not allow for this, and in the eyes of the NRA leadership support by the Soviet Union also seemed unlikely (Rukikaire 2019, 197). According to its head, Matthew Rukikaire, there was no expectation on the side of the NRA that it would gain immediate recognition by any foreign government anyway:

> We did not get recognition from anybody, not even from African countries. We were looking for sympathisers and for restraining them from supporting Obote. We were not looking for recognition, we knew we would not get it.[8]

[5] Interview with Matthew Rukikaire, 17 February 2023. According to him, neither the Burundian nor the Rwandan governments knew about this delivery. Another former leading NRA commander and later general, however, talked about the tacit agreement of both governments for using their countries as transit (Interview with general of the Ugandan People's Defense Force, 19 February 2023). The delivery as such was revealed by Yoweri Museveni (1997, 141).

[6] Interview with Ugandan general, 19 February 2023, Western Uganda.

[7] Interview with Matthew Rukikaire, Kampala, 17 February 2023.

[8] Interview with Matthew Rukikaire, Kampala, 17 February 2023.

The external relations of the NRA were thus rather weak and shaky. Even in Nairobi, the 'External Committee' was not entirely safe. Rukikaire's house was often searched by Kenyan police as the Obote government tried to push Kenyan authorities to act against Ugandan exiles (Rukikaire 2019, 212).[9] The real diplomacy unfolded only in the last year of the war, from early 1985 to January 1986. Even after the victory, Museveni immediately visited neighbouring governments to calm any worries about the war spilling over.

As it turned out, once the NRA had taken power, the ground had been well prepared in the region for its recognition. Museveni as new head of state travelled to meet Zaire's Joseph Mobutu and Rwanda's Juvenal Habyarimana in Goma the day after his *pronunciamento* in January 1986 in order to assure them that the NRA had no intention of exporting its anti-regime war into these neighbouring countries. Relations with Kenya and Tanzania had already been formalized de facto during the peace negotiations that lasted from August to December 1985.

In earlier years of the war, the reaction of foreign governments to the NRA representatives was usually the same, according to Rukikaire: '"We have listened, you have a cause, but really, we are not going to support you."'[10] Only in the third phase of the war (1984–1985) did the NRA become more active in different foreign capitals. Based on diplomatic archives, three of these capitals, London, Stockholm, and Washington DC, can be described in more detail. They confirm the quote from the title of this chapter, namely that the NRA got more attention when it became militarily successful.

Great Britain's late reaction

The NRA was not able to achieve officialization of relations by the British government until the very last phase of the war, after Tanzania and Kenya had already engaged with the NRA in a series of peace talks that these two governments mediated. Yet London was relevant for the NRA as an important site of the Ugandan diaspora. In 1981, after his visit to Libya, Museveni and other NRA leaders travelled to London to sensitize the Ugandan community and to calm worries about the connection to Libya (Rukikaire 2019, 203). According to the head of the External Committee, these efforts were necessary as the NRA was not yet well-known among diaspora members. In particular, when contacted by state officials in the United Kingdom and in the US, such knowledge and possibly even sympathy among the diaspora was required in order to achieve at least tolerance or neutrality by these governments (Kategeya 2020, 94).[11]

[9] The German ambassador in Nairobi wrote to the German Foreign Office that several Ugandan oppositionists had been abducted in Kenya, telex from embassy, 14 November 1984, PA AA BAV 106 KAMP 21175. Here, and in further footnotes, I refer to unpublished sources, both from the 'Political Archive of the German Foreign Office' (starting with the acronym PA AA), and 'The National Archive' in Kew, London (starting with the acronym TNA).

[10] Interview with Matthew Rukikaire, Kampala, 17 February 2023.

[11] Interview with Matthew Rukikaire, Kampala, 17 February 2023.

In the course of the war, however, the NRA figured that they needed to work on their image in the UK, as the British government continued to support the Obote regime in Kampala. Eriya Kategeya presided over the external wing of the London branch. It was he who organized a decisive step in 1984 when he invited William Pike, a British freelance journalist, to travel to a rebel-held area and write—in the NRA's interest—about the war in Uganda. In August 1984 Pike published his article in *The Observer* (London).[12] It was widely quoted and translated and appeared the next day as the front page story in the French daily *Libération*. Also, it led to reporting in the BBC's Africa programme in which interviews with Museveni were quoted. The massive human rights violations of the Obote regime became a public issue in Great Britain, yet without any change in Britain's policy towards the regime.[13]

Stockholm—a safe haven

Two leading figures in the NRA's External Committee, Amama Mbabazi and Ruhakana Rugunda, were among those who found refuge in Sweden during the war (Rukikaire 2019, 205). Despite the secret support of the NRA by Kenya's president Daniel Arap Moi, Kenyan security agencies occasionally acted against NRA members and other oppositionists who had sought refuge there. The most prominent case of such danger concerned Museveni's wife Janet who, alerted by a Swedish diplomat working for UNHCR, went into exile in Sweden, which became the most important hub of the NRA in continental Europe (Museveni 2011, 119). While the Swedish government did not support the NRA with resources, it helped exiled individuals, and it allowed the NRA to agitate: in summer 1984, for example, the British Embassy in Stockholm reported about a prominent NRA member shouting in front of the embassy 'Obote is a murderer, Thatcher is a murderer'.[14] Museveni himself finally contacted Swedish diplomats in April 1985, and they reacted positively to the idea of talks, but ruled out recognition.[15]

Washington DC—divided attitudes

With regard to the US administration under Ronald Reagan, the NRA applied a similar diplomatic strategy as in the UK. It was essential to win over at least parts of the

[12] The fact that *The Observer* was owned by Lonrho—a London-based conglomerate, triggered speculations that Pike's article was 'propaganda' in order to tarnish the Obote regime's image while presenting an inaccurate account of the democratic and human rights protection record of the NRA/M (Otunnu 2017, 169). Rumours persist that Lonrho was rewarded with business licenses in Uganda for having supported the NRA during the war (cf. Tandon 1987).

[13] On similar cases of how denouncing the sitting government's human rights violations would help to legitimize a rebellion, see the chapters by Andrea Oñate-Madrazo on the FMLN and Eline van Ommen on the FSLN in this volume.

[14] Letter from the British Embassy in Stockholm to EAD, 31 August 1984; TNA FCO 31/4467 Uganda Human Rights, 1984.

[15] Telegram from the Embassy in Stockholm to FCO, 17 April 1985, TNA FCO 31/4731, JEU 015/1, Uganda: Dissidents.

exiled community, yet it was also important to establish unofficial contacts with US diplomats. Here, the former minister and Ugandan ambassador to the UN, Grace Ibingira, was instrumental (Kategeya 2020, 93). Yet the NRA had only mixed success in getting acknowledgement in the State Department, and this seemingly had to do with divided attitudes within the agency. In September 1984, Assistant Secretary of State for Africa Chester A. Crocker met with President Obote and assured continued US support despite allegations of mass murder and torture by the Ugandan army. This was in reaction to a report that had brought Uganda in the limelight of US foreign policy. In June 1984, the assistant secretary of state for human rights, Elliot Abrams, had told Congress that 200,000 civilians had been massacred by government forces ('Assistant Secretary of State' 1984). The background of this report was the work of a humanitarian freelancer, Robert Gersony, hired by the State Department to collect 'what we thought were hard facts' on the human rights situation in the Luwero triangle, as then US ambassador to Uganda, Allen C. Davis (1998), recalled.[16] Museveni's guerrilla war was mentioned in the *New York Times* in reports on Crocker's visit ('U.S. Aide raises rights issue' 1984). This was a huge success for the NRA, as these discussions were internationally noticed and tarnished the image of the Obote government considerably. The official US administration's attitude towards Uganda, however, did not change, as Chester A. Crocker's statement indicated.

Turning points: dynamics in 1985

The international acknowledgement of the NRA changed only in the course of 1985, but then assumed a rapid pace. Not only were the rebels able to conquer large swathes of territory in Western Uganda and approach Kampala with several thousands of fighters, they also enhanced their diplomatic efforts and had relations officialized by neighbouring governments through their participation in the peace negotiations starting in Nairobi in August 1985. These had become possible through the collapse of the Obote regime. In spring 1985, the decay of the government army was evident, and clandestine talks between government members and the NRA took place. Museveni met the then prime minister Paul Muwango in the hospital of the German armed forces in Koblenz, Germany. Muwango had already contacted Matthew Rukikaire to ensure his survival after a possible regime change.[17] A split in the armed forces along ethnic lines was appearing, dissolving the central command and leading to mutinies in June 1985 (cf. Kasozi 1994, 173). These culminated in a military coup d'état on 27 July 1985, which opened the avenue for peace negotiations. Yet the NRA had also discovered the need for a more programmatic stance. In early 1985,

[16] Gersony has repeatedly worked as a freelancer for the US State Department and has been lauded for his information gathering, which has corrected US diplomatic knowledge production. His report on the Resistência Nacional Moçambicana (RENAMO) in Mozambique, for example, also led to a policy change by the US (cf. Kaplan 2023). On the problems of US foreign policy to assess African contexts, see also Shackelford (2020).

[17] Interview with Matthew Rukikaire, Kampala, 17 February 2023.

a group of leading NRM/A figures met in Austria to agree a 'Ten Point Programme' that from then on served as the political programme that was to be communicated to foreign governments. Finally, the NRA used the negotiations in Nairobi from August to December 1985 to produce a number of programmatic statements while simultaneously doubling its work to create diplomatic contacts.

To sum up, the NRA efforts to gain legitimacy in the international realm display all three main claim-making practices outlined in the Introduction of this book. They brought military force against the central government (fighting), they ensured that they were heard by the international community by inviting journalists to rebel areas so they could write about the conflict (voicing), and finally, they established committees that mirrored the central government and worked on a political programme to present to potential international allies (mimicking).

Yet beyond that, the case offers a series of other insights: First, it shows that at times at least, regional political constellations can overrule global constellations. The regional political relations are of utmost importance in order to understand the trajectory and political success of the NRA. A number of NRA leaders, first and foremost Yoweri Museveni himself, were known politicians or student leaders before the war started. Numerous contacts, in particular in Tanzania and Mozambique, existed, including a personal acquaintance with Julius Nyerere, the then president of Tanzania who acted as the conductor of East African regional relations, who maintained at times contentious relations with Daniel Arap Moi, Kenya's head of state between 1978 and 2002.

In the years of its armed fight, the NRA was therefore at first less concerned with global political relations or what great powers thought about it than with the attitude of Uganda's immediate neighbouring governments. It was in particular the government of Tanzania, certainly not a global player, that had a key role in the inner Ugandan power battles. Up until the 1990s, it was Julius Nyerere who acted as the pivot of a left-leaning Pan-Africanist position that many East and Southern African politicians and intellectuals shared and that has remained very influential in the continent's conflicts and alliances (cf. Roessler and Verhoeven 2016). Other relations, with Mobutu in Zaire or with the government of Sudan, were equally important in order to contain counter-movements that partly recruited former regime soldiers.

A second observation is connected to this: 'I don't think the Cold War mattered at all' noted the former head of the External Committee, Matthew Rukikaire, in an interview on 17 February 2023. In a way, the war in Uganda from 1981 to 1985 remained unnoticed by the superpowers. At first sight this is all the more astonishing as at the same time both superpowers were heavily engaged in a number of other African conflicts, like in Ethiopia, Mozambique, and Angola, and in particular in the Apartheid system in South Africa. Yet the war in Uganda was not 'charged' with the Cold War, although several Ugandan actors tried to solicit superpower support. Obote's government, for example, tried to tarnish the image of the NRA by labelling them as 'socialists', although Obote himself had once subscribed to 'African socialism'. Yet a considerable part of the NRA leadership was suspicious of socialist countries and their influence.[18] In fact, as will be shown below, socialist governments in Europe, such as that of the German Democratic Republic (GDR), remained very reluctant with regard to support for the NRA and stuck with the government in power, which, due to its links to Tanzania, had elements of African socialist traditions in its self-representation and was furthermore benefiting from North Korean military.

That the conflict in Uganda did not become an episode of the Cold War can also be explained by the NRA's strategy. From what is known, the rebels never tried to engage with the Soviet Union or any of its allies. Such a step could have engaged the US on Obote's side, as under Reagan the 'need to block Soviet adventurism' was the highest priority of US foreign policy.[19]

Archived files from the East German foreign ministry confirm this abstention of the Soviet Union and its allies from engaging in Uganda. East German diplomats were approached by both sides and coordinated their reaction to the Soviet Union and Cuba with the maxim: 'any partial orientation should be avoided, political balancing should be maintained'.[20] Even in December 1985, during the last phase of peace negotiations, an NRM representative's demand for support from the GDR was rejected: the GDR could not deliver medicine and medical equipment to the NRA due to the GDR's formal relations with the Republic of Uganda.[21]

An explanation for this 'neglect' of the war in Uganda by parties who were otherwise willing to give support to various sides in global conflicts might draw on this conflict's particular global historical timing. Britain's role in Uganda was still considerable, so post-colonial continuities still structured the insertion of Uganda in its relations with the West. The NRA also had a clear sense that denouncing its opponents' disastrous human rights record was more rewarding diplomatically than joining the already weakened socialist camp.

'The need for an accommodation with Museveni is clear': British and German reactions to the rebellion

The interpretation of the NRA's diplomacy given above is supported by the analysis of files from two diplomatic archives—the British National Archives in Kew, London and the 'Political Archive' of the German Foreign Office (Auswärtiges Amt) which contains the former GDR foreign office's files as well. All three countries' diplomatic correspondence supports the impression that regional dynamics mattered more than great power politics when it came to validating the legitimacy claims of the NRM/A. In fact, the British policy towards Uganda was 'anti-NRA' almost until the very last moment, when the power balance had already tipped. While the US did not develop

[18] Interview with Matthew Rukikaire, Kampala, 17 February 2023.

[19] Cf. Interview with the then undersecretary of state for Africa, Chester A. Crocker (ADST 2011).

[20] MFAA (Foreign Office of East Germany), instruction for a conversation of comrade König with comrade Bruno May, ambassador of the GDR in Uganda, 24 September 1985, PA AA M 95/20388.

[21] Note to Dr Makumbi Kidza, 18 December 1985, PA AA M 95/20388.

any particular interest in Uganda, both governments and their African departments in foreign service were at the same time much busier dealing with South Africa's Apartheid regime and the wars in Southern Africa (cf. Schraeder 1992, 572). The West German Foreign Office showed activity only very late in the process, even though the Germans had a more neutral stance on the question. The attitude of not recognizing the NRA until it became inevitable was also shared by all other foreign governments. Acknowledgement, support, and finally recognition was thus first granted by East African governments, triggering a chain reaction.

Great Britain—last minute acknowledgement

For a long time, the British government remained quite conservative in its attitude towards the NRA. It stood with Obote's regime and did not change its position even when Obote was overthrown by his generals in July 1985. Post-colonial ties mattered for relations. In the early 1980s, over 570 British expatriates lived in the country, and Britain was still Uganda's largest non-African trading partner. Britain's exports to Uganda totalled 32 million pounds, with France and Germany being the main competitors. Uganda's armed forces received military training from the UK while arms exports were banned, with the exception of armoured vehicles. This support was intended to serve Britain's primary aims of stabilizing the country and disciplining the army, as a 'country assessment paper' from July 1983 stated.[22]

Even engaging in informal contact with the NRA was considered to be dangerous, as it would weaken Britain's position vis-à-vis Obote's regime. The British high commissioner in Kampala in particular repeatedly emphasized 'progress' in the observation of human rights in the army and police, both of whom benefitted from UK aid till the very last moments of the old regime. The high commissioner expressed his support, however, in the face of the growing concern being voiced by the British public and also in his own headquarters, the Foreign and Commonwealth Office (FCO) in London. It is plausible to assume that his continuous plea to maintain support of the regime was also a result of his own institutional interests, avoiding the delegitimization of his own work.

This attitude only began to crumble when international pressure increased, in particular in the US due to the Elliot report (see above) and when the NRA achieved massive military advances. In January 1985, after the war had gone on for four years, High Commissioner McLean admitted that 'Obote's regime did not do well' and that the NRA had become 'increasingly effective'. A government victory would be beyond its 'immediate reach'. Yet, British military training and 'obscure' North Korean military trainers had 'boosted UNLA performance'. McLean also reported that France had suspended its aid to Uganda, and Germany had reduced theirs.[23] In September 1984 European diplomats had already stated in a joint declaration that 'Obote had

[22] TNA FCO 31/4095, JEU 014/5.
[23] Annual Review for 1984, TNA FCO 160/225, DD 1985/151–219.

no control of the army'.[24] Despite this, none of these events endangered the support of the British government, including providing military aid.

The overall turning point in the international perception of the NRA's war and the problems of Uganda's government were the massacres in Namugongo, committed by government soldiers at Kampala's outskirts in June 1984. A Ugandan human rights group in London immediately reacted with a protest march around Trafalgar Square. It was around that time that the NRA organized the trip of William Pike, a British journalist, to feature the bush war in British mass media (see above). Later on, in June 1985, William Pike contacted British diplomats in London, offering to establish contact with Museveni, who at the time was in Sweden. The diplomats replied that they saw 'no reason to respond to this approach or in any way to modify our carefully considered policy of avoiding contact with Museveni'.[25]

Change was in the air, though. With the coup in July 1985, British military training of Ugandan forces came to a halt.[26] Seemingly in late August 1985 the FCO revised its attitude. Until then, contacts with the NRM had remained spurious and unofficial. From the sources it seems that it was only when the military situation around Kampala had become so clearly in favour of the rebellion that the UK government saw the need for negotiations with the NRM. With this decision, the UK government swung towards a development that governments in the East African region had already prepared. Since about that same time, the negotiations mediated by Kenya's president Moi had already taken shape, supported by Tanzania's government. The officialization of relations with the NRA by East African governments thus preceded the acknowledgement by the former colonial power and permanent member of the UN Security Council, Great Britain.

In the summer of 1985, sometime between the end of August and the end of September, the lever was thrown. Colin McLean, the high commissioner in Kampala, was overruled by the FCO. The British foreign secretary Geoffrey Howe intervened and wanted to reach a solution through negotiations with tripartite mediation (UK, Kenya, Tanzania). At the same time, the military successes of the NRA, approaching Kampala in September, also made its acknowledgement unavoidable. It was seemingly the NRA's military might that made the FCO change its position. In the archived correspondence, a handwritten remark on a confidential report on the situation in Uganda from 2 August 1985 looks like the tipping point of the internal deliberation:

> If the army is really down to the strength of 12,000 grouped in brigades around the country, they might be hard put to it to defend Kampala from a determined attack by the NRA. The need for an accommodation with Museveni is clear.
>
> (William Marden, Assessment of East Africa Department, 2 August 1985, TNA FCO 31/4726)

[24] Telegram of Foreign Mission in Ireland to FCO, 5 September 1984, TNA FCO 31/4467, and TNA JEU 240/1, Uganda Human Rights, 1984.

[25] Letter to Wenba-Smith, EAD, 7 June 1985; FCO 31/4731,TNA JEU 015/1, Uganda: Dissidents.

[26] Loose minute of minister of armed forces, 30 October 1985, TNA DEFE 13/1931, Uganda.

The same report, however, had also stated that '... stability will depend on neutralising the political opposition, notably Museveni's NRA, ...' (ibid.). Yet Britain's minister of state, Malcom Rifkind, suggested engaging with the governments of Kenya and Tanzania, who had already asked the UK government to join a tripartite diplomatic initiative in order to shore up the faltering mediation efforts of the Kenyan government.[27] With that step, the British government's officialization of relations with the NRA as a legitimate actor was accomplished. Yet the lead had already been taken by regional neighbours, not by the former colonial power of Uganda.

Once the situation in Uganda had received enough attention in the upper echelons of the ministry, the decision also led to diplomatic action. In September 1985, Foreign Secretary Geoffrey Howe authorized a change in Britain's foreign policy towards Uganda. From this moment onwards, the British High Commission in Nairobi, where the negotiations took place, was more important than the one in Kampala. The High Commission in Kampala had still considered Okello's 'Military Council' to be more legitimate than the NRA, probably not least because General Okello brought long-standing civilian politicians like Paul Ssemogerere and Olara Otunnu into his government, both well-known to British diplomats. British military and police training aid was continued until August 1985.

Similar to Germany, the final decision about officializing relations with the NRA came from the head office, that is, the FCO in London. What mattered here in addition to the military situation was publicly available reports and an entire network of contacts and private initiatives that had been building pressure since shortly after the beginning of the war.

Within the domestic political field in the UK, the High Commission in Kampala had no monopoly on information about Uganda. The respective folders in the National Archive contain numerous letters, documentations, and interventions from private individuals and prominent politicians. Since 1982, church members, Amnesty International, individual exiled Ugandans, etc., had written to Members of Parliament about human rights violations in Uganda, often with extended documentation. Typically, ties from colonial times mattered here, even twenty years after Uganda had gained independence. For example, the former dean of Makerere University in Kampala, Hugh Dinwiddy, wrote to the foreign minister in the summer of 1985 and lamented the intransigence of the High Commission in Uganda. Furthermore, in the debates of the House of Lords at the beginning of the 1980s, the upper classes, most of whom had an imperial family history, spoke up because of human rights violations happening in Uganda ('Uganda: Human Rights Violations' 1984).

It is not so easy to explain why years of strong and public protest against the support of an openly repressive regime did not trigger any action, though. Institutional inertia might be one explanation, and a reason for the sudden change might have been the evolving international situation. In March 1985, Mikhail Gorbachev had become the eighth leader of the Soviet Union, and in November 1985 the first summit between US president Reagan and Gorbachev took place in Geneva, where they

[27] Rifkind, Uganda, 19 September 1985, TNA OD 53/144.

agreed on a 50 per cent reduction of nuclear arms. Détente was in the air and perhaps it made many things possible, both in Europe and in Africa (cf. Schraeder 1992).

In the US and UK, diplomacy with Africa during the early 1980s was presumably much more preoccupied with developments in Southern Africa. Wars were ongoing in Angola, Mozambique, and Namibia in which Cuba and other socialist states stood against an alliance of Western governments with the Apartheid regime in South Africa, creating numerous dilemmas for the US and UK governments. In 1985, the anti-Apartheid movement in the US reached new heights and posed a serious challenge for the Reagan administration (cf. Rich 1988, 190). Once there was a decision at the top of the ministry, the British were no longer any obstacle, and once the NRA had taken Kampala and Museveni was sworn in as the new head of state, recognition followed swiftly. The new regime had already started to receive British military aid in February 1986.

The supreme position of Great Britain as the most influential Western actor in Uganda was also acknowledged by the US administration. On the whole issue of the war in Uganda, the US had left the field to the British. At the end of July 1985, after the coup, all but twelve American diplomats left Uganda. The US ambassador in Uganda told his British colleague that Britain is regarded as the 'senior Western partner in any moves to exert pressure on the present regime here'.[28] The US, according to the British, did not even have a problem with North Korea's training of Ugandan government forces 'as long as human rights are observed by North Korea'.[29] In the summer of 1985, US foreign policy was probably sufficiently busy with Iran and Nicaragua and considered relations with South Africa to be the supreme African issue. At least, a particular role taken by the US was mentioned neither in British or German diplomatic archives nor in interviews with leading NRA members or in their published accounts. The US Embassy had already been deserted in June 1985 after the coup d'état (cf. Davis 1998).

Only reactive: West Germany's recognition of the NRA

In opposition to British diplomats dispatched in Kampala, even early on the West German ambassador was quite critical of human rights violations committed by Uganda's government forces. In September 1982, when Amnesty International had published a report with respect to the allegations, the ambassador cabled back to Bonn that President Obote had requested the NRA to give up, and he just repeated the characterization of the NRA as 'bandits'.[30] Two years later, the same ambassador had a different assessment and reported about a one-to-one conversation with President Obote in which it was said that the regime would not be able to arrange a

[28] High Commissioner McLean to East African Department, 16 August 1985, Uganda relations with USA, TNA JFCO 31/473.

[29] High Commissioner McLean to East Africa Department, 5 March 1985, TNA JFCO 31/473.

[30] Telex from AV Kampala to 320, 14 September 1982, PA AA BAV 106 KAMP 21169.

turn-around of the critical situation and that Obote knew 'that sooner or later he himself could become a victim of the spreading violence'.[31] Yet another year later, in June 1985, shortly before the putsch, the West German diplomats reported that the NRA had a firm stand in Uganda and that government forces were meanwhile 'unable to fight, despite their North Korean trainers'.[32]

With the start of negotiations in Nairobi, mediated by Tanzania and Kenya, the German diplomats also got into closer contact with the NRA. In a three-page telex, the diplomats describe Museveni as a 'waterproof Marxist, yet with a distance towards Eastern European socialist states'. That the NRA had brought peace to Western Uganda was acknowledged, and also that its troops were much more disciplined than the government forces.[33] The lengthy text of an NRA press release from August 1985 in Nairobi was cabled in its entirety to Bonn as an indication of the programmatic stance of the NRA.[34]

Mirroring the British FCO's assessment from 2 August 1985, German diplomats also reported in August 1985 on how the situation in Kampala and surrounding areas had changed. There were rumours that the NRA was marching towards Kampala from three different directions and was by then only 15 km away from the capital. With the army being too weak to resist, there was no alternative to the negotiations. If these failed, the ambassador argued, Museveni would most certainly take over Kampala directly.[35]

It was only during the negotiations in Nairobi that German diplomacy became active in the conflict. The actual officialization of relations between the NRA and Germany took place when the German diplomats in Kampala and Nairobi suggested inviting Museveni to Germany in early December 1985. He was to participate in a workshop in Königswinter near Bonn and be received by the German Foreign Office 'at a high level'. The NRA's request to get support in the form of medicine and health equipment was, however, made dependent on the junta's agreement to come to the meeting.[36] In the days of 9 to 12 December, Museveni indeed visited Germany. On 9 December he was received by a third-level diplomat and secured an agreement to receive large shipments of medicine which were dispatched the very next day via Kigali and the NGO Cap Anamur. The Germans also reported that two British advisers were in contact with Museveni to counsel him on political action after the peace deal which was signed on 18 December 1985.[37]

The reception of Museveni, a de facto officialization of relations by the West German government, was thus rather late, four months after Tanzania's president Nyerere and Kenya's president Moi had officialized their countries' relations with the

[31] Telex from AV Kampala to 320, 5 July 1984, PA AA BAV 106 KAMP 21169.

[32] Ambassador Held to Foreign Office, 'Innenpolitik Ugandas', 18 June 1985, PA AA BAV 106 KAMP 21172.

[33] Telex from AV Kampala to 320, 16 August 1985, PA AA BAV 106 KAMP 21172).

[34] *Uganda Resistance News*, Nairobi, vol. iii, no. 3, August 1985, PA AA BAV 106 KAMP 21172.

[35] Telex from AV Kampala to Ref. 320 und AV Nairobi und Dar, 24 August 1985, PA AA BAV 106 KAMP 21172.

[36] Telex of 22 November 1985 from Nairobi to 320: PA AA BAV 106 Kamp 21173.

[37] Telex from AV Kampala to 320, 29 November 1985, PA AA BAV 106 Kamp 21173.

NRA. Only after the encounter in Bonn did the German diplomats come to the conclusion that Museveni would be a reliable partner, that he would not drive Uganda into the socialist camp, and that he would stick to a market economy at least in a 'mixed economy version'. In another cable from the German Embassy in Rwanda, further confirmation of Museveni's status was produced: The NRA, it was said, was considered a reliable partner by Rwanda and Burundi.[38]

But it was not only acceptance and de facto recognition by others, it was also military might that pushed the Germans further towards final recognition of the NRA. In late December 1985, marauding government soldiers plundered Kampala, and it was only after the NRA marched into the city in late January that the shooting and fighting ended. The NRA pushed the former government forces eastwards out of the city, and open violence ended. 'White persons, as far as it is known, have not suffered', a handwritten note from the ambassador reported back to Berlin.[39]

Part of the explanation for West Germany's reluctant approach was certainly the desire to act in line with legal procedures with a view to the policies of Western allies like the UK and US. The preoccupation with legality produced the second decision point in German diplomacy on 18 February 1986, three weeks after Museveni had been inaugurated. Up until then, the question had not been explicitly answered as to whether Germany would now recognize him as head of state. The wish of the federal president to transmit congratulations directly after the inauguration in January was put on hold until mid-February as the German ambassador did not find this appropriate, rightly assuming that the Foreign Office would still need to find a position first. This position was defined in a paper that summarized the events of the last year and came to the conclusion that the NRM was a reliable actor and also without any viable alternative. Given that both the US and UK had already signalled their acceptance of the NRM as the new government, this was actually an easy step. It is probably typical of German diplomacy that the international law aspect made a long reflection on the situation and the appropriateness of official recognition necessary. With the traditional 'monopoly of lawyers' in German higher administrative echelons generally (Dahrendorf 1965, 251) and in the Foreign Office as well, Jelinek's 'normative power of the factual' unfolded much more slowly.

Conclusions—a war at the thresholds of historical times?

Why did the NRA succeed? As it was argued in this chapter, it was ultimately the combination of military strength, the generation of a minimum of international support, and the indirect effects of its warfare, namely the erosion of the governing regime, that allowed for the victory of the NRA. The successful delegitimation of its opponent, the Obote government, by denouncing its human rights violations heralded the global historical time of humanitarianism.[40]

[38] Letter of German Embassy Kigali to FO, 20 December 1985; PA AA BAV 106 Kamp 21173.
[39] Telex from AV Kampala to 320, 30 January 1986, PA AA BAV 106 Kamp 21173.
[40] Cf. the section of the Introduction to this volume on the historical time of humanitarianism.

All six practices of claim-making and claim validation outlined in the Introduction can be observed in this case, but that is not all we can learn from it. Another remarkable observation is the insertion of leading NRA personnel in what can be labelled as the regional political class of post-colonial East Africa was a major factor in gaining informal support, acknowledgement, and finally recognition by other governments in the region. One primary result of this case analysis is thus that it was less global constellations and global powers that explain the international dynamic surrounding the NRA, but rather the much less studied regional political constellation.

A second core result of this case analysis concerns the question of global historical timing (cf. Introduction to this book). The NRA seems to have operated at a double threshold in the history of international politics. First, the case shows clear post-colonial features. Great Britain was still the most important Western political arena and the most important non-African actor in this conflict. This changed after the war. As the role of the World Bank and IMF after the war showed, multilateral agencies set the terms for the reconstruction of the country after 1986. In addition, relations to the US intensified considerably, not least because Uganda became a strategic partner for the US in the region of East Africa. In that sense, the NRA operated—and was itself an actor—in a shift of international politics from the post-colonial and Cold War constellation to the 'unilateral moment' and liberal hegemony led by the US.

The NRA operated at another threshold insofar as more or less unconditional support for allies in the Cold War was slowly superseded by a discourse on human rights in the 1980s. The long traces of the Second World War and the use of human rights for delegitimizing socialist states transformed human rights from a weapon into an autonomous discursive field that affected Western foreign policy making as well, as the cutting of aid for Uganda, based on the 'Foreign Assistance Act' of 1973 in the US, indicated (cf. Dezalay and Garth 2006, 240). The case of the NRA was thus a portent of the global historical time of humanitarianism sketched in the Introduction to this volume.

The case finally shows how uneven and divided the world was when this war took place. Although information about the war and about massive human rights violations was accessible and globally distributed, Western governments did not make any cohesive effort to change the situation in Uganda. A generalized expectation that concerted action was needed to end humanitarian disasters did not yet exist in this global historical time. All Western foreign services turned out to be rather conservative and not without internal cleavages. Another lesson from the story is thus that the cohesion and coherence of the Western foreign policy apparatus might at times be over-rationalized in interpretation and theory.

References

Association for Diplomatic Studies and Training (ADST). 2011. Interview with Assistant Secretary Chester Arthur Crocker, 5 June 2006. https://adst.org/OH%20TOCs/Crocker,%20Chester%20Arthur.toc.pdf. Accessed 25 January 2025.

'Assistant Secretary of State Chester Crocker Met with President'. 1984. *United Press International*, 4 September. Accessed 30 January 2022.

Dahrendorf, Ralph. 1965. *Gesellschaft und Demokratie in Deutschland.* München: Piper.

Davis, Allen C. 1998. Interview, 26 June. *Association for Diplomatic Studies and Training, Foreign Affairs Oral History Project.* Accessed 3 April 2023.

Dezalay, Yves and Bryant Garth. 2006. 'From the Cold War to Kosovo: The Rise and Renewal of the Field of International Human Rights'. *Annual Review of Law and Social Science* 2: pp. 231–255.

Golooba-Mutebi, Frederick. 2008. *Collapse, War and Reconstruction in Uganda: An Analytical Narrative on State-making.* Crisis States Working Papers no. 2, London: London School of Economics and Political Science.

Jellinek, Georg. 1905. *Allgemeine Staatslehre.* Berlin: Häring Verlag.

Kaplan, Robert. 2023. 'The Good American: The Epic Life of Bob Gersony, the U.S. Government's Greatest Humanitarian'. Carnegie Council Podcast. Accessed 8 April 2023.

Kasfir, Nelson. 2005. 'Guerrillas and Civilian Participation: The National Resistance Army in Uganda, 1981–86'. *Journal of Modern African Studies* 43(2): pp. 271–296.

Kasozi, A. B. K. 1994. *The Social Origins of Violence in Uganda, 1964–1985.* Kampala: Fountain Publishers.

Kategeya, Eriya Tukahirwa. 2020. *Impassioned for Freedom.* 2nd edn. (1st edn. 2006). Kampala: Wavah Books.

Kretschmer, Fabian. 2010. 'Ugandischer Befreiungskampf im Weinviertel'. *Die Presse*, 8 July. Accessed 9 July 2022.

Museveni, Janet Kataaha. 2011. *My Life's Journey.* Kampala: Fountain Publishers.

Museveni, Yoweri Kaguta. 1997. *Sowing the Mustard Seed: The Struggle for Freedom and Democracy in Uganda.* London: Macmillan.

Ngoga, Pascal. 1999. 'Uganda: The National Resistance Army'. In *African Guerillas*, edited by Christopher Clapham, pp. 91–106. Kampala: Fountain Publishers.

Otunnu, Ogenga. 2017. *Crisis of Legitimacy and Political Violence in Uganda, 1979 to 2016.* London: Palgrave Macmillan.

Rich, Paul. 1988. 'United States Containment Policy, South Africa and the Apartheid Dilemma'. *Review of International Studies* 14(3): pp. 179–194.

Roessler, Philip and Harry Verhoeven. 2016. *Why Comrades Go to War: Liberation Politics and the Outbreak of Africa's Deadliest Conflict.* Oxford: Oxford University Press.

Rukikaire, Matthew. 2019. *70 Years a Witness: From Colonialism to Resistance and Beyond.* Kampala: Dominant Seven Publishers.

Schlichte, Klaus. 2009. *In the Shadow of Violence: The Politics of Armed Groups.* Frankfurt a.M.: Campus; Chicago: Chicago University Press.

Schraeder, Peter J. 1992. 'The Horn of Africa: US Foreign Policy in an Altered Cold War Environment'. *Middle East Journal* 46(4): pp. 571–593.

Shackelford, Elizabeth. 2020. *The Dissent Channel: American Diplomacy in a Dishonest Age*. New York: Public Affairs.

Tandon, Yash. 1987. 'Elements of Continuity and Change Between Obote and Museveni: Some Lessons from the Obote's Rule for Museveni's Government'. *UFAHAMU* 15(3), Winter 1986/87: pp. 79–97.

'Uganda: Human Rights Violations'. 1984. *The Hansard*, Volume 455, debated on Tuesday 16 October.

'U.S. Aide Raises Rights Issue During His Talks in Uganda'. 1984. *The New York Times*, 5 September.

7
The Politics of Mutual Legitimation

UNITA, Lobbyists, and Washington in the Angolan Civil War

Reyko Huang

As soon as Angola gained independence from Portugal in 1975, it was at war within its own borders. Two of the erstwhile anti-colonial resistance movements—the Movimento Popular de Libertação de Angola (People's Movement for the Liberation of Angola, MPLA) and the União Nacional para a Independência Total de Angola (National Union for the Total Independence of Angola, UNITA)—turned against each other in a violent contest for control of the Angolan state. The MPLA seized control of the capital, Luanda, while UNITA established bases in the Central Highlands and a headquarters in Jamba in the remote southeast. If national liberation had been the rallying cry during the fight against Portuguese colonialism, in independent Angola the MPLA and UNITA quickly became embroiled in opposing sides of the Cold War and its regional manifestations. The MPLA government embraced the ideology—or at least the rhetoric—of Marxism-Leninism, while UNITA, led by its founder Jonas Savimbi, declared itself an anti-communist movement. The Angolan civil war was thus internationalized from the outset, with the Soviet Union, Cuba, and other Eastern bloc countries backing the MPLA while South Africa and later the United States supported UNITA.[1]

For UNITA, the politics of international legitimation was central to the war effort. The organization needed to appeal to sympathetic states for political and military support to sustain its insurgent army. But its quest for sovereign control of Angola had an additional dimension: the fact remained that while a number of African and European governments had formally recognized the MPLA as the legitimate government of Angola soon after independence, the United States had refrained from doing so, due mainly to its opposition to the presence of Cuban troops in Angola. For UNITA, then, the ultimate prize in the legitimation struggle was nothing short of diplomatic recognition by a superpower.

From the early 1980s onwards, UNITA thus attached considerable importance to its outreach to the US government. While the armed group had other friends and

[1] For historical analysis of the Angolan conflict, see, among others, Minter (1991), James (2011), and Pearce (2015).

Reyko Huang, *The Politics of Mutual Legitimation*. In: *Armed Groups and the Politics of International Legitimation*. Edited by: Stephan Hensell and Klaus Schlichte, Oxford University Press. © Reyko Huang (2025).
DOI: 10.1093/9780198974529.003.0007

allies in Africa, the Middle East, and Europe (James 2011, 180), it saw aid from the Reagan administration as indispensable, both materially and symbolically, for countering Soviet and Cuban support for the MPLA. Furthermore, despite the 'Reagan Doctrine' elevating support for anti-communist movements to a dominant US foreign policy issue, Washington's support for UNITA was far from guaranteed; indeed, it was a deeply controversial policy both within the Reagan administration and among lawmakers on Capitol Hill. The Clark Amendment of 1976 had prohibited all US military aid to insurgent groups in Angola.[2] When Congress lifted it a decade later in 1986, opponents of US aid for UNITA pressed hard for a reinstatement of the ban, arguing the United States ought not support a rebel group propped up by apartheid South Africa. In contrast, proponents of support doubled down on the necessity of fighting the Soviet and Cuban threat in the region.[3] UNITA, through its multiple contacts with the US government, media, and civil society, was inserting itself with its own messages into this fraught domestic policy debate.

This chapter examines the ground-level tactics of UNITA's diplomacy and lobbying in the United States during the Angolan war. If armed groups' broad interests in legitimacy-seeking are now well-established in the literature (e.g. Caspersen 2012; Jo 2015; Duyvesteyn 2017; Griffiths 2021; Geis et al. 2021), we arguably have a less clear sense of the processes of legitimacy-seeking by armed groups. What does legitimacy-seeking entail, and what bureaucratic, financial, and social considerations feed into it? What are the relevant practices and publics in armed groups' politics of legitimation? Who are the rebel diplomats who lead the work of international legitimacy-seeking, what credentials and skills do they bring to bear, and what do they actually *do* as rebel representatives? Given their non-state status, how do they 'set up shop' in foreign capitals? Who do they reach out to, how do they gain access to them, and what messages do they bring to the table? In this chapter I identify the individuals and organizations that do the work of international public diplomacy and the ways they go about doing so.[4]

While I limit the scope to UNITA's diplomatic efforts in Washington, I also propose that many of the tactics UNITA employed in Washington are now routine practices among rebel diplomats working in advanced democracies.[5] In this sense, UNITA diplomacy in 1980s Washington is representative of a phenomenon I call 'the rebel lobby', defined as targeted policy advocacy in external states by armed non-state organizations (Huang 2023). This chapter shows that UNITA's legitimacy-seeking in the United States entailed fielding a sustained presence in Washington through extensive network-building, face-to-face interactions with the political elite, and public outreach to American civil society, combined with strategic narrative

[2] Even so, it was known that the US indirectly provided aid to UNITA prior to 1986.

[3] Of the proponents of aid to UNITA, Minter (1991, 137) writes, 'This group derived political force not only from true believers among the Reaganauts, but also from important sectors of the Democratic party seeking convenient ways to prove their anti-communist credentials.'

[4] On the importance of studying the practices of diplomacy in IR, see Adler-Nissen and Pouliot (2014).

[5] Another case in that regard is the Polisario that is routinely lobbying EU institutions in Brussels. See the chapter by Stephan Hensell in this volume.

programmes to reassure allies of its commitment to shared values. Furthermore, much of UNITA's success in deftly managing the art of diplomacy can be attributed to the role of another key set of actors, namely paid professional lobbyists. Part of UNITA's armed campaign to take over Angola, in other words, involved paying American PR firms to do the legitimacy-building on the armed group's behalf.

For UNITA, diplomatic operations in Washington were a major component of its armed campaign to topple the MPLA regime in Luanda. This argument stands in stark contrast to the notion that external sponsorship of armed groups is driven primarily by geopolitics (e.g. Cold War imperatives) and/or by the sponsor's political interests. In such accounts, armed groups' diplomatic activism is epiphenomenal—external actors will provide support to an armed group if doing so is in their national interest in a given context, regardless of how ardently rebels invest in diplomacy. The wider political context and external actors' interests are certainly important, but a focus on these factors alone gives short shrift to the politics and tactics of armed groups' claim-making and the dynamic interplay between rebel diplomats and decision-makers of target states. In the politics of international legitimation, those with the power to confer legitimacy must be convinced that a party is deserving of it. The case of UNITA suggests that it mattered to American officials and lawmakers that UNITA representatives easily intermingled with those in Washington's highest political circles, were foreign educated and spoke English, delivered compelling speeches without the need for interpreters, and could be counted on to tout the ideology of 'anti-Marxism', 'freedom', and 'democracy'. These qualities made UNITA more appealing and acceptable—or at least could be instrumentalized to portray the group as such—amidst mounting reports of the group's violence and brutality in Angola. UNITA leaders, in turn, were more than willing to play the part, and did so with a savvy melding of diplomatic professionalism and performative politics.

In the conceptual vocabulary of this book, the politics of legitimation in the case of UNITA can be sketched as follows: In order to make a claim to legitimacy, UNITA engaged in *mimicking* by adopting state-like formal behaviour through the procedures and customs of diplomacy. UNITA also practiced *voicing* in a commercialized form, employing communications professionals who helped frame UNITA's vision and actions in ways that resonated with its Washington constituency. The US government, in turn, validated these claims by *officializing* UNITA in the framework of diplomatic protocol, *acknowledging* its leader as a 'patriot' and 'loyal friend', and *supporting* UNITA with money and weapons.[6] As a result, UNITA and its constituency in the US developed a tight symbiotic relationship in which claim-making and claim-validation and, thus, the legitimation of UNITA became a mutually beneficial, collusive effort.

One of the advantages of studying UNITA's wartime lobby in the United States is that it is exceptionally well documented, for several reasons. First, US involvement in

[6] On these practices of claim-making and claim-validation, see the introduction in this volume.

the Angolan war, and especially its support for UNITA, was controversial within the US government from the outset and hence generated intense debate. Consequently, there is a large amount of archival records on UNITA from both executive and legislative offices, especially from 1986 until the withdrawal of US support following the end of the Cold War in the early 1990s. Second, as discussed below, UNITA itself was highly active in Washington, generating a voluminous paper trail of its activities. UNITA leaders made speeches, toured US cities, produced Washington-based newsletters, met with American policymakers, and rallied support among churches and other civil society organizations in the United States, leaving troves of documents, media reports, and relics of rebel diplomacy for later scrutiny by researchers. Finally, and related to the second point, US lobbying disclosure laws required that UNITA—and any other foreign entity engaged in political lobbying in the United States—regularly report their lobbying activities to the US Department of Justice. These lobbying disclosure files reveal not only how the organization tried to portray its activities, but also which individuals its leaders met with, what propaganda they disseminated, and how much they spent on various lobbying activities, and essentially serve as a rarely-used historical archive of their routine diplomatic engagements during the height of the Angolan Civil War.

I use these primary and secondary sources to trace the tactics of UNITA's public diplomacy in Washington. Throughout its efforts, the UNITA lobby focused squarely on securing and maintaining US support—for the war effort at home, and for recognition abroad.

The UNITA lobby

UNITA's lobbying efforts in Washington were as high-profile, choreographed, and costly as they come. The organization in many ways epitomizes armed non-state actors' lobbying in foreign capitals as it pursued essentially every avenue of government and civil society lobbying allowed by law in democracies. At the same time, UNITA was in a tight symbiotic relationship with the US government under the administration of President Ronald Reagan—UNITA needed US support for its war in Angola, but Reagan and his political base in Congress also needed UNITA for their political pursuits both domestically and internationally. UNITA certainly sought legitimacy and support from the US government, but in this case, the latter also played the former's legitimacy politics for its own political ends. This dynamic of mutual benefit both defines this particular case and sets it apart from cases where an armed group makes appeals to a disinterested foreign government for support.

I conceive of the UNITA lobby as comprised of two related strategies—the first focused on establishing a physical presence in Washington and building up its networks therein, and the second geared towards substantiating its politics of legitimation through tactful showmanship and claim-making.

The structures of legitimation: local embeddedness in Washington

Structurally, UNITA's international legitimation project in the United States consisted of three main pillars, each of which mutually reinforced the effective operation of the others and jointly enabled a professional, polished, and ultimately compelling endeavour—so long as political circumstances remained favourable. The three pillars were the opening of a Washington representation office and stationing a 'foreign secretary' to lead it; hiring high-profile lobbyists and PR professionals; and running an active local propaganda outlet. The UNITA lobby, in short, had a face, a strategy, and a message.

Opening a representation office in strategically important capitals is a time-tested tactic of modern rebel diplomacy (Huang 2016b). Rebel offices abroad are as much a simulacrum of state diplomacy as they are an embodiment of the group's physical presence and symbolic investment in a particular locale, signalling its 'open for business' status. UNITA exemplified this phenomenon as it launched 'UNITA USA.' as a foreign representation in 1980 (Leonard 1989, 13).[7] By 1986, UNITA had formally incorporated its Washington office as a 'charitable, literary and educational' organization called Free Angola Foundation (and later Free Angola Information Service)—in other words, the office was not a mere outpost, but had legal standing in the United States as a non-profit organization (US Department of Justice 1986b). With three Angolan nationals led by Tito Chingunji representing UNITA in Washington, the office served as the official lobbying clearinghouse for UNITA's political activities in the United States (US Department of Justice 1987). By all accounts, Chingunji—a name that will reappear towards the end of this chapter—was a respected face of UNITA in Washington; indeed, much of UNITA's successful image-making could be attributed to Chingunji and his popularity among Republican politicians. He was a seasoned UNITA spokesperson, having earlier represented the organization in Western Europe. In addition to managing UNITA affairs in Washington, Chingunji also toured cities and towns across the United States to relay UNITA's official message to the American public (Brittain 1998, 46–47; Sly 1992; Nesbitt 1988).

Maintaining an office in Washington is a costly affair, and more so if it is located in the city's political and diplomatic hub. A 1989 disclosure document shows UNITA expending over $300,000 in a six-month period for office rent, personnel, and supplies, while a 1991 disclosure reveals an office address in Washington's Dupont Circle neighbourhood, located just minutes from the White House and coveted by embassies, think tanks, and political offices alike (US Department of Justice 1989a). Somewhat curiously, UNITA's Washington representatives were 'reimbursed for

[7] Though this study's focus is on the UNITA lobby in Washington, note that UNITA also opened representation offices in several European and African cities including Lisbon, Munich, Geneva, London, Paris, Dakar, and Rabat (James 2011, 119).

living and out of pocket expenses, but receive[d] no additional compensation' (US Department of Justice 1986b).

If UNITA's expenditures were not salarying its foreign emissaries, they were plentifully spent contracting professional lobbyists, another common feature of rebel diplomacy (Huang 2016b). Throughout the latter half of the 1980s UNITA relied heavily on hired American lobbyists to strategize and implement its influence activities in the United States. Lobbying firms are invariably well connected to politicians and government officials, as they typically recruit those retiring from public office who have insider knowledge of congressional and bureaucratic politics and the interpersonal connections needed to navigate it.[8] In March 1986, for example, UNITA hired TKC International for an annual retainer of $198,000 (over $540,000 in 2023 dollars) for services including 'providing advice, analyses, and subjective counseling' to enhance UNITA's relations with the US government (US Department of Justice 1986c). In a six-month period in 1986 UNITA spent over $400,000 for its Washington operations (US Department of Justice 1986d), including $100,000 paid to the PR firm Keefe Company International and nearly $100,000 spent arranging a UNITA delegation's visit to the United States (US Department of Justice 1986e). Likewise, in 1988–1989 the organization expended over $1.2 million (nearly $3 million today), including the usual fees to TKC and sizable sums to various other lobbyists (though a significant portion of the spending also went to prosaic expenses like rent and other costs of running its Washington operations) (US Department of Justice 1989a, 1989b). The most widely reported lobbying episode surrounds UNITA leader Jonas Savimbi's 1986 visit to the United States, for which the organization contracted the firm Black, Stone, Manafort and Kelly—then deemed 'the capital's hottest power broker' thanks to its close links with the Reagan administration (Leonard 1989, 14)—for a retainer of $600,000 (Apple 1986).[9] The source of these large sums remains unclear, with suggestions ranging from UNITA-controlled diamond mines in Angola and South African support, to covert aid from the CIA and donations from church organizations in the United States (Apple 1986; Leonard 1989, 13; Heywood 1989, 64).

Given UNITA leaders were learning to navigate a foreign political terrain, their lobbyists helped strategize their outreach to US officials, lawmakers, and the public, arrange the contacts, and craft tailored messages and speeches. Savimbi himself was 'meticulously coached on everything from how to answer his critics to how to compliment his patrons' prior to his 1986 visit to Washington (Tyler and Ottaway 1986). His lobbyists understood that the performative details of claim-making were not mere accessories, but consequential, to UNITA diplomacy and efforts at state mimicry. Savimbi's carefully orchestrated Washington visit thus included rides in stretch limousines, escorts by federal security agents at the State Department's

[8] Indeed, so close are some lobbyists' ties to the government that the lobbying industry has come under intense scrutiny in the United States in recent years for its possible roles in corruption. See, for instance, Foer (2018) on Paul Manafort (who appears in this chapter as a lobbyist with Black, Manafort, Stone and Kelly).

[9] For a fuller list of lobbyists and PR firms that worked as UNITA agents in the United States, see Nesbitt (1988, 122–124) and Leonard (1989, 51–58).

diplomatic entrance, a banquet at the Capital Hilton, meetings not only with Reagan but also with a lineup of political notables including Secretary of State George Schulz, the Defense Secretary, the Senate Majority Leader, and the chairman of the Senate Foreign Relations Committee, among many others, and a final departure in a private jet—all captured on camera for media blasting (Apple 1986; Carlson 1990). With this visit, the officialization of UNITA by the US government and the consecration of Savimbi as a loyal partner appeared complete.

Working with lobbyists was important for another reason, which is that the MPLA itself was working with large Washington lobbying firms to coax the US government to its side (Carlson 1990). With Chester Crocker, the Assistant Secretary of State for African Affairs and the face of US diplomacy in southern Africa, meeting multiple times with Angolan president Eduardo Dos Santos to negotiate a war settlement, the political and military situation in the Angolan war in the late 1980s was in flux. In this context, for UNITA and MPLA, Washington had become a key battleground fought with diplomatic tactics. By most accounts, however, the MPLA lobby was no match for UNITA's glossy operations; its lobbying efforts were 'clumsy and uncoordinated' and 'hampered by having little knowledge of the US political scene' (Minter 1991, 137).

As for the voicing, armed groups today are active netizens on platforms such as Twitter and Facebook, where they run external outreach operations through regular, strategic messaging. Though lacking the convenience of digital media in the 1980s, UNITA managed overseas propaganda with targeted messages for various audiences.[10] In fact, UNITA's claim-making in the United States was, like its strategic planning, nearly entirely outsourced to private firms. For instance, in Washington UNITA hired the PR firm Griffin Communications to publish a newsletter called *Kwacha News*, an English-language outlet featuring multiple essays and commentary voicing unwavering praise for UNITA. As Griffin Communications disclosed in a 1986 lobbying file, UNITA 'asked us to write their newsletter, *Kwacha News*, which will be published every 2 to 3 months. The newsletter is basically an information piece. However, UNITA is shown in a favorable light, and any bills favoring UNITA in the US House or Senate are listed and editorialized about' (US Department of Justice 1986f). One report published in 1984 by 'Kwacha UNITA Press' is a 93-page adulation for the organization replete with full-page photos depicting Savimbi, UNITA forces, and village life in Angola's 'liberated' zones. Featured essays liken Savimbi to other 'remarkable statesmen' such as Woodrow Wilson, Winston Churchill, John F. Kennedy, and Charles de Gaulle, and tout his personal qualities including his education, knowledge of the Bible, experience with guerrilla warfare, 'impeccable' attire, and penchant for poetry (Döhning 1984, 7–9). If UNITA claim-making borders on the theatrical, its close resemblance to the politics of public dissimulation and rhetorical manipulation employed by authoritarian regimes is worth noting (Wedeen 1999).

[10] Within Angola, UNITA ran a radio station called Voice of the Resistance of the Black Cockerel.

The substance of legitimation: tailored appearances and rhetorical tactics

The structures of the UNITA lobby—its foreign offices and representatives, lobbyists, and propaganda outlets—enabled its strategic operations abroad. These centred on forging close relations with key individuals and political offices within the US government and civil society and advancing a carefully crafted image of the organization through speeches, media appearances, and displays of its organizational capacity.

Thanks largely to the work of their lobbyists, throughout the 1980s UNITA representatives had innumerable exchanges with US officials, lawmakers, the media, and civic leaders. Their long list of contacts and meetings included State Department and Defense Department officials, offices of the Senate and House of Representatives, the US ambassador to the United Nations, White House officials, retired cabinet members, and individuals representing numerous think tanks such as the Center for Strategic and International Studies, Freedom House, Heritage Foundation, the Center for Ethics and Policy Studies, and the New York Council on Foreign Affairs. Their records also document a series of media appearances including interviews and radio segments with such prominent outlets as *Los Angeles Times*, *American Spectator*, *Voice of America*, CBS's '60 Minutes', and Radio Martí, a US government-funded radio programme broadcast in Cuba in Spanish. UNITA representatives also delivered a number of briefings, speeches, and lectures before audiences at the State Department, think tanks, government committees, universities, and legislative offices (US Department of Justice 1986b). During Savimbi's 1986 visit to the United States he was received by President Reagan in the Oval Office and was given treatment typically reserved for heads of state (Apple 1986)—rendering, perhaps, the $600,000 lobbying bill value for money.

All evidence points to an easy recognition by UNITA that a certain social facility on the part of its representatives would have large payoffs in the distinct milieu of Washington's political sector. For one thing, UNITA representatives had no need for interpreters, which greatly facilitated its claim-making practices; the organization made sure those dispatched to Washington had the requisite backgrounds, experiences, and skills. Jeremias Chitunda, UNITA's US spokesperson for a time, held an engineering degree from the University of Arizona and spoke fluent English. Lobbying disclosure files show Chitunda gave many speeches and wrote letters directly to US politicians.[11] He would go on live television on the public affairs channel C-SPAN to participate in a call-in programme and take questions directly from the American public (C-SPAN 1990). His successor Jardo Muekalia did much the same, having served as UNITA representative to the United Kingdom prior to his post in Washington (C-SPAN 1991; James 2004, 108). Jonas Savimbi, UNITA's top leader, was himself known for being a Swiss-educated polyglot (who reportedly spoke seven languages) and easily conversed in English with American officials and the media during his

[11] Though, granted, they may have been written by his lobbyists. See US Department of Justice (1986a).

US tours.[12] So comfortable was Savimbi among his supporters in Washington that he told an audience at The Heritage Foundation, a conservative think tank, that coming to the Foundation was 'like coming back home' (Savimbi 1989).

Indeed, he not only conversed, he captivated and charmed his American audiences in the process. In turn, US officials engaged in claim validation by acknowledging Savimbi as a great leader. At a gathering of the Conservative Political Action Conference in 1986, former US ambassador to the United Nations, Jeane Kirkpatrick, introduced Savimbi to a large crowd, stating: 'Linguist, philosopher, poet, politician, warrior ... Savimbi has admirers the world over, and I have long been one of them'. Calling him 'one of the few authentic heroes of our time', she called on the US government to support Savimbi with 'real weapons', adding, 'Whether that help is overt or covert is a bureaucratic detail'—to which 'the crowd stood and went wild with applause', according to news coverage of the event (McCombs 1986). In public speeches Savimbi regularly highlighted personal attributes he believed would please his American constituency, including his upbringing by his preacher father and pious mother (see, for instance, Tooley 2017). Addressing the oft-heard charge that Savimbi was once a Maoist (in the 1960s he received guerrilla training in China), he insisted to a conservative gathering in 1985: 'I am not a communist. I am a Christian. I am a Protestant', to much applause. Even a critic of Savimbi noted in a 1986 op-ed (Ungar 1986):

> I have met Jonas Savimbi during his previous visits to the United States, and I have been greatly impressed with his political and rhetorical skills. He is a genuinely charismatic figure whose eyes seem alive with fire when he talks of the struggle in Angola. He speaks superb English.

US media chimed in, with a 1987 *Reader's Digest* article declaring that Savimbi, the 'charismatic, swashbuckling general', had 'stopped Soviet imperialism dead in its tracks in Angola' (quoted in Minter 1988, 1). President Reagan and Vice President George H. W. Bush were similarly rapt. 'He is a very impressive figure', noted Reagan in his diary after receiving Savimbi in the Oval Office in January 1986 (White House Diaries 1986). Calling Savimbi 'a true patriot', Bush stated in 1988 that not providing aid to UNITA would be 'an immoral sellout of a loyal friend' (McAllister and Ottaway 1988). Accordingly, the following year Bush, this time as US president, again received Savimbi in the White House (Gerstenzang 1989). On this occasion, Savimbi presented Bush with a gift of two wood-carved face sculptures[13] while Bush offered the guerrilla leader promises of further US support.

Such was the devotion of his American sympathizers that for some, the line between the professional and the personal became blurred. When Savimbi was killed

[12] See, for instance, Savimbi's 1986 interview with *The MacNeil/Lehrer NewsHour* show, at American Archive of Public Broadcasting (1986).

[13] Displayed at the George H.W. Bush Presidential Library and Museum in College Station, Texas, as observed by the author. This is an instance of armed group 'gift-giving,' as it were, alongside claim-making. This, too, can be seen as a mimetic act, given the practice of gift exchanges among state leaders during foreign visits is a routine part of international diplomacy.

in combat in 2002, his longtime American allies, including former ambassador Jeane Kirkpatrick, members of the political-religious right, and associates of the PR firm Griffin Communications organized a dinner memorial in Washington in his honour[14]—notwithstanding the fact that by then, the mass atrocities and human rights abuses committed by UNITA forces had long been exposed to the American public, leading to a total abandonment of the group by the US foreign policy establishment (Krauss 1992).

The substantive content of the UNITA lobby in the United States was targeted and consistent: it aimed to cast Savimbi as the sole promise of democracy and stability in southern Africa and UNITA as a bulwark against a communist takeover of the region, all to bolster the case for US political and weapons support in the Angolan war. UNITA statements frequently warned Washington of the threat of Soviet domination in the region, asserting that 'the Soviet-Cuban presence is far more oppressive than colonialism under Portugal' (Chitunda 1986a). UNITA, in contrast, was 'committed to the defense of human liberties, political tolerance, religious freedom, free trade unions and a socioeconomic system in which the individual is the main factor for progress and wellbeing'—a claim obviously manufactured to hit all the key terms in one breath. UNITA also sought to assuage concerns about the support it received from South Africa. Its leadership claimed Soviet imperialism was 'far more dangerous' than apartheid and that its receipt of Pretoria's support was driven by pragmatism and did not indicate UNITA's endorsement of apartheid (Chitunda 1986b). Savimbi himself addressed this charge head-on, stating to its Washington audience: 'I don't approve of apartheid. How can I, as a black leader, approve of apartheid? ... On one side I am branded as a communist, on the other side I am a puppet of South Africa!' (McCombs 1986). In a 1985 letter typed on UNITA letterhead and addressed to the House Foreign Affairs Committee, UNITA Foreign Secretary Jeremias Chitunda systematically rejected several other 'misperceptions' about UNITA while advancing another one of its refrains—that 'Dr. Savimbi would easily win a free and fair election' in Angola (US Department of Justice 1986a).

UNITA's claim-making practices of mimicking the state and voicing the achievements of the rebel group also included constructing a state-like apparatus within Angola. Amidst the intense war-fighting and the operation of a repressive and violent organization that often used terror tactics (James 2011, 108–109; Brittain 1998), UNITA also ran a robust administrative and governance structure in Angola's southeast interior where it exercised near-complete control. Many rebel organizations develop alternative governance structures during armed conflict (Mampilly 2011; Huang 2016a; Albert 2022), but UNITA went a step further: it ensured that the outside world, particularly its constituents in sympathetic states, observed its civilian governance accomplishments first-hand and reported on them for broader audiences. Throughout the 1980s UNITA thus encouraged and assisted numerous foreign journalists, politicians, and other observers to fly into its capital, Jamba. There, they

[14] http://www.griffnews.com/images/savimbi-page2.jpg; http://www.griffnews.com/images/savimbi-page1.jpg.

could observe UNITA's internal congresses, interview soldiers and civilians, and participate in guided tours of its 'liberated zones'. Lavish praises followed such visits. A reporter for the *Christian Science Monitor*, a prominent US newspaper, wrote in 1988 that Savimbi had established 'a new government—replete with ministries of education, health, information, natural resources and such to rule the three million people in his country' (quoted in Heywood 1989, 60). Others reported on UNITA's investments in education, with primary and secondary schools; its several sports fields; 250 hospital beds; hot and cold running water; electricity; roads and avenues; traffic lights; 'an airport with UNITA immigration facilities'; publications and radio stations operating in several Angolan languages; its 'modern bureaucratic institutions' linking a president (Savimbi) to a congress, a political bureau, and regional and village-level committees along with several ministries, all of which were governed by a UNITA constitution; and even its own postage stamps, in what one Angola scholar called a 'state within a state' (James 2011, 110; see also Brittain 1998; Heywood 1989). One journalist, having spent an extended time in UNITA territory, remarked in a US news outlet that 'Unitaland in southern Angola is perhaps the most impressive guerilla zone anywhere' (Sikorski 1989, 35–36). Part of UNITA's external legitimation efforts thus consisted of showcasing its internal legitimation and governance projects to the outside world. The flip side was also true—UNITA boasted of its overseas diplomatic achievements to its local supporters. Following Savimbi's high-profile visits to Washington, photos of Savimbi with Reagan and Bush would be 'blown to enormous proportions' and 'cover virtually every village entrance, community center, hospital and workshop' in UNITA territory (Windrich 1992; Zucchino 1988). In this way, UNITA linked its operations in the bush with its diplomacy in Washington to run a mutually reinforcing legitimation project on two fronts (Huang 2016b, 120).

That Savimbi's supporters in the United States extolled his virtues by repeating and amplifying his narratives in a public echo chamber is clear. In this case, claim-making and claim-validation were a collusive effort. More critical analysis attributed this chorus of praise and fanfare surrounding Savimbi to a paucity of accurate information on the Angolan war and a lack of journalistic due diligence, including a failure to consult Portuguese-language material (which itself offered only patchy information) and a recycling of the same available sources (Minter 1988, 1–2). The US government itself acknowledged the challenges of collecting information on Angola in the absence of diplomatic representation in the country (US Department of State Cable via Wikileaks, 1979). Others went further, charging US media with relying primarily on US government and UNITA statements for information, leading to overtly biased, superficial, and sensationalist coverage (Windrich 1992). More recent analysis reflects on the tight professional and social relations between Washington lobbyists and journalists—'[e]ach side was using the other'—and how they created a feedback loop of bias between UNITA propaganda and US news reports (Kitchener 2018). It could, nevertheless, also signal two concurrent influences: the sheer power of a strategic public influence campaign run by UNITA's Washington lobbyists, and the seemingly dogmatic devotion on the part of a segment of the

US government to defending UNITA and its leader—notwithstanding the growing evidence from Angolan battlefields that contradicted UNITA's self-branding as a champion of freedom and democracy.

Outcomes

The big irony of the UNITA lobby in the United States was that it turned out to be a project far removed from political reality at best, and of tremendous deception with severe consequences for the conflict at worst. The main outcome of this high-profile, high-cost lobbying effort was that UNITA secured covert aid from the Reagan administration from 1986 onwards, starting at $15 million a year and expanding to $50 to $60 million by 1989 (Minter 1991, 136, 139). The Reagan administration even granted UNITA one of its top demands, the delivery of Stinger anti-aircraft missiles (Lewis 1987). Overall, US covert aid for UNITA totaled about $250 million between 1986 and 1991, making it the largest US covert aid program of the period after US support for the mujahideen in Afghanistan (Human Rights Watch 1999). Supporting UNITA to such an extent can also be seen as a validation of the group's claim to legitimacy. UNITA's American patrons obliged the wishes of its proxy and the war continued—or, more critically, the war became prolonged thanks to the fresh infusions of US military aid.

Given the perceived stakes of the Cold War, it is possible that Washington under the Reagan administration would have backed UNITA irrespective of the extent, content, or quality of the armed group's lobbying efforts. While counterfactuals are difficult to assess, the foregoing discussion nonetheless provides strong support to the notion that the UNITA lobby had a notable influence on individual and collective perceptions of the organization and its leadership. Even if UNITA appealed solely to its conservative constituency in the US and won no new sympathizers, it was a powerful bloc within the government with the ability to steer policymaking. Even if US support for UNITA was a certainty given the Cold War context, the assurances of a skilled and impressive leader in Savimbi and a reliable ally in UNITA likely not only boosted policymakers' confidence that their support was well-placed, but also helped them to justify their position to more sceptical audiences. To the extent that public and especially elite perceptions ultimately shape policy, the UNITA lobby likely had its intended effect. As described above, US domestic politics was such that support for UNITA was also shakier than the consistent support from 1986 onwards might suggest in hindsight. At a minimum, UNITA's behaviour demonstrates that the organization itself believed investing in US lobbying was of utmost importance for securing and maintaining US support. Regardless of any effects on US policy, then, the fact remains that UNITA placed a heavy emphasis on—and poured millions of dollars into—diplomatic and lobbying activism in the United States during the Angolan war and continued to do so even as the Reagan administration ramped up its aid to the organization. US lawmakers responded favourably, affording UNITA all the elements of claim validation.

The Angolan Civil War and US support for UNITA can only be understood in the context of the Cold War superpower rivalry.[15] Likewise, the broader historical context directly shaped the total US policy reversal towards Angola in the early 1990s. Specifically, around 1989 and 1990 even the most ardent supporters of UNITA were losing steam. As Cuban forces withdrew from Angola, the Soviet Union fell, and the Cold War ended, the dogged belief in the Reagan Doctrine dissipated as US policymakers turned to new issues in a radically new political environment. Supporting UNITA was simply no longer a priority by the early 1990s. Instead, the State Department ramped up efforts to bring a negotiated settlement to the Angolan war.

Concomitantly, the overall tone of media coverage of UNITA began to change. In 1989, US media began to report on UNITA's brutal killings of its internal rivals in Angola. These reports came to a head when in 1992 news reports revealed the assassination of a series of UNITA defectors and their families stretching over 15 years. The most shocking of these cases was the murder of Tito Chingunji, who had served as UNITA's Washington representative until just a few years prior, alongside many of his family members (Wren 1992; Bridgl 1992; Brittain 1998). Savimbi's support base in the United States showed its first big cracks, while US officials began to doubt UNITA's commitments to participating in a democratic process.

The 1992 national elections in Angola—the first free elections in its history, held in accordance with the Bicesse Accords signed by the belligerent parties the previous year—became the culminating event marking Washington's repudiation of UNITA. When the elections results favoured the MPLA, UNITA rebuffed the outcome and resumed fighting, signalling a complete rejection of all its democratic rhetoric of the previous 17 years. So much for the big-ticket lobbyists of the 1980s: Savimbi, now exposed as a ruthless betrayer of democracy, quickly became 'something of an embarrassment' to his erstwhile supporters in the United States (Krauss 1992). In 1993 the United States formally recognized the MPLA as the government of Angola—but not before UNITA made a last-ditch attempt to dissuade Washington from doing so. UNITA's representative to the UN, Marcos Samondo, issued a letter to *The New York Times* on the eve of US recognition, threatening further destabilization should Washington use that 'most potent policy tool' to legitimate the MPLA (Samondo 1993). But the train had already left the station. With the MPLA government formally recognized, the United States established formal diplomatic relations with Luanda, bringing the long-running contest for US recognition to a close. The UN Security Council imposed its first set of sanctions on UNITA that same year. By then, an estimated 450,000 Angolans had died in the war ('UN Sanctions Hit Angolan Rebels', 1993). The UNITA lobby in Washington was over. Nevertheless, the Angolan war continued for another decade as UNITA pivoted from courting US support to gaining near total control of Angola's lucrative diamond trade in the 1990s (Bakonyi and Stuvøy 2005, 370–371). It took the death of Savimbi in 2002 for the war to come to an end.

[15] On the Cold War context and its role in the legitimation of armed groups, see the introduction to this volume.

Conclusion: legitimation is politics

UNITA in the 1980s was something of a trailblazer in its Washington lobbying efforts. It engaged in the gamut of activities that have since become fairly routine among armed groups seeking to establish long-term relations with the US government. In opening a representation office, working with lobbying firms, making public appearances, reaching out to local communities, and maintaining a propaganda project—in pursuing every available avenue to make itself known to government and civil society—UNITA played its legitimation politics to the full.

If armed groups seek to act and be treated like a state, UNITA accomplished the goal. It received the official reception, TV prime time coverage, lethal weapons, political backing, and public fanfare fit for any US state ally. In anticipation of claim-validation by Washington, UNITA ensured that its claim-making checked all the right boxes in a dynamic that was insular, circular, and mutually beneficial. The United States, too, was engaged in a politics of legitimation within its own government, with UNITA doing its part to boost the claim-making of one camp within a divided policy arena. Legitimation was thus a tandem effort. UNITA created the trappings of a state, from local social service provision in Jamba to the stationing of representatives at its US offices. These efforts ensured that the United States could justify its officialization of UNITA and thus maintain the momentum of US support for its proxy.

The politics of the Cold War pervaded every aspect of this mutual legitimation dynamic; UNITA's legitimation practices could not be examined apart from the broader political context in which it took place. And in a remarkable display of the fickleness of legitimacy politics, Washington performed a complete policy flip when the Cold War ended. Almost overnight, Savimbi went from being the darling of American conservatives to a persona non grata, and UNITA from a loyal friend to a criminal organization punished with international sanctions.

References

Adler-Nissen, Rebecca and Vincent Pouliot. 2014. 'Power in Practice: Negotiating the International Intervention in Libya.' *European Journal of International Relations* 20(4): pp. 889–911.

Albert, Karen E. 2022. 'What is Rebel Governance?: Introducing a New Dataset on Rebel Institutions, 1945–2012.' *Journal of Peace Research* 59(4): pp. 622–630.

Apple Jr., R. W. 1986. 'Red Carpet for a Rebel, Or How a Star Is Born.' *The New York Times*, 7 February. https://www.nytimes.com/1986/02/07/us/red-carpet-for-a-rebel-or-how-a-star-is-born.html.

American Archive of Public Broadcasting. 1986. The MacNeil/Lehrer NewsHour Episode. 30 January. https://americanarchive.org/catalog/cpb-aacip-507-wd3pv6c30t?start=1691.66&end=2029.48.

Bakonyi, Jutta and Kirsti Stuvøy. 2005. 'Violence and Social Order beyond the State: Somalia and Angola'. *Review of African Political Economy* 32(104–105): pp. 359–382.

Bridgl, Fred. 1992. 'Angola's Secret Bloodbath'. *The Washington Post*, 29 March. https://www.washingtonpost.com/archive/opinions/1992/03/29/angolas-secret-bloodbath/c2bf6ed2-7dc2-4568-8ba5-23bc15165fc2/.

Brittain, Victoria. 1998. *Death of Dignity: Angola's Civil War*. London: Pluto Press.

Carlson, Peter. 1990. 'The Image Makers'. *The Washington Post*, 11 February. https://www.washingtonpost.com/archive/lifestyle/magazine/1990/02/11/the-image-makers/ccf6936e-4f50-4292-8224-fee6db3b582d/.

Caspersen, Nina. 2012. *Unrecognized States: The Struggle for Sovereignty in the Modern International System*. Cambridge, Malden: Polity.

Chitunda, Jeremias. 1986a. 'Statement Delivered to the Council for National Policy at The Homestead, Hot Springs, Virginia, September 9, 1983'. In *Amendment to Registration Statement Pursuant to the Foreign Agents Registration Act of 1938*, as amended: National Union for the Total Independence of Angola ('UNITA') USA, 1 March. US Department of Justice.

Chitunda, Jeremias. 1986b. 'Statement, July 2, 1985'. In *Amendment to Registration Statement Pursuant to the Foreign Agents Registration Act of 1938*, as amended: National Union for the Total Independence of Angola ('UNITA') USA, 1 March. US Department of Justice.

C-SPAN, 1990. 'Angola'. 13 March. https://www.c-span.org/video/?11463-1/angola.

C-SPAN, 1991. 'Transition Toward Democracy in Angola'. 28 May. https://www.c-span.org/video/?18153-1/transition-democracy-angola.

Döhning, W. 1984. *UNITA*. Kwacha UNITA Press.

Duyvesteyn, Isabelle. 2017. 'Rebels & Legitimacy: An Introduction'. *Small Wars & Insurgencies* 28(4–5): pp. 669–685.

Foer, Franklin. 2018. 'Paul Manafort, American Hustler'. *The Atlantic*, March: pp. 62–78.

Geis, Anna, Maéva Clément, and Hanna Pfeifer (eds). 2021. *Armed Non-State Actors and the Politics of Recognition*. Manchester: Manchester University Press.

Gerstenzang, James. 1989. 'Savimbi Agrees to Return to Talks on Angolan Peace'. *Los Angeles Times*, 6 October.

Griffiths, Ryan D. 2021. 'Secessionist Strategy and Tactical Variation in the Pursuit of Independence'. *Journal of Global Security Studies* 6(1): pp. 1–19.

Huang, Reyko. 2016a. *The Wartime Origins of Democratization: Civil War, Rebel Governance, and Political Regimes*. Cambridge: Cambridge University Press.

Huang, Reyko. 2016b. 'Rebel Diplomacy in Civil War'. *International Security* 40(4): pp. 89–126.

Huang, Reyko. 2023. 'The Rebel Lobby: Understanding Policy Advocacy by Foreign Armed Groups'. Paper presented at the Annual Conference of the International Studies Association, Montreal, Canada, 15–18 March.

Heywood, Linda M. 1989. 'Unita and Ethnic Nationalism in Angola'. *The Journal of Modern African Studies* 27(1): pp. 47–66.

Human Rights Watch. 1999. *Angola Unravels: The Rise and Fall of the Lusaka Peace Process*. New York: Human Rights Watch.

James, W. Martin. 2004. *Historical Dictionary of Angola*. Lanham, MD: Scarecrow Press.

James, W. Martin. 2011. *A Political History of the Civil War in Angola, 1974–1990*. New Brunswick: Transaction Publishers.

Jo, Hyeran. 2015. *Compliant Rebels: Rebel Groups and International Law in World Politics*, 1st edn. Cambridge: Cambridge University Press.

Kitchener, Caroline. 2018. 'How Paul Manafort Influenced the Media'. *The Atlantic*, 20 February. https://www.theatlantic.com/membership/archive/2018/02/how-paul-manafort-influenced-the-media/553529/

Krauss, Clifford. 1992. 'Angolan Rebel Lays Killings to a C.I.A. Plot'. *The New York Times*, 5 May, Section A, p. 7.

Leonard, Richard. 1989. *Apartheid Whitewash: South Africa Propaganda in the United States*. New York: The Africa Fund.

Lewis, Neil A. 1987. 'Administration Decides that Aid to Angolan Rebels Will Continue'. *The New York Times*, 11 June, Section A, p. 1.

Mampilly, Zachariah Cherian. 2011. *Rebel Rulers: Insurgent Governance and Civilian Life during War*. Ithaca, NY; London: Cornell University Press.

McAllister, Bill and David B. Ottaway. 1988. 'Reagan Vows Support of Angola Rebels'. *The Washington Post*, 1 July. https://www.washingtonpost.com/archive/politics/1988/07/01/reagan-vows-support-of-angola-rebels/4f58496c-a622-4336-93b7-82fb098790cf/.

McCombs, Phil. 1986. 'The Salute to Savimbi'. *The Washington Post*, 1 February. https://www.washingtonpost.com/archive/lifestyle/1986/02/01/the-salute-to-savimbi/b37b8411-a9fe-4cf3-8040-bb189e936d4a/.

Minter, William. 1988. *Operation Timber: Pages from the Savimbi Dossier*. Trenton: Africa World Press.

Minter, William. 1991. 'The US and the War in Angola'. *Review of African Political Economy* 18(50): pp. 135–144.

Nesbitt, Prexy. 1988. 'Terminators, Crusaders and Gladiators: Western (Private & Public) Support for Renamo & Unita'. *Review of African Political Economy* 43: pp. 111–124.

Pearce, Justin. 2015. *Political Identity and Conflict in Central Angola, 1975–2002*. Cambridge: Cambridge University Press.

Samondo, Marcos. 1993. 'Don't Reward Angola With US Recognition'. *The New York Times*, Section 1, p. 18, 20 February.

Savimbi, Jonas. 1989. 'The Coming Winds of Democracy in Angola'. *The Heritage Lectures* No. 217. Washington, DC: The Heritage Foundation.

Sikorski, Radek. 1989. 'The Mystique of Savimbi' *National Review*, 18 August.

Sly, Liz. 1992. 'Angola's Savimbi Brings Bloody Past to Election'. *Chicago Tribune*, 29 September. https://www.chicagotribune.com/news/ct-xpm-1992-09-29-9203280630-story.html.

Tooley, Mark. 2017. 'Angolan Dinner with George H. W. Bush & Jerry Falwell'. *Providence*, 28 August. https://providencemag.com/2017/08/angola-dinner-bush-falwell-savimbi/.

Tyler, Patrick and David B. Ottaway. 1986. 'The Selling of Jonas Savimbi: Success and a $600,000 Tab'. *The Washington Post*, 9 February. https://www.washingtonpost.com/archive/politics/1986/02/09/the-selling-of-jonas-savimbi-success-and-a-600000-tab/d9fd8686-8f8d-497b-a3b4-7b636fec9b69/.

'U.N. Sanctions Hit Angolan Rebels'. 1993. *The Washington Post*, 26 September. https://www.washingtonpost.com/archive/politics/1993/09/27/un-sanctions-hit-angolan-rebels/bd077625-7f08-42dd-8775-9e78dacdadf6/.

Ungar, Sanford J. 1986. 'Jonas Savimbi: Big Welcome For a Bad Bet'. *The Washington Post*, 26 January.

US Department of Justice. 1986a. *Amendment to Registration Statement Pursuant to the Foreign Agents Registration Act of 1938*, as amended: National Union for the Total Independence of Angola ('UNITA') USA. 1 March.

US Department of Justice. 1986b. *Exhibit A and B to Registration Statement Under the Foreign Agents Registration Act of 1938*, as amended: National Union for the Total Independence of Angola ('UNITA') USA. 1 March.

US Department of Justice. 1986c. *Exhibit A and B to Registration Statement Under the Foreign Agents Registration Act of 1938*, as amended: TKC International, Inc. 1 March.

US Department of Justice. 1986d. *Supplemental Statement Pursuant to Section 2 of the Foreign Agents Registration Act of 1938*, as amended: National Union for the Total Independence of Angola ('UNITA') USA/dba/Free Angola Information Service, Inc. and Free Angola Foundation, Inc. 1 September.

US Department of Justice. 1986e. *Amendment to Registration Statement Pursuant to the Foreign Agents Registration Act of 1938*, as amended: UNITA, USA/dba/Free Angola Information Service. 1 October.

US Department of Justice. 1986f. *Exhibit A and B to Registration Statement Pursuant to the Foreign Agents Registration Act of 1938*, as amended: Griffin Communications. 1 December.

US Department of Justice. 1987. *Supplemental Statement Pursuant to Section 2 of the Foreign Agents Registration Act of 1938*, as amended: National Union for the Total Independence of Angola ('UNITA') USA/dba/Free Angola Information Service, Inc. 1 March.

US Department of Justice. 1989a. *Supplemental Statement Pursuant to Section 2 of the Foreign Agents Registration Act of 1938*, as amended: National Union for the Total Independence of Angola ('UNITA') USA. 1 March.

US Department of Justice. 1989b. *Supplemental Statement Pursuant to Section 2 of the Foreign Agents Registration Act of 1938*, as amended: National Union for the Total Independence of Angola ('UNITA') USA/dba/Free Angola Information Service, Inc. 1 September.

US Department of State Cable from Secretary of State, Washington, DC to US Embassies, 1979, 'Exchange of Information on Angola: Savimbi Visit to US, Internal MPLA Situation', 18 December, WikiLeaks, https://wikileaks.org/plusd/cables/1979STATE326028_e.html.

Wedeen, Lisa. 1999. *Ambiguities of Domination*. Chicago, IL: The University of Chicago Press.

White House Diaries. 1986. *Entry for Thursday, January 30*. Ronald Reagan Presidential Foundation & Institute. https://www.reaganfoundation.org/ronald-reagan/white-house-diaries/diary-entry-01301986/.

Windrich, Elaine. 1992. 'Media Coverage of the Angolan War'. *Africa Today* 39(1/2): pp. 89–99.

Wren, Christopher. 1992. 'Deaths and Desertions Undermine Tenacious Angolan Rebel'. *The New York Times*, 7 April, Section A, p. 17.

Zucchino, David. 1988. 'Savimbi: A Revolutionary Who's Full of Contradictions'. *Inquirer*, 1 July.

8 Between the 'Dome of the Rock' and a Hard Place

Fatah and the Trade-off between Armed Struggle and International Legitimacy

Christian Olsson

Introduction

This chapter focuses on Fatah, a Palestinian armed 'movement' created in 1959 that was long embodied by its charismatic leader and chairman, Yasser Arafat (1929–2004). It considers the historical trajectory of Fatah from its foundational engagement in the Palestinian Liberation Organization (henceforth PLO) in 1967 to the control by its chairman of the Palestinian National Authority (henceforth PA) in 1994, until today. Through the PA, successive Fatah leaders—Arafat and Mahmud Abbas—have been exercising a limited Palestinian 'self-governance' in parts of the Occupied Palestinian Territories (OPT: Gaza, East Jerusalem, and West Bank): Gaza until 2007, parts of the West Bank, but not occupied East Jerusalem.

This contribution adopts a long-term perspective, focusing on the relation between armed force and political legitimacy. The main and central question is how the pursuit of political legitimacy with regards to diverse publics affects the use of force in the quest for Palestinian statehood; and vice versa: how the use of armed force against diverse publics affects Fatah's claims to political legitimacy. In the latter case, the notion of 'public' serves to highlight that the use of armed force has a communicative function (Wieviorka 2009; Olsson and Malešević 2022): beyond the immediate victims, it generally targets the state whose territory, or whose citizens, are being attacked. However, here as well, diverse publics must be considered.

In this context, the distinction between Fatah's national constituencies ('the Palestinian people') and their external or international environment will prove analytically important. In a nutshell, this chapter's central argument is that Fatah long maintained a delicate balance between armed struggle (that earned it Palestinian popular support and hence internal legitimacy, but had a paradoxical effect on external legitimacy) and the quest for diplomatic recognition (for which 'internal' and 'international' legitimacy were important). However, in the 1990s outright contradictions between external and internal legitimacies took centre stage

Christian Olsson, *Between the 'Dome of the Rock' and a Hard Place*. In: *Armed Groups and the Politics of International Legitimation*. Edited by: Stephan Hensell and Klaus Schlichte, Oxford University Press.
DOI: 10.1093/9780198974529.003.0008

in Fatah's trajectory. This became clear when Yasser Arafat officially renounced armed struggle in 1993 and recognized Israel as a state. From this moment on he would be held to account by 'the international community'—and see his international legitimacy contested—should violence erupt in the parts of the OPT under his control. At the same time, he would now be held to account by his Palestinian constituencies should he stay idle, and refrain from taking decisive action, in the face of obstacles on the way towards Palestinian independence. Progressively, Fatah lost much of its 'domestic' legitimacy, in particular in favour of Hamas, in the absence of concrete steps to end Israel's military occupation in the OPT. The internationally celebrated Oslo Accords came to seen by Palestinians as offering no perspectives with regards to Fatah's ultimate goal, not even mentioning Palestinian statehood.

The case of Fatah highlights that 'national liberation movements' are simultaneously dependent on many publics; and their practices simultaneously impact different publics. The use of armed force against one public might produce very different effects with regards to political legitimacy depending on the 'public' one is considering. The use of all-out violence against the enemy state(s)' interests in third countries most likely will undermine the group's claim to international legitimacy. It may even reinforce its stigma as a 'terrorist group'. It might however simultaneously boost the organization's legitimacy with regards to its 'national constituencies', for example by showing its determination to live up to what it claims to stand for. Inversely, the use of armed force by an armed group or any authority against its own 'national constituencies', if considered as excessive by the latter, might undermine its 'internal legitimacy'. Such 'excessive force' indeed highlights that this authority is not supported by the population that it administrates (Olsson 2013). 'Excessive force' exposes the arbitrariness of the powerful (Bourdieu 2000, 105). This has been highlighted by the unpopularity of the PA under Abbas, since 2004, in the West Bank. Many Palestinians increasingly accuse the PA of arbitrarily repressing any Palestinian resistance to the illegal military occupation of the OPT. At the same time, this repression might legitimate this authority with regards to the 'international community' if the latter sees it as an attempt to enforce unpopular international engagements; as signalling a commitment on the part of this authority to international agreements. This is clearly what motivates the PA's increasingly frequent 'security crackdowns' on Palestinian opposition to the Oslo Accords, as well as the current PA's principled renunciation of Fatah's original armed struggle (as opposed to Arafat's pragmatic renunciation).

Given the long-term perspective adopted, Fatah is an obvious study case among the Palestinian groups. Prior to 1993, it was the main organizational embodiment of the wider Palestinian national movement. It was the main group to join the PLO in 1967, de facto controlling it from 1969 until today (Miller 1983; Parsons 2005; Sayigh 1997b). The PLO can be described as a Palestinian 'proto-state', an organization mimicking statehood in many ways (Charountaki and Havlová 2022). It theoretically represents Palestinians from all the political factions embracing the struggle for 'national liberation', at least prior to the creation of Islamic Jihad (1981) and Hamas

(1987). As a secular organization fighting for a free and democratic Palestine, the PLO claims to represent Palestinians of all ethnicities, confessions (Muslim, Christian, and Jewish), and of all walks of life: from the Palestinian camps in Lebanon, Jordan, and Syria, the OPT, the diaspora, the Palestinians that stayed when Israel was created in 1948. It can be described as an 'umbrella organization', a confederation of Palestinian factions embracing the struggle for national independence.

In reality, the PLO allows for Fatah's core leaders to formally unify the (secular) Palestinian national movement under their leadership. In this sense, the PLO can be seen as an extension of Fatah's project of gathering all Palestinians yearning for national independence, irrespective of political inclination, under its authority. This project also explains why Fatah wants to describe itself as a movement rather than as a party. In reality, however, Fatah is both very centralized and centralizing at the level of its leadership, yet very factionalized and fragmented when looking at mid-rank positions. Fatah's chairman has since 1969 always been chairman of the PLO and since 1994 the president of the PA. In what follows, it will hence be impossible to consider Fatah without many overlaps with the PLO and the PA.

The PLO is especially important for Fatah's struggle for national independence. Since 1969, the PLO has allowed for Fatah's chairman (first Arafat then Abbas) to claim international legitimacy for the Palestinian national movement. Fatah made a claim to legitimacy, first through *fighting* Israeli occupation, later through *mimicking* the state with the PA as the core institution of the future Palestinian state. An increasing number of states and international organizations reacted to these claims by *acknowledging*, *supporting*, and *officializing* the PLO as 'the representative of the Palestinian people'.[1] In October 1974 both the Arab League and the UN recognized the PLO as respectively 'sole representative' and 'the representative' of the Palestinian people. In 1988, Arafat officially proclaimed the State of Palestine. It has been recognized by 146 out of 193 members of the UN, mostly—but not exclusively—in the 'Global South'. The entity that is thus legally recognized is the PLO. From the Oslo Accords onwards, it is however the PA that manages most of the everyday external relations of the Palestinian state. The chairman of Fatah has since 1988 always been president of the state of Palestine (in addition to chairman of the PLO and president of the PA). In spite of its significant (but partial) recognition, the Palestinian state is actively deprived of the most important empirical attributes associated with statehood: a relatively unified territory, clear borders, an ability to effectively administrate its population. Therefore, it needs not to be addressed separately in the context of this contribution. For the moment being, it is a symbolic attribute of the PLO, concretely managed by the PA.

Among the original Palestinian armed factions, Fatah is the one that remained the most committed to the process initiated by the Oslo Accords of 1993 and 1995. The latter established the PA to exercise limited Palestinian 'self-governance' in parts of the OPT[2] in 1994. The PA has no power over Palestinians beyond the OPT.

[1] On these practices of making and validating a claim to legitimacy see the Introduction in this volume.

[2] In particular in the 'Area A' in the West Bank. The PA has however no control in the 'Area C' of the West Bank (that is under Israeli control by virtue of the Oslo accords), nor in East Jerusalem (that has

This as opposed to Fatah and the PLO, which are also present in the wider diaspora, including in the refugee camps in Lebanon, Syria and Jordan. In sum, Fatah is historically the main embodiment of the Palestinian struggle for statehood and also the de facto caretaker of the PLO, the PA, and the state of Palestine.

The aim of this chapter is not to retrace the convoluted history of Fatah. This would be impossible given that the time frame considered spans from the 1960s until today. Rather, the first section deals with Fatah's pre-Oslo history, highlighting the role of armed struggle in mobilizing Palestinians in favour of 'national liberation'. In the second section, I highlight how the relation between armed force and political legitimacy played out in the OPT after the Oslo Accords, showing how, in the context of a stalling peace process, Fatah's renouncement to armed struggle undermined its claim to 'domestic legitimacy'. The last and third section explores how the post-Oslo 'transnational government' of the West Bank, borne out of the quest for international recognition, currently fuels the emergence of new armed groups and challenges Fatah's leadership.

Fatah's armed struggle and the evolution of the PLO from regional tutelage to international recognition

Fatah's trajectory from the 1960s to today is of particular interest when considering the international legitimation of armed groups. Indeed, it did not follow the typical sequential evolution that is often assumed when dealing with the internationalization of armed groups. It did not start as a grassroots 'local' organization that progressively sought to legitimize its agenda internationally. From the outset, it was highly transnationalized and linked to the Palestinian diaspora.

The Palestinian National Liberation Movement (Fatah): From the diaspora to the Palestinian refugee camps

Fatah is the inverted acronym of ***Harakat at-Tahrir i-Filastini*** (The Palestinian National Liberation Movement) and means 'opening', referring historically to the Islamic conquest. Fatah finds its roots in the surge in Palestinian nationalism during the British mandate on Palestine granted by the League of Nations in 1921, and that lasted until 1947. The Palestinian national movement progressively became more organized after the creation of Israel in 1948 (Khalidi 1989; Swedenburg 2003): small and autonomous groups of armed *fedayeen* (partisans; literally: 'those who sacrifice themselves') emerged. They were particularly active in the All-Palestinian protectorate that ruled in Gaza from 1948 to 1959, with the exception of a first Israeli occupation in 1956–1957. In reality, the protectorate's Palestinian government was at

been unilaterally annexed by Israel) and Gaza (that has been under Palestinian Hamas rule since 2007). Under internal law, however, the OPT include the whole of the West Bank, East Jerusalem and Gaza.

times a mere client state of Egypt. It however enjoyed a limited international recognition, in particular by most of the Arab League, as the Palestinian state. The *fedayeen* were present in the Jordanian-occupied (1948–1950), then annexed (1950–1967), West Bank and East Jerusalem until 1967; this in spite of Jordan's best efforts to root them out. They were also well implanted amongst the Palestinian refugees in scattered camps in Jordan, Lebanon, and Syria. These combatants were initially locally embedded groups with very little interconnection or coordination: 'Youth for Revenge', 'Heroes of Return', the 'Palestinian section of the Nasserite Arab Nationalist movement', etc. Some of them collaborated with neighbouring countries' security services, thus limiting their autonomous 'action power' (Popitz 2017).

Fatah emerged from these early *fedayeen* groups, in particular when their Gazan leaders were forced to flee after the first Israeli occupation of the Gaza Strip in 1956–1957. The Gaza strip had in 1948–1949, after having been seized by Egypt in the first Arab-Israeli war, become a refuge for nearly a quarter of the Palestinian population of Mandatory Palestine. Most had fled the massacres and the destruction of hundreds of villages by the Zionist paramilitary Haganah, then the Israeli armed forces. As a result, Gaza was a hotbed for Palestinian nationalists wanting to return to what remained of their villages and cities.

Fatah was created in the years 1957–1959 in exile. It was led by Yasser Arafat from 1959 to his death in 2004. In addition to the exiled Gazan leaders and *fedayeen*, the transnational microcosm that served as the recruitment pool was diverse: Palestinian refugees in the Levant, Palestinian students in the Arab world and beyond, young professionals established in the Persian Gulf, Kuwait in particular. Their transnational circulation forged Fatah as a movement: it managed to break with the political and geographical parochialism of some of the original *fedayeen* (for a similar argument, see Olsson 2022). Their original members were also less constrained by the states—Israel, Jordan, Egypt—now controlling the territory of Mandatory Palestine. Fatah presented itself as a broad, 'transpolitical' movement representing all Palestinians provided they embraced the struggle for national liberation. It progressively took root in the West Bank and liaised with the local *fedayeen*.

In 1965 Fatah officially set up its own armed branch 'al Asifa' (the storm). This move consolidated its support base, and hence its Palestinian constituency, first in the Palestinian camps of Lebanon, Syria, and Jordan, and then progressively in 'Occupied Palestine'. The Algerian Front de Libération Nationale (FLN) was a main source of inspiration. The way it had put armed struggle at the service of political power and diplomatic pressure to achieve national independence (Connelly 2002) served as a model. It is not for nothing that Fatah's first success in establishing official relations with an Arab state was with Algeria in 1962. Syria was soon to follow in 1963. Fatah increasingly received financial support from Arab states and other Arab constituencies.

Fatah and its guerilla operations against Israel increased the pressure on Arab states to be seen to show resolve on the Palestinian question. The Palestinian cause, as the main Arab cause, enjoyed deep-seated popular support and regional legitimacy throughout the Middle East (Barnett 1998). The Arab governments' domestic legitimacy in part hinged on their ability to credibly portray themselves as supporters of the Palestinian national cause. They at the same time wanted to avoid conflict with Israel. In this context, the decision to create the PLO was taken by the Arab states during the January 1964 Arab League meeting in Cairo. This decision showed Arab resolve. At the same time, the PLO was designed to have very limited autonomy (Norton and Greenberg 1989). It was a highly hierarchic, and therefore easily controllable, state-like structure. It formally had an armed branch, the Palestinian Liberation Army (PLA). Initially, the PLA was largely subordinated to the armed forces of Egypt, Syria, and to a lesser extent Jordan (Frisch 2008; Sayigh 1997b). In sum, the Arab states had little interest in seeing the PLO effectively engaging in armed struggle outside of their control. It was therefore not the source of any independent Palestinian 'action power' (Popitz 2017). This was consistent with the claim of many Pan-Arabists that Palestinian liberation and Arab unification were inseparably linked: the objective of Palestinian self-determination was considered as subordinated to the creation of a unified Arab state (Barnett 1998).

The PLO, however, in part because it mimicked state-structures and had strong support in the Arab League, was an ideal site from which to make claims to international legitimacy. International audiences and states sympathetic towards Pan-Arabism, the anti-colonial principles proclaimed by the 1955 Bandung Conference, or Marxism, were prone to validate Fatah's struggle. Moreover, the PLO's organizational backbone had the ability to generate cohesion among what was in the 1960s still a loose network of poorly structured armed Palestinian groups, Fatah being a relative exception in this regard. Whomever controlled the PLO, could potentially extend its authority over all Palestinian groups. The PLO, as already mentioned, presented Fatah with a stark choice between diplomatic legitimacy under regional control or an autonomous 'action power' (Popitz 2017) outside the PLO. Fatah in 1969 chose both.

An autonomous capacity to represent the 'Palestinian People' earned through armed struggle

The regional legitimacy crisis triggered by the Six Days War of 1967 solved this dilemma for Fatah. Among the Palestinian territories, Egypt lost Gaza and Jordan lost East Jerusalem and the West Bank to Israel. According to international law, these three territories were now occupied by Israel. The latter however claimed the annexation of East Jersulem in 1980. In the face of Egypt's, Jordan's, and Syria's abject failure against Israel, it was clear that state-led Pan-Arabism would not lead the struggle for Palestinian statehood. In 1967 Fatah now joined the PLO and in 1968, leaders of the Palestinian armed factions (including Fatah) gained representation in the PLO's legislative assembly, the Palestinian National Coucil. Arafat became the PLO's chairman in 1969 thanks to the broad Palestinian support earned by Fatah's guerilla operations. In particular, the 1968 battle of Karameh in Jordan, in spite of Fatah suffering heavy losses (along with the Jordanian army), shattered the myth of the invincible Israeli army. The battle led to a surge in Fatah popularity among Palestinians. Since the PLO had now become an umbrella organization, Fatah gained—through the chairmanship of the PLO—de facto authority over the other Palestinian organizations.

For Fatah, armed struggle was primarily a way to mobilize the Palestinian masses into recognizing the group as a political force able to represent their interests and aspirations (Sayigh 1986). Fighting as a practice of claim-making served to express independent 'action power' (Popitz 2017) both to Palestinian and Arab constituencies. In this sense, although Fatah's operations were far from coercing Israel into relinquishing any of the Palestinian territories it controlled, they conveyed a sense of Palestinian agency to Fatah's constituencies. While the Pan-Arabist slogan had been that Arab unity would pave the way to the liberation of Palestine, Fatah now could reverse the slogan: Arab unity could only be achieved through the liberation of Palestine. Although the *fedayeen* operations had very little military utility with regards to the Israeli army, they had a constitutive and political dimension (Sayigh 1997b). In this sense, Fatah's early trajectory confirmed the Weberian idea of the origins of political authority in the independent decision-making capacity created through armed force (Mann 1984; Weber 1978). Fatah carved out an autonomous power base for itself in the Palestinian refugee camps in the countries neighbouring Israel, both recruiting willing fighters and claiming an exclusive authority over their inhabitants. When gaining leadership over the PLO, Fatah conferred relative autonomy to the PLO itself (Miller 1983; Sayigh 1997b).

Armed violence also played a symbolic role in defining a Palestinian 'Self' in relation to an 'Other' (Sayigh 1986). The Other was primarily Israel. Although Israeli leaders, such as Golda Meir in 1969, had claimed that 'there was no such thing as Palestinians' (Cattan 2000, 21), spectacular attacks seemed to prove the contrary. In relation to Arab rulers, armed violence had a more subtle role. It placed them in a potentially difficult position: they risked either war with Israel if they supported such violence, or corroborating the claim to political agency on the part of Fatah if they disavowed these actions. In this sense, the performance-based practice of fighting expressed a claim to independent nationhood. It is what concretely constituted the capacity of 'external representation of a people', on which diplomacy is based (Mitzen 2015). At this point, there was hence no contradiction between armed violence and the claim to be a legitimate diplomatic interlocutor of the 'international community'. This claim, however, still needed to be fully validated by the main regional and international actors before the PLO's status as a 'diplomatic actor' could be established.

From ideological legitimation to diplomatic recognition of the PLO

While the ideological goals of the PLO as well as of the main Palestinian armed factions had been couched in the language of Pan-Arabism and decolonization, the Six Day War led to an ideological transformation. This change reflected a wider shift in world 'historic times' from claims to international legitimacy formulated in the language of decolonization to claims formulated in terms of one out of the two Cold War phraseologies, in this case Marxism-Leninism.[3] The Palestinian section of the Arab Nationalist Movement transformed into the Popular Front for the Liberation of Palestine (PFLP) to mark its independence from Nasserism that had lost its appeal as a source of international and internal legitimacy. It split in 1969 between the PFLP and the Democratic Front for the Liberation of Palestine (DFLP). Both factions adopted an explicit Marxist-Leninist ideology. These groups gained representation in the PLO at the same time as Fatah.

Marxism-Leninism seen as inseparable from Third World national liberation was widely shared internationally and gave increased visibility to the PLO. This was highlighted by Arafat's diplomatic successes in the communist bloc and amongst the increasingly left-leaning non-aligned states. Whereas the USSR had initially been sceptical towards Fatah, the Arab defeat of 1967 marked a shift towards a marked pro-Fatah/pro-PLO stance (Sayigh and Shlaim 1997). Arafat visited Moscow in 1968, 1971, and 1978. Throughout the 1970s the USSR provided weapons and training to Fatah guerillas (Reppert 1989). Fatah as such however remained ideologically non-aligned in its pursuit of domestic legitimacy. Its symbols were largely linked to anti-colonial wars of national liberation and imposed no ideology other than nationalism. Its members ranged from pro-Soviet Marxist-Leninists, Maoists, to Arab nationalists and Islamists (Lefèvre 2021).

The turn towards Marxist-Leninism of significant parts of the Palestinian national movement had three implications that Arafat sought to avoid within Fatah and the PLO. The first is that the groups embracing revolutionary socialism, like the DFLP and PFLP, polarized positions internationally in a context in which Yasser Arafat pursued a broad legitimacy transcending ideological cleavages. Arafat claimed international recognition rather than a necessarily fragile ideological legitimacy. The idea was that purely ideological solidarities could hamper not only national unity but also a broad-based diplomatic recognition. Indeed, many of the 'leftists' within Fatah (and other PLO factions) had trained like-minded activists from third countries. This military training took place in Palestinian camps in Lebanon during the civil war. It was, for example, provided to the Armenian Secret Army for the Liberation of Armenia and the Partiya Karkeren Kurdistan (PKK), both fighting against Turkey. This dissuaded Turkey, a potentially important ally, from validating the PLO's claims to international legitimacy in the 1970s and 1980s (Norton 1989, 6).

[3] On the historical times of decolonization and the Cold War and their role for the legitimation of armed groups see the Introduction of this volume.

The second is that the openly Marxist-Leninist groups such as the PFLP and DFLP quite logically advocated revolution in their Middle Eastern host states. This put them on a crash course with local authorities. The most dramatic event in this regard was 'Black September' in 1970 when the Jordanian army crushed PLO forces established in Jordan. Most of the latter had fled to Jordan when the West Bank was occupied. After the Jordanian military crackdown, they were forced to relocate to Lebanon. Fatah was often criticized in host states for creating parallel power structures in Palestinian refugee camps, which it did. It however generally did not actively seek to overthrow the government. By publicly recognizing the sovereignty of host states, Fatah was mimicking state-behaviour and seeking validation of its own claim to legitimate statehood. The hope was that, by virtue of the principle of reciprocity, host states would validate Fatah's claim to sovereignty.

The third implication is that these 'leftist' groups were coveted by regional states such as Iraq, Libya, and Syria because their hard-line ideological stance made them relatively less preoccupied by issues of organizational autonomy (Muslih 1976). Their quest for ideological validation by patron states contradicted the unity that Arafat wanted to maintain in the PLO and Fatah. It also went against Arafat's quest for broad-based recognition across the Cold War divide. While the PLO's initial diplomatic successes were in the Arab world, the Eastern bloc, and the non-aligned states, Arafat was aiming for broader support. This quest for external support was not limited to states and international organizations. Fatah's legitimacy claims were also addressed to intellectual circles in the West. By attracting high-level Palestinian public intellectuals such as Edward Saïd or Mahmoud Darwish, Fatah's claims were validated by many Western intellectuals. This cultural dimension of Fatah's external relations is not to be underestimated.[4]

In June 1974, the Palestinian National Council of the PLO adopted the 'Ten Point Programme' that called for the establishment of a national authority in all liberated Palestinian territories rather than in the whole of historical (mandatory) Palestine. It was a move away from the demands previously made. It actively voiced a readiness to make concessions, a will to negotiate. It foremost was a call for 'the international community' to engage diplomatically with the PLO. This programme led several 'leftist' groups to break with the PLO, at the time largely encouraged by Iraq (Muslih 1976). Fatah's strategy would however prove successful at the end of the same year. On 14 October 1974, Yasser Arafat made an official speech at the United Nations General Assembly (UNGA). On the same day, the PLO was recognized and thus officialized as 'the representative of the Palestinian people' by UNGA resolution 3210 (UNGA 1974). One month thereafter, the UNGA recognized the Palestinian people's right to autonomy, independence, and sovereignty (res. 3236) and accorded the PLO observer status in the UN (res. 3237). The claim of a right to statehood was thereby at least indirectly validated. This paved the way for the PLO's recognition as the (sole)

[4] A similar role of public intellectuals in the legitimation of armed groups can be observed in the case of the FSLN. See the chapter by Eline van Ommen on Nicaragua in this volume.

'legitimate representative of the Palestinian people' by more than 100 countries, starting with the Arab states. The PLO gained full membership in the Arab League in 1976.

The success of this strategy came at a cost (Kursani 2023). Fatah had to water down its demands and curtail some of its previous armed activities to avoid being labelled a terrorist organization. This risk was constantly present after the attack of the 'Black September Organization'—a group associated with Fatah—during the 1972 Olympic Games in Munich (Gruen 1988). In 1974, Arafat accordingly rejected transnational political violence, hijackings, and high-profile bombings outside of the Levant. The promise was difficult to hold, however, given the level of factionalization of the PLO and even of Fatah (Stern 2022). The issue of terrorism reverberated in 1987 when the PLO was designated a terrorist organization by the US (Gerber 1990). In 1988, Yasser Arafat finally explicitly denounced terrorism at the UNGA, largely to satisfy US demands.

The 1980s saw the PLO redoubling its diplomatic efforts. This was at least in part due to its realization that there was no chance of establishing a favourable power balance with Israel in the absence of reliable external allies (Sayigh 1986). The PLO proclaimed the State of Palestine in 1988 in Algiers. In less than two years, more than 100 states recognized Palestine as a fully-fledged state. The UN changed the designation of the PLO to 'Palestine' (Stokes 2019). This was also a period during which smaller but significant steps were achieved in some countries in which Palestine was not recognized as a state. In France, the PLO's Bureau of Liaison and Communication, a mostly informal and non-recognized office, was upgraded to become the General Delegation of Palestine. Many Western European countries followed suit. In 1989, François Mitterand even received Yasser Arafat on an official visit. This was the first time he officially visited a West European country, with the exception of a visit to Austria in 1979 as the party leader of Fatah at a congress convened by Chancellor Kreisky (Wood 1993).

Fatah can be credited with two moves that paved the way to broad international recognition of the PLO. First, it built up the organizational capacity to emancipate the PLO from the regional patronage structures that had until then prevented it from being considered an autonomous political force. Armed struggle played a significant role in this regard. Second, its legitimacy claims were increasingly steeped in the language of diplomacy—non-intervention in the political affairs of host states, a willingness to compromise, the renunciation of transnational political violence (beyond the Levant), insistence on the function of external representation of the Palestinian people—rather than in a polarizing ideological language.

Mimicking the state after the Oslo Accords: the legitimacy predicament of the PA

The delicate balance between armed struggle and diplomatic recognition could be maintained by Fatah and the PLO as long as a substantial number of states supported national liberation struggles. As the Cold War drew to a close, however, it became clear that the armed struggle was becoming an obstacle to the international validation of the PLO's legitimacy claims. The US was now the most important actor to convince. This became clear when Arafat failed to show support for Operation Desert Storm against Iraq in 1990. He was temporarily marginalized by the 'international community' as a result. In 1991, the Madrid Peace Conference on the Israel–Palestinian conflict confirmed this US hegemony: Russia was not even invited. It is in this context that the Oslo Accords were signed, and that the PLO recognized the state of Israel (Doxey 1995).

The broad-based recognition of the PLO as credible interlocutor: opportunities and constraints

The process leading up to the Oslo Accords of 1993 and 1995 challenged the PLO's strategy which until then had been based on both political-military struggle and diplomatic negotiations. A few days before the Oslo I Accord, Arafat signed an agreement of mutual recognition with Israel. In reality, this mutual recognition was highly asymmetric.[5] Moreover, it highlighted what Kursani has called the 'taming effect of recognition' (Kursani 2023): the agreement was based on Arafat's prior renunciation of armed struggle. It also implied that Arafat would be held to account, and even blamed, by both Israel and the wider 'international community', should violence erupt in the OPT. Arafat's options were now severely limited. He was walking on eggshells: Palestinians would also hold him to account if his new strategy would not satisfy their demand for an end to Israeli military occupation in the OPT and Palestinian independence.

Now that the PLO had been formally recognized as a 'partner in peace' (or at least in the 'peace process'), even by Israel, armed struggle entered in direct contradiction with international legitimacy. This 'taming effect of recognition' (Kursani 2023) is consistent with Weber's observation that the institutionalization of political power generally mitigates its use of violence (Weber 1978, 314–315). Historical context also played an important role. In the post-Cold War world-historical time, beliefs in the emancipatory virtues of armed struggle were markedly on the decline (Wieviorka 2009).

The combination of political contention and limited armed action had, however, been crucial in creating conditions favourable to the Oslo Accords. The first Intifada (1987–1993) had put the PLO in a relatively favourable position with regards to Israel (Frisch 2015). With this uprising, contentious politics had erupted from the heart of the West Bank and Gaza, rather than from Palestinian exiles abroad. The Palestinian quest for statehood was internationally acknowledged as being totally independent from regional powers: the Palestinians were now counting only on

[5] Israel was recognized by the PLO as a state, the PLO was recognized by Israel as 'the representative of the Palestinian people.'

themselves. The West Bank and Gaza had, during the first Intifada, also proven to be ungovernable for Israeli military authorities. Fatah appeared as the main force able to create Palestinian institutions in spite of its headquarters still being abroad. Before the first Intifada, Fatah was mainly a transnational movement, having its strongest roots in Palestinian refugee camps in neighbouring states. During the first Intifada, Fatah progressively came to play a central role in coordinating throughout the OPT what initially had been a spontaneous uprising (Frisch 2009). Finally, the Intifada had demonstrated Fatah's willingness to limit attacks to soldiers and settlers, and refrain from targeting Israeli civilians outside of the OPT (Frisch 2015). This self-restraint contributed to its international legitimacy.

The Oslo Accords, as conceived largely by the US on the basis of the 1978 Camp David Accords, gave priority to the management of the violence generated by the conflict over the resolution of the underlying issues. The latter issues were left for future negotiations: the absence of effective Palestinian statehood, the Palestinian refugees, the status of Jerusalem, the settlements in the OPT, etc. In this regard, the Accords launched a 'peace process' that aimed at resolving the conflict through negotiations between Israel and the PLO over a period of 5 years (1995–2000). The Accords were embraced by the international community as an embodiment of its vocal post-Cold War commitment to world peace. They led to Arafat's shared Nobel Peace Prize with Peres and Rabin in 1994. A certain number of Palestinian diplomatic successes also accompanied the Accords, starting with the Israeli and American recognition of the PLO as 'the representative of the Palestinian people' and therefore the sole Palestinian interlocutor of Israel.

The Oslo Accords led to the creation of the PA, an interim self-governance structure in parts of the West Bank and Gaza. The armed groups affiliated to Fatah, that had proliferated during the first Intifada, such as Fatah's Falcons or the Black Panthers, were disbanded and integrated into the PA's security forces. The objective was both to end the armed struggle and put these forces at the service of the 'peace process' (Frisch 2008). The creation of the PA had the advantage for Fatah's leaders to allow imposing Fatah rule in the Occupied Territories in a political-military context, that of the first Intifada, that had seen the involvement of groups that were not represented in the PLO such as Islamic Jihad and Hamas (Stern 2022). The fervour of the uprising was indeed impossible to fully control, especially as Arafat was in exile in Tunis at the time.

Theoretically, the PLO remained the authority representing all Palestinians, while the newly created PA, originally an emanation of the PLO, in particular of Fatah, only governed over part of the Occupied Territories (Parsons 2005). While treaties with the Palestinian state are in reality signed with the PLO, the PA is charge of the external relations of the State of Palestine through its Ministry of Foreign Affairs. Relations with Arab countries are, however, managed by the PLO's political department (Gürbey et al. 2023). In reality, these institutions are all tightly conjoined through Fatah control, thus limiting power struggles between them.

In spite of the initial hopes, the Oslo process failed to produce any agreement on a two-state solution. It was soon to be followed by a more militarized 'second Intifada'

(2000–2005). The 'roadmap for peace' put in place during the latter stalled as well. The expansion of settlements continued in the West Bank. There have indeed been very few incentives for Israel to reach an agreement with the PLO on a two-state solution. Israel's main preoccupation after the first Intifada was for a Palestinian administration to contain the Palestinians in the Occupied Territories (Leech-Ngo 2014; Turner 2014; Wildeman and Tartir 2014). This it achieved. Indeed, the Oslo Accords compelled the PA to cooperate with Israel on security matters to prevent threats towards the latter.

It is clear that international legitimacy and recognition (formally of the PLO) is the absolute priority for Mahmud Abbas, after succeeding Yasser Arafat in 2004. This recognition, however, does not allow for the PA to exercise state powers given the conditions of military occupation. It does not translate into pressure on Israel, at least as long as the PA manages to enforce peace. In a way, therefore, the Oslo Accords deprived Fatah and the PLO of the repertoire of political-military contention that had laid the foundation for these very accords. The PLO's quest for international recognition—mainly as State of Palestine, now that the status of 'sole representative of the Palestinian people' was broadly recognized—did, however, achieve some symbolic successes under Mahmud Abbas. In 2011, Palestine became a fully-fledged member state of UNESCO. This was not the first time the PLO had vied for membership to UN specialized agencies since 1988. It was, though, the first time that the US had not succeeded in aborting the procedure. In November 2012 Palestine was recognized as a 'non-member observer state' by the UNGA through Resolution 67/19 (UNGA 2012). These symbolic successes showed that Mahmud Abbas's strategy of rejecting armed struggle and calling for legal recognition had some potential. They could however not create the material conditions for a (not only nominal) State of Palestine, one that would effectively be sovereign on the entirety of the OPT.

Moreover, these symbolic achievements hardly made up for Mahmud Abbas's failure in the campaign to get Palestine recognized as a full UN member state: the 'Palestine 194' campaign. This campaign to make Palestine the 194th member of the UN was formally launched in 2011 and is theoretically still ongoing in spite of having lost momentum. This failure created an awkward reality. Indeed, some countries, like France or Belgium, that voted in favour of Palestine becoming a member state of UNESCO and a UN non-member observer state, nevertheless refused to recognize Palestine as a state. What is more, some Central and Eastern European countries—such as Hungary or the Czech Republic—abstained from voting on Resolution 67/19 in the UNGA although they had recognized Palestinian statehood in 1989 while under communist rule. France, Belgium, and other Western European countries, however, used the momentum to upgrade the status of their Palestinian delegations to Mission of Palestine with a director enjoying the status of ambassador. In this context, it is worth noting that the US did everything in its power to undermine 'Palestine 194', like most of Mahmud Abbas's legal initiatives, thus undermining his policy of principled non-violence in the eyes of Palestinian constituencies.

The PA and the contradictions between internal and external legitimacy

On the domestic front, after Oslo, the main issue for Fatah leaders was how to transform the PA's limited right to self-rule into an exercise in state-building on its scattered areas of self-government. The Oslo Accords were to highlight the PA's ability to exercise governmental responsibilities in parts of the OPT. The strategy was to progressively assemble all characteristics of statehood until it would become impossible to deny that there was indeed a Palestinian state (Frisch 2012). Such a strategy depended on Israeli goodwill or on international pressure, both of which failed to materialize.

Peculiarly, the Oslo Accords did not even mention a future state of Palestine. Yet implicitly it defined the achievement of statehood as the norm by which the PA would be judged. The Oslo Accords did, however, explicitly condition the creation of the PA on its commitment to the peace process and to non-violence with regards to Israel. In practice, this meant containing the Palestinian population under conditions of military occupation (Leech-Ngo 2014). As the peace process rapidly proved structurally incapable of advancing towards a two-state solution, the maintenance of peace (in fact, the *status quo*) became the only remaining objective, an aim itself, one that made real negotiations redundant and yet that was morally irreproachable.

This reality was readily accepted by the new Palestinian governing elite. Indeed, after Oslo, international aid and donor money flowed freely into the OPT. This gave international donors immense political power in shaping the post-Oslo political order, and donors wanted peace at all cost (Turner 2014; Wildeman and Tartir 2014). International legitimacy was no longer a means of advancing the cause of Palestinian statehood, but a crucial source of enrichment for Palestinian elites (Leech-Ngo 2014). The fact that the aid economy did not favour development and facilitated corruption in the territories did not draw much criticism as long as it favoured peace (Wildeman and Tartir 2014). In sum, the situation amounted, and still amounts, to what Gilles Dorronsoro has analysed in terms of a 'transnational government' in the case of Afghanistan between 2001 and 2021 (Dorronsoro 2021).

This situation predictably exacerbated the tension between internal legitimation of Palestinian institutions and their external legitimation. The PA's authority over its Palestinian constituencies was contingent, as it had been from the start, on its ability to respond to Palestinian demands for statehood. As the two-state solution rapidly gave way to a one-state reality, that is, a state of Israeli occupation, the PA appeared as both a guardian of the *status quo* and a convenient scapegoat for Israel rather than as a representative of the Palestinian people. The PA was designed to maintain order in the OPT (Turner 2014), and yet it could be conveniently blamed by Israel when violence erupted. In fact, the PA was presented with a particular acute version of the predicament of many Third World states: engaging in potentially violent state-building processes, yet needing to conform to international norms created by fully consolidated states that limit the use of violence (Ayoob 1995).

This predicament left two potential solutions. The first was for Palestinian organizations to resume political-military contention at the risk of losing international legitimacy. This was to a limited extent what happened during the second Intifada between 2000 and 2005. During this period, Fatah maintained ambiguous relations to some armed groups that claimed to be organically linked to Fatah: neither rejecting them nor condoning them. Indeed, what was at stake for Fatah was how not to lose out to Hamas and the Islamic Jihad in the symbolic battle to embody the struggle against Israeli occupation (Bloom 2004).[6] Mahmud Abbas, however, seemed to be convinced that violence is always counter-productive. And, indeed, the second Intifada did not bring much to the Palestinians; at least if one excepts the unilateral Israeli withdrawal from Gaza in 2005 (Frisch 2015).

The second possible solution was for the PA to focus on its security obligations under the Oslo Accords at the expense of its 'internal' legitimacy. This 'scenario' would see Fatah and the PLO lose much of their revolutionary legitimacy to more radical organizations (Bloom 2004), as highlighted by the seizure of Gaza by Hamas in 2007.[7] Moreover, this 'scenario' increasingly engaged the PA on the path of authoritarianism to compensate for lacking popular support (Ouchaklian 2020). In April 2021 legislative elections were 'postponed'. Human rights violations are rife and the independence of the judiciary is seriously in doubt in the West Bank (Cheong 2012; Sayigh 2011). Mahmud Abbas is increasingly seen by Palestinians as serving the interests of Israel and the US (Leech-Ngo 2014; Wildeman and Tartir 2014), rather than the Palestinian people. The commitment to the *status quo* is, however, clearly privileged by the international quartet (the US, the UN, Russia, and the EU) supervising what little remains of the 'peace process'. The democratic credentials of the PA are a lesser concern. On the other hand, among international human rights organizations, the PA's behaviour weighs negatively on its international legitimation. These two scenarios, while distinct and seemingly contradictory, coexist to a certain extent currently in the West Bank as I will try to show.

The return of armed struggle in the West Bank?

As the conditions of military occupation remain draconian in the West Bank, and the perspective of a two-state solution becomes more and more unrealistic, Fatah's promise to refrain from armed violence increasingly put it at odds with the constituencies it claims to represent. As a result, since at least 2015, the presence of independent and informal armed groups is on the rise in the West Bank. Skirmishes with the Israeli army, settlers, and the PA's security forces are becoming more frequent. The bigger refugee camps in the West Bank are largely beyond PA control, even those dominated by Fatah-affiliated groups. This trend highlights the PA's and the PLO's failure to embody a credible two-state solution, or even to ease the hardships of life under occupation. It confronts Palestinian institutions with the stark

[6] On the role of Hamas here see the chapter by Leila Seurat in this volume.
[7] See the chapter by Leila Seurat in this volume.

choice to either seek to reclaim their historical legacy of political-military resistance or to entrench their international legitimacy as the authoritarian guarantor of the post-Oslo *status quo*.

The PA's loss of control in the West Bank

According to the Oslo II Accords, Area A of the West Bank is under the administrative and police control of the PA (18% of the West Bank) and Area B is under shared PA–Israeli control (22% of the West Bank). Area C (60% of the West Bank) is under Israeli control. The PA is, however, hardly able to exercise full police powers in Area A. Israeli military incursions are routine and many Palestinian refugee camps are beyond PA control, being controlled by a burgeoning microcosm of armed groups. Pressure is mounting on these armed groups from both Israeli troops and PA security forces.

In a context of frequent killings and arrests of their members, either by the Israeli army or the PA's security forces, the Palestinian armed groups are largely clandestine in the West Bank. An overwhelming majority of the fighters claim affiliation to the armed branch of a political party: Ezzedine el Qassam Brigades for Hamas, *Saraya al Qods* for Islamic Jihad, Abu Ali Mustapha Brigades for PFLP, and National Resistance Brigades for the DFLP. These armed groups are however generally highly independent from their respective parties. The case of the al-Aqsa Martyrs' Brigades (AMB) is particularly interesting in this regard. It encompasses a vast array of small groups, that all have in common that they claim loyalty towards Fatah's legacy and the memory of Yasser Arafat. The official Fatah today denies any association with the AMB. However, the AMB were presumably secretly created by Arafat in 2000 during the second Intifada (Frisch 2008). Indirect channels of communication between Fatah's leadership and the AMB have been claimed to exist under Arafat's chairmanship of Fatah (Cheong 2012).

Since 2022, there has been a trend of trans-political cooperation between armed groups on a local basis in the West Bank. The armed groups mainly operate in the refugee camps of Nablus, Jenin, Tulkarem, Hebron, Tubas, and Bethlehem. Since 2022, they often collaborate through camp-specific, or city-specific, 'joint brigades' (*Katibat Jenin, Katibat Nablus*, Lion's Den, etc.). Most of the time these 'joint brigades' are a small group of 30–40 young men that welcome combatants from all political parties. As such, these 'joint brigades' do not claim any party-based identity, although their individual members often do. As a group they are bound by a very strong sense of local identity that transcends party affiliation or ideological persuasion. Attachment to the neighbourhood by far overrides political allegiances following a pattern that can be encountered in the impoverished areas of other Levantine conflict cities (Seurat 1985). During exploratory fieldwork carried out in diverse Palestinian camps in the West Bank in 2021, most interlocutors proclaimed the legitimacy of the struggle pursued by the armed groups and celebrated the solidarity and cooperation between them beyond ideological or party affiliation. This evolution is important, as it reflects a return to the type of fragmented camp-based grassroots activism the PLO had historically managed to transform into a political force on the international arena but that is currently unfolding outside of its purview.

The emergence of these 'joint brigades' signals a rejection of official parties and institutional politics on the part of the disaffected youth. All armed groups are fiercely opposed to the Oslo Accords and to the current PLO, PA, and Fatah leadership, personified by Mahmud Abbas. This leadership is largely seen to be linked to foreign interests and as coming from the Palestinian diaspora rather than from the West Bank. From their inception, the AMB criticized official Fatah leaders for their corruption and incompetence. Interestingly, they also castigate this leadership for its engagement in international diplomacy, seeing it as turning its back on the ongoing struggle for national liberation (Cheong 2012). The AMB's Islamist outlook arguably symbolizes a rejection of the secular-nationalist narrative of previous Fatah leaders. It has been argued that the secular-nationalist narrative is generally favoured by organizations seeking international recognition. Following this argument, the AMB's mimicry of Hamas-affiliated groups in the West Bank would indicate a relative indifference to questions of international legitimacy or diplomatic recognition (Frisch 2005).

The AMB pose an interesting puzzle. How can a political movement, such as Fatah, coexist with clandestine armed groups that claim to be a branch of the 'real Fatah', rather than of the current 'official Fatah'? How can the PA fight against these groups through its security forces, while at the same time celebrating half-heartedly some of their combatants in order not to appear as opposing what many Palestinians see as 'legitimate resistance'? Moreover, if the armed groups increasingly reject official parties and organize in 'joint brigades', why does the AMB still proclaim allegiance to Fatah?

The fragmentation of Fatah by the post-Oslo political order

A first reading is that the PA is losing control in certain areas. The attempts on the part of the PA to rein in the armed groups by offering their members positions in the PA security forces have until now largely failed. This seems to confirm a loss of internal legitimacy on the part of Fatah. The situation can, however, also be analysed in terms of a PA that, because of international constraints, secretively delegates the armed struggle to independent groups, while at the same trying to capitalize on their struggle. This is consistent with the idea that the AMB were originally created by Arafat but were structured in cells that did not answer to the Fatah-PA leadership (Cheong 2012). Some evidence can be found to corroborate such an interpretation. For example, when the armed groups threaten the status quo, this sometimes leads the international community to push Israel for limited concessions (with few results). Moreover, the very presence of these armed groups highlights a lack of political alternatives other than the PA for an 'international community' eager to find Palestinian interlocutors sharing its vision of 'peace' as *status quo*. The presence of armed groups is also likely to make the PA's 'security cooperation' more important for Israel, thus indirectly giving the PA some leverage in the context of highly asymmetric bilateral relations. In this sense, although these new armed movements are opposed to the prioritization of diplomacy, their violence inevitably fuels PA diplomacy.

The interactions are, however, more complex and must be analysed in the context of a triadic game between the PA, the armed groups, and Israel. In this 'game', convergence and conflict of interests coexist between the PA and the armed groups with regards to Israel as just highlighted. This is also the case between Israel and the PA with regards to the armed groups: Israel blames the PA for the armed groups, yet at the same time it works through the PA to get their members arrested or 'eliminated'. At the same time, Israel offers political cover for the PA: PA officials can claim some arrests to be attempts to protect the armed militants from being killed by the Israeli army. The PA then shifts from being the 'punisher' to the 'protector of resistance'. In the West Bank, PA officials are, however, seen to use this argument in order to limit the popular outrage at arrests that are perceived as being carried out by the PA on behalf of Israel.

It is highly unlikely that the PA leadership as such is actively fanning the flames of insurgency in the West Bank. Even during the second Intifada, Mahmud Abbas was staunchly opposed to any 'armed struggle', and disagreed with Arafat's unwillingness to condemn Palestinian activists. While the armed groups in the West Bank weaken the PA's authority, Israel uses this visible weakness to de-legitimize it in international arenas. The frequent claim on the part of Israeli authorities that the PA is complicit in the low-level insurgency in the West Bank must be interpreted as an attempt to push the PA to rein in the armed groups. The puzzle must therefore be analysed in the context of an increasingly fragmented Fatah undermined by infighting.

The current situation highlights two things. The first is that Fatah and the PA are not the same thing. One needs to distinguish at least three levels: a) Fatah's leadership, which partly overlaps with the PA leadership; b) the wider official Fatah movement (including local authorities in the West Bank); c) the informal Fatah (including the AMB) that has openly broken with the leadership. The second is that Fatah is becoming increasingly factionalized in spite of its centralized leadership (Cheong 2012; Stern 2022). Multiple factions compete for power. Some of the more peripheral factions have broken with the official movement, at least in part in opposition to the Oslo Accords (Stern 2022). They seek to be reintegrated in a better position and demand a change in Fatah's leadership. This explains why AMB members continue to signal their commitment to Fatah in spite of their organizational disconnections from the 'official Fatah'. At the same time, the AMB allow for the wider Fatah movement to claim continuity with its revolutionary legitimacy (Cheong 2012). This is consistent with the observation that many Palestinians distinguish the 'local Fatah' from the 'Fatah of the PA' (Larzillière 2003), one being credited with some revolutionary legitimacy and the other castigated for its security cooperation with Israel (Ouchaklian 2020).

Nevertheless, there are connections between the 'Fatah of the PA' and groups like the AMB. It is, for example, noticeable that several members of the AMB that have been killed or arrested were found also to be members of the PA security service. Family connections between members of the AMB and officials at the PA are not rare. This is, in fact, not astonishing. Indeed, the PA manages to maintain its rule in the West Bank largely through the co-optation into its plethoric security forces of large sectors of Palestinian society (Wildeman and Tartir 2014). This includes Fatah loyalists that could otherwise turn against the PA and that sometimes do so in spite of their co-optation. As a consequence, there is a continuum between formal and informal Fatah. The fragmented political order created by the Oslo Accords thrives on the tension between a popular support based on connections to the 'popular resistance' and a formal recognition based on international norms promoting a vision of peace as *status quo*.

The PA leadership between extraversion and grassroots mobilization

The armed groups are the expression of a wider social struggle within Palestinian society. Sociologically, the PA leadership is entirely different from the members of the armed groups in the camps: on the one hand a rich elite, on the other the unemployed poor; on the one hand the Palestinians living outside of the camps and, on the other hand, the camp dwellers; the generational difference between Fatah cadres and barely 20-year-old youths (Larzillière 2003); the high level of internationalization of the PA leadership (Frisch 2009) and the 'local embeddedness' of the members of armed groups, etc. (Bucaille 2004; Cheong 2012). This last line of division is not to be underestimated. The Oslo Accords allowed for Fatah's leaders in the diaspora to return to the territory they claimed, to assume formal power; this in spite of local militants of the OPT having borne the brunt of successive Israeli crackdowns (Frisch 2009).

Armed struggle is not only a way of resisting settler-colonial expansion in the West Bank, but is also a way of contesting the PA leadership from within the very political order that it has created (Larzillière 2003; Ouchaklian 2020). The armed struggle mobilizes the very same 'symbolic capital'—provided by military engagement—that constituted the current leadership but that the latter exchanged for institutional positions after the Oslo Accords. In a sense, the armed struggle offers an example of a 'weapon of the weak' (Scott 1985) by which the subalterns oppose social elites in the name of the very principles that the elites originally claimed to embody (Swedenburg 2003). The resistance to PA rule emerging from within the larger Fatah movement creates a complex power balance that constrains Mahmud Abbas's options.

As such, this situation is not unprecedented. From the times of the British mandate, Palestinian elites were caught between their economic and political extraversion on the one hand, and on the other their need to mobilize at the grassroot level (Swedenburg 2003). Today, transactions simultaneously involving international sponsors and grassroot militancy are at the foundations of their authority. This leads them to balance between international demands for accountability and popular demands for grassroots mobilization. As such, these elites bear many resemblances to the 'transnational brokers' analysed by Dezalay and Garth (2002): they act as go-betweens; on the one hand allowing for diverse international and Israeli actors to shape the political order in the West Bank; on the other hand, voicing the concerns of their Palestinian constituencies and redistributing some of the economic advantages that come with international recognition.

This reminds of the fact that international legitimation is not only instrumental to the 'liberation struggle' pursued by armed groups. It is also constitutive of the social hierarchies that emerge from the political orders that they create. It is in great part Mahmud Abbas's international support, rather than his 'domestic' popularity, that singled him out for succession to Yasser Arafat. More generally, access to international resources is a crucial asset for the Palestinian political elites and, reversely, the ability of these elites to regulate and control political-military militancy makes them a crucial asset for external powers.

Conclusion

This contribution on Fatah illustrates the dilemmas entailed by the struggle for statehood in a modern international system in which the distinctive means through which states have historically been established—armed struggle—is reserved to already established states. In the case of Fatah's leadership, even a distant hope of entry into the select club of internationally recognized states comes at very high costs in terms of popular support.

Indeed, the extension of the illegal Jewish settlements in the OPT and the frequent Israeli military incursions into territory theoretically held by the PA portrays any progress towards statehood as illusory. Moreover, the PA–PLO's international legitimacy now hinges upon the PA containing and controlling the Palestinian population, the *status quo* being more or less acceptable to both Israel and the 'international community'.

The consequence is an increasingly factionalized Fatah and a Palestinian leadership that is less able to represent its national constituency in diplomatic settings. At the same time, armed groups are on the rise in the West Bank. The splintering of these armed groups increases the risk of armed violence (Bloom 2004; Stern 2022). It was against this type of splintering that Arafat had tried to centralize both Fatah and the PLO. In a way it is not only Fatah's strategy that is failing, but also the international strategy of 'taming by recognizing': to the extent that the international politics of recognition of the PLO was a means of pacifying the (secular) Palestinian national movement, it is increasingly failing.

The current situation of the PLO/PA in the West Bank bears many resemblances to the situation of the PLO before 1967. While the armed struggle on the part of 'local' independent armed groups is gaining traction in the refugee camps, the PLO and the PA are severely limited by their international sponsors in their ability to embody any Palestinian 'action power' (Popitz 2017). The PA's power today arises from its international recognition rather than from the ballot box.

References

Ayoob, Mohammed. 1995. *The Third World Security Predicament: State Making, Regional Conflict, and the International System*. Boulder, CO: Lynne Rienner Publishers.

Barnett, Michael N. 1998. *Dialogues in Arab Politics: Negotiations in Regional Order*. New York: Columbia University Press.

Bloom, Mia M. 2004. 'Palestinian Suicide Bombing: Public Support, Market Share, and Outbidding'. *Political Science Quarterly* 119(1): pp. 61–88.

Bourdieu, Pierre. 2000. *Pascalian Meditations*. Stanford, CA: Stanford University Press.

Bucaille, Laetitia. 2004. *Growing Up Palestinian: Israeli Occupation and the Intifada Generation*. Princeton. NY: Princeton University Press.

Cattan, Henry. 2000. *The Palestine Question*. New York: Taylor & Francis.

Charountaki, Marianna and Radka Havlová. 2022. 'States-to-Be as Foreign Policy Actors'. In *Mapping Non-State Actors in International Relations*, edited by Marianna Charountaki and Daniela Irrera, pp. 95–110. New York: Springer.

Cheong, Damien D. 2012. 'Fatah's al-Aqsa Martyrs' Brigades: Armed Struggle as a Response to a Crisis of Legitimacy'. *Melbourne Journal of Politics* 35: pp. 39–59.

Connelly, Matthew. 2002. *A Diplomatic Revolution: Algeria's Fight for Independence and the Origins of the Post-Cold War Era*. New York: Oxford University Press.

Dezalay, Yves and Bryant G. Garth. 2002. *The Internationalization of Palace Wars: Lawyers, Economists, and the Contest to Transform Latin American States*. Chicago, IL; London: University of Chicago Press.

Dorronsoro, Gilles. 2021. *Le gouvernement transnational de l'Afghanistan: Une si prévisible défaite*. Paris: KARTHALA Editions.

Doxey, Margaret. 1995. '"Something Old, Something New": The Politics of Recognition in Post-Cold-War Europe'. *Diplomacy & Statecraft* 6(2): pp. 303–322.

Frisch, Hillel. 2005. 'Has the Israeli–Palestinian Conflict Become Islamic?: Fatah, Islam, and the Al-Aqsa Martyrs' Brigades'. *Terrorism and Political Violence* 17(3): pp. 391–406.

Frisch, Hillel. 2008. *The Palestinian Military: Between Militias and Armies*. London: Routledge.

Frisch, Hillel. 2009. 'The Death of the PLO'. *Middle Eastern Studies* 45(2): pp. 243–261.

Frisch, Hillel. 2012. *Countdown to Statehood: Palestinian State Formation in the West Bank and Gaza*. New York: State University of New York Press.

Frisch, Hillel. 2015. 'Choosing the Right Strategy: Why the Palestinians Were More Successful in the First Intifada than in the Second'. *Contemporary Review of the Middle East* 2(3): pp. 176–191.

Gerber, Michael. 1990. 'The Anti-Terrorism Act of 1987: Sabotaging the United Nations and Holding the Constitution Hostage'. *New York University Law Review* 65(2): pp. 364–402.

Gruen, George E. 1988. 'The Anti-terrorism Act of 1987—An Update on the Controversy Surrounding Efforts to Close down PLO Offices in the US'. *Terrorism* 11(3): pp. 235–239.

Gürbey, Gülistan, Sabine Hofmann, and Ferhad Ibrahim Seyder. 2023. *Between Diplomacy and Non-Diplomacy: Foreign Relations of Kurdistan-Iraq and Palestine.* London: Palgrave Macmillan.

Khalidi, Rashid. 1989. 'The PLO as Representative of the Palestinian People'. In *The International Relations of the Palestine Liberation Organization*, edited by Agustus R. Norton and Martin H. Greenberg, pp. 59–73. Carbondale, Edwardsville, IL: Southern Illinois University Press.

Kursani, Shpend. 2023. 'Costs of International Recognition: Palestine's and Kosovo's Struggle with Negotiated Statehood'. *Geopolitics* 29(1): pp. 174–202.

Larzillière, Pénélope. 2003. 'Construction Nationale et Construction de Soi: L'évolution de l'engagement Politique Des Jeunes Palestiniens de La Première à La Deuxième Intifâda'. *Égypte/Monde arabe* 6: pp. 19–39.

Leech-Ngo, Philip. 2014. 'After "Security First": An Analysis of Security Transition and "Statebuilding" in the West Bank 2007–2011'. *New Middle Eastern Studies* 4(1).

Lefèvre, Raphaël. 2021. *Jihad in the City: Militant Islam and Contentious Politics in Tripoli.* Cambridge: Cambridge University Press.

Mann, Michael. 1984. 'The Autonomous Power of the State: Its Origins, Mechanisms and Results'. *European Journal of Sociology* 25(2): pp. 185–213.

Miller, Aaron David. 1983. *The PLO and the Politics of Survival.* Westport, CT: Greenwood Press.

Mitzen, Jennifer. 2015. 'From Representation to Governing: Diplomacy and the Constitution of International Public Power'. *Cambridge Studies in International Relations* 136(1): pp. 111–139.

Muslih, Muhammad Y. 1976. 'Moderates and Rejectionists within the Palestine Liberation Organization'. *Middle East Journal* 30(2): pp. 127–140.

Norton, Agustus Richard. 1989. 'Introduction'. In *The International Relations of the Palestine Liberation Organization*, edited by Agustus R. Norton and Martin H. Greenberg, pp. 1–11. Carbondale, Edwardsville, IL: Southern Illinois University Press.

Norton, Agustus Richard and Martin H. Greenberg (eds.). 1989. *The International Relations of the Palestine Liberation Organization.* Carbondale, Edwardsville, IL: Southern Illinois University Press.

Olsson, Christian. 2013. '"Legitimate Violence" in the Prose of Counterinsurgency: An Impossible Necessity?' *Alternatives* 38(2): pp. 155–171.

Olsson, Christian. 2022. 'No One Is Prophet in His Own Land?: Hezbollah and the Transnational Constitution of Non-State Armed Organizations'. *Partecipazione e Conflitto* 15(1): pp. 55–71.

Olsson, Christian and Siniša Malešević. 2022 'Historical Sociology of Organized Violence'. In *Palgrave Handbook of the History of Human Sciences*, edited by David McCallum, pp. 225–250. New York: Springer.

Ouchaklian, Minas. 2020. 'Le Gouvernement Sécuritaire Palestinien Au Prisme de l'émeute Retour Sur Une Enquête de Terrain Dans Le Camp de Dheisheh'. *Revue des mondes musulmans et de la Méditerranée* 147(1): pp. 89–105.

Parsons, Nigel. 2005. *The Politics of the Palestinian Authority: From Oslo to Al-Aqsa*. New York: Routledge.

Popitz, Heinrich. 2017. *Phenomena of Power: Authority, Domination, and Violence*. New York: Columbia University Press.

Reppert, John C. 1989. 'The Soviets and the PLO: The Convenience of Politics'. In *The International Relations of the Palestine Liberation Organization*, edited by Agustus R. Norton and Martin H. Greenberg, pp. 109–137. Carbondale, Edwardsville, IL: Southern Illinois University Press.

Sayigh, Yezid. 1986. 'Palestinian Armed Struggle: Means and Ends'. *Journal of Palestine Studies* 16(1): pp. 95–112.

Sayigh, Yezid. 1997b. *Armed Struggle and the Search for State: The Palestinian National Movement, 1949–1993*. Oxford, New York: Clarendon Press.

Sayigh, Yezid. 2011. *Policing the People, Building the State: Authoritarian Transformation in the West Bank and Gaza*. Washington, DC: Carnegie Endowment for International Peace.

Sayigh, Yezid and Avi Shlaim. 1997. *The Cold War and the Middle East*. Oxford: Clarendon Press.

Scott, James C. 1985. *Weapons of the Weak: Everyday Forms of Peasant Resistance*. New Haven, CT: Yale University Press.

Seurat, Michel. 1985. 'Le Quartier de Bâb Tebbâné à Tripoli (Liban)'. In *Mouvements Communautaires et Espaces Urbains Au Machreq*, edited by Mona Zakaria and Bachchâr Chbarou, pp. 45–86. Beirut: Presses de l'IFPO.

Stern, Moran. 2022. 'Factionalisation from Below: The Case of Palestinian Fatah'. *Civil Wars* 24(1): pp. 27–72.

Stokes, DaShanne. 2019. 'Political Opportunities and the Quest for Political Recognition in Tibet, Taiwan, and Palestine'. *International Review of Sociology* 29(1): pp. 102–124.

Swedenburg, Ted. 2003. *Memories of Revolt: The 1936–1939 Rebellion and the Palestinian National Past*. Fayetteville, NC: University of Arkansas Press.

Turner, Mandy. 2014. 'The Political Economy of Western Aid in the Occupied Palestinian Territory Since 1993'. In *Decolonizing Palestinian Political Economy: Rethinking Peace and Conflict Studies*, edited by Mandy Turner and Omar Shweiki, pp. 32–52. London: Palgrave Macmillan.

UN General Assembly. 1974. 'Invitation to the Palestine Liberation Organization'. (29th sess.: 1974–1975). United Nations Digital Library.

UN General Assembly. 2012. 'Resolution 67/19 on the Status of Palestine in the United Nations'. United Nations Digital Library.

Weber, Max. 1978. *Economy and Society: An Outline of Interpretive Sociology*. Berkeley, CA: University of California Press.

Wieviorka, Michel. 2009. *Violence: A New Approach*. London: SAGE.

Wildeman, Jeremy and Alaa Tartir. 2014. 'Unwilling to Change, Determined to Fail: Donor Aid in Occupied Palestine in the Aftermath of the Arab Uprisings'. *Mediterranean Politics* 19(3): pp. 431–449.

Wood, Pia. 1993. 'France and the Israeli–Palestinian Conflict: The Mitterrand Policies, 1981–1992'. *Middle East Journal* 47(1): pp. 21–40.

9

From Government to Armed Struggle

Hamas's Legitimation Politics

Leila Seurat

Hamas, the acronym of Ḥarakah al-Muqāwamah al-ʾIslāmiyyah, the Islamic resistance movement which also means 'enthusiasm' in Arabic, emerged from the Palestinian branch of the Egyptian Muslim Brotherhood. It was founded in December 1987 during the First Intifada, the popular uprising of Palestinians against Israel's occupation of the West Bank and Gaza. After publishing its own charter in August 1988,[1] the movement established itself as an alternative to the Palestinian Liberation Organization (PLO), which had been dominant in the Occupied Palestinian Territories until then. Made up of several Palestinian factions with Fatah as the most important one, the PLO has been recognized by Arab states as the sole legitimate representative of the Palestinian people since 1974.[2] The PLO officially renounced its armed struggle and declared the independence of Palestine in 1988. In the course of the Oslo Accords of 1993–95, it recognized Israel's right to exist while Israel recognized only the PLO as a representative of the Palestinians. With these treaties, the Palestinian Authority (PA) had also been created as a civil administration of Gaza and the West Bank with an ultimately military rule by Israel. Israel along with the US and most of its allies did not recognize Palestine as a state, despite the fact that around one hundred UN member states had done so by 1990. While the international status of Palestine as a state has remained contested, since its inception Hamas has competed with the PLO and other groups for the representation of the people of Palestine.

Hamas received global public attention with its surprise attacks on 7 October 2023 against Israel in which it killed more than a thousand Israelis and foreign nationals and took 251 hostages and prisoners. Following the attack, Israel brutally retaliated with a military offensive against the Gaza strip.[3] The offensive was officially aimed at destroying Hamas and freeing the hostages and prisoners, but the extremely

[1] The Charter was oftentimes a source for judging Hamas and criticizing its affiliation and overall objectives. Some of these critics questioned whether Hamas actually belonged in the national Palestinian struggle and accused the movement of belonging to trans-boundary projects, while also characterizing Hamas as anti-Semitic. Cf. Seurat (2022, 13–19).

[2] See the chapter on Fatah and the PLO by Christian Olsson in this volume.

[3] At the time of writing the military conflict is still ongoing. This text has been written in 2022–2023 and covers the most recent Gaza war only to a limited extent. It has been last updated in July 2025.

Leila Seurat, *From Government to Armed Struggle*. In: *Armed Groups and the Politics of International Legitimation*. Edited by: Stephan Hensell and Klaus Schlichte, Oxford University Press.
DOI: 10.1093/9780198974529.003.0009

high number of killed Palestinian civilians (70% are women and children according to the UN)[4], as well as the targeted bombings of schools and hospitals, show that the Israeli government of Benjamin Netanyahu is pursuing other objectives.[5] Nearly 59,000 Gazans have been directly killed by the Israeli army, but the number of indirect deaths is much higher reaching more than ten per cent of the total population.[6] This war of annihilation caused unprecedented destruction and sparked a major humanitarian crisis in Gaza including the forceful displacement of nearly all of the population. It has also provoked a broader international crisis. The International Court of Justice is reviewing a case accusing Israel of committing genocide and the International Criminal Court, following an investigation of war crimes and crimes against humanity, has issued arrest warrants for Israeli and Hamas leaders. Various states have cut their diplomatic relations with Israel while massive collective mobilizations have expressed solidarity with the Palestinian people.

Armed struggle has not always been central to Hamas's strategy. Like other Palestinian organizations, Hamas interprets resistance (*Muqawama*) in a broader sense than just armed struggle (*al-kifah al-musallah*), including popular and peaceful forms (Hroub 2000; Signoles 2010). For instance, Hamas considered its participation in the 2006 election of the Palestinian Legislative Council, the legislature of the PA, to be an integral part of its resistance programme. Yet, the election of Yahya Sinwar in the Political Bureau of Hamas in 2013 has greatly contributed to a strategic renewal: due to Sinwar's belief that armed struggle is a central tool that will allow Palestinians to regain a strong position in the negotiations with Israel to achieve a political settlement, he had no hesitation to confront Israel with violent means. Sinwar's ascension to the top of Hamas's leadership in the Gaza Strip in February 2017 simultaneously marked another strategic shift, namely Hamas's willingness to detach itself from the executive structures of the PA.

This chapter highlights Hamas's quest for political legitimacy and it puts Hamas's politics of self-legitimation centre stage. Having won a landslide victory in the 2006 Palestinian legislative elections, the movement has de facto been recognized as a legitimate actor by the Palestinian constituency. However, its victory at the ballot box was denied by international actors although the US and the EU had strongly encouraged the holding of elections. This contribution argues that Hamas has, since 2006, experimented with different practices for making a claim to legitimacy: Hamas initially relied on *mimicking* by duplicating the institutions of the PA in the Gaza Strip as a resource to achieve recognition and as a claim to legitimacy. Later, however, it

[4] https://news.un.org/fr/story/2024/11/1150406, accessed 23 July 2025.

[5] According to the Middle East Scholar Barometer, a majority of Middle East scholars see a primary Israeli objective of the war as making Gaza uninhabitable so as to force Palestinians out. A wide majority also describe Israeli actions as constituting genocide or major war crimes akin to genocide. See the Middle East Scholar Barometer online: https://www.brookings.edu/articles/gloom-about-the-day-after-the-gaza-war-pervasive-among-mideast-scholars/ (accessed 21 February 2025).

[6] According to renowned British medical journal *The Lancet* 'it is not implausible to estimate that up to 186,000 or even more deaths could be attributable to the current conflict in Gaza'. See online https://www.thelancet.com/journals/lancet/article/PIIS0140-6736(24)01169-3/fulltext (accessed 21 February 2025). The current death figures for both sides have been taken from the UN: https://www.ochaopt.org/content/reported-impact-snapshot-gaza-strip-16-july-2025, accessed 22 July 2025.

counted more on armed struggle to confront Israel and *fighting* as a form of claim-making.[7] In doing so, Hamas has gradually favoured popular support over the quest for international legitimacy. In the Palestinian colonial context, winning the support of the Palestinians by showing itself capable of confronting Israel became the core of Hamas's politics of legitimation. Thus, the evolution of Hamas's legitimation from 2006 until the present can be understood as a trajectory from government to armed struggle.

As an object of research, Hamas has been approached from various angles. A first generation of researchers focused on the movement's opposition to the Oslo Accords, viewing Hamas as a social movement. In the process from armed struggle to negotiation, Hamas's politics have been largely understood as the result of internal rivalry among various Palestinian factions (Milton-Edwards and Farrel 2010; Gunning 2007; Hroub 2000; Mishal and Sela 2000). Following Hamas's electoral victory in 2006, a second generation of researchers highlighted the precarious balance between exercising power and resistance (Signoles 2006; Hamed 2023). For some, governance overshadowed armed struggle (Bao 2016), while others insisted on the primacy of armed struggle over governance (Berti 2015). Hamas's experience in power in the Gaza Strip since 2007 subsequently opened up other works that considered Gaza to be a laboratory for analysing Islamism in power (Brenner 2017; Berti 2015; Alijla 2022; Tuastad 2021). Among these works, however, few combined an analysis on Hamas's politics of legitimation with a reflection of the organization's evolving positions regarding both the PA and the PLO.

Understanding Hamas's politics of legitimation requires one to disentangle the contradictory relations between Hamas on the one hand and the PLO as well as the PA on the other.[8] Having first opposed the PLO, Hamas then oscillated between a desire to build an alternative to the PLO and an attempt to negotiate its entry into this organization. Shortly after its creation, Hamas challenged the PLO's role in leading the popular uprising (Klein 1996). By offering a religious definition of armed struggle (Awad 2021) and a vast charitable network, Hamas had the social resources to resist the symbolic hegemony of the PLO and started presenting itself as an alternative, publishing its own charter and communiqués in its own name (Abou Amer 2017). While Article 27 of Hamas's charter stipulates its entry into the PLO on the condition that the latter adopts Islam as a rule of life (Legrain 2020), Hamas's opposition to the PLO is far from strictly religious. Khaled Hroub (2000) has shown that Hamas fought the PLO mainly because Hamas was opposed to Fatah, a faction that had dominated the PLO and which had embarked on a path of negotiated solution with Israel, accepting the two-state solution. However, Hamas's opposition to the PLO has evolved over time. The death of Arafat in 2004 and the election of Mahmoud Abbas as president of the PA in 2005 brought about a new time of cooperation

[7] I use the terms mimicking and fighting according to the Introduction of this volume in the sense of practices of claim-making.

[8] On the PLO and the PA see the chapter by Christian Olsson in this volume.

between the two organizations. In the Cairo Declaration of 2005, Hamas recognized the PLO as 'the sole legitimate representative of the Palestinian people'. This was reiterated one year later in the so-called Prisoners' Document signed by Hamas and other Palestinian factions.[9]

As for the PA, Hamas had refused to participate in its institutions in the 1990s, considering them to be a betrayal because they validated the Oslo Accords. Then in 2005, Hamas decided that participation in the PA was part of its resistance programme, and it took part in the Palestinian legislative election in 2006. This shift was explained by Hamas's leaders as a reflection of the changing situation on the ground: having exceeded the five-year interim period that was to lead to the building of a Palestinian state according to the Oslo Accords, they considered that their participation in the PA would no longer be considered a validation of these Accords since they had actually expired. After winning the elections, Hamas failed in its attempt to bring the Palestinians together through a government of national unity, resulting in violent clashes between Fatah and Hamas in 2007. Since then, Palestine has been divided into two administrative territories. One government was established in Ramallah led by Fatah, the other in the Gaza Strip led by Hamas. Hamas's politics of legitimation have thus unfolded in a setting of ambiguous relations with the other Palestinian actors and institutions.

This chapter is structured in two parts. First, we will see how the victory of Hamas in the 2006 Palestinian legislative elections inaugurated a period in which Hamas presented itself as the embodiment of the institutions of the PA and sought to acquire legitimacy as a government. By heading the PA, Hamas conducted the official government business of Palestine including its foreign relations. After Hamas took over the Gaza Strip in June 2007 and the establishment of a second internationally recognized government in Ramallah, Hamas decided to duplicate the institutions of the PA. Its practice of institutional reforms was aimed at appearing to be an authority that maintained order in the Gaza Strip. In the second part we will see how Hamas has progressively detached itself from the structures of the PA and reoriented the movement's strategy. Hamas began to stress the importance of armed struggle to unify the Palestinian ranks and gain the support of even more Palestinians. This strategy culminated in the attack of 7 October 2023 against Israel which provoked the most recent Gaza war (2023–2025). A short conclusion briefly assesses Hamas's strategy, the search for a sustainable ceasefire, and the future of Hamas. This chapter relies on primary and secondary sources including interviews with the movement's leaders inside and outside Palestine. These interviews were conducted between 2010 and 2015 and more recently also in Lebanon and Qatar from 2017 to 2023. Reflecting an insider's perspective, these sources allow us to highlight Hamas's strong desire for self-legitimation and the image its leaders wish to convey to their constituency and wider audiences.

[9] For the Cairo Declaration see http://www.palestine-pmc.com/details.asp?cat=2&id=849. For the National Conciliation Document of the Prisoners see https://www.un.org/unispal/document/auto-insert-208621/, accessed 18 February 2025.

Embodying the PA under blockade

Hamas's victory in the Palestinian legislative elections of 2006 marked a major shift for the movement as it suddenly had to exercise power. Having already enjoyed recognition as a movement, Hamas now also sought to gain recognition as a government. After seizing power in June 2007 and having duplicated the institutions of the PA in the Gaza Strip, Hamas engaged in claim-making practices that were oriented towards institutional reforms. It presented itself as capable of maintaining order and exercising authority over a restricted territory.

Conducting governmental diplomacy

Hamas formed the 10th government of the PA without Fatah and was faced with a siege imposed by both Israel and Egypt, and an economic blockade imposed by the Quartet.[10] In order to prevent Hamas from governing, international aid was transferred directly to the presidency in Ramallah, making Hamas dependent on aid from other external states such as Iran to run its government. Faced with these difficulties, Hamas made every effort to get the 2006 election results recognized. Hamas representatives' statements were all focused on respecting electoral legality. In his inaugural speech in front of Parliament on 27 March 2006, Hamas leader and prime minister of the PA, Ismail Haniyeh, requested that the international community respect the choices of Palestinians who voted for Hamas and to reconsider its negative stance on the Hamas government: 'The Palestinian people should not be punished for exercising their right and choosing their direction through free and democratic elections (...) Those who think that economic pressure will bend our government or undermine the determination of our people are mistaken (...) We call on the international community to review its positions towards the new government and to respect the democratic choice of the Palestinian people.'[11]

Having de facto obtained the PA's Ministry of Foreign Affairs, Hamas now represented Palestinians outside Palestine and spoke on behalf of the PA. It thus exercised, in parallel with its own partisan diplomacy,[12] governmental diplomacy embodied by both Prime Minister Ismail Haniyeh and Foreign Minister Mahmoud al-Zahhar. The

[10] The Quartet, which includes the United States, the European Union, Russia, and the United Nations, established three conditions for Hamas to be considered a legitimate negotiating partner. These conditions, known as the Quartet Principles, require Hamas to renounce violence, recognize the state of Israel, and accept previous agreements made between Israel and the PLO.

[11] « Les principaux points du discours d'investiture d'Ismaïl Haniyeh», in: *Le Monde*, 27 March 2006, online available: https://www.lemonde.fr/proche-orient/article/2006/03/27/les-principaux-points-du-discours-d-ismail-haniyeh_755207_3218.html (accessed 21 February 2025). All quotes have been translated by the author.

[12] Members of Hamas have engaged in international diplomacy, including a notable visit to Moscow in March 2006. This visit was significant as it marked one of the few instances where Hamas was engaged in high-level discussions with a major international power. Russia, a member of the Quartet on the Middle East along with the United States, the European Union, and the United Nations, did not adhere strictly to the Quartet's conditions for engaging with Hamas.

former, who left the Gaza Strip on 28 November 2006 for a diplomatic tour, expressed himself as follows: 'This is the first time since the formation of the government that I am leaving my homeland. The purpose of these visits is to deepen relations between the "tenth government" and Arab and Muslim governments, and to do everything necessary to enable the Palestinian people to face the unjust blockade imposed on it.'[13] Al-Zahhar conducted two diplomatic tours: the first in Arab countries, the second in Asia where he met with the presidents of Indonesia and Malaysia, as well as the sultan of Brunei. In Pakistan, he met the vice president, the prime minister, and the foreign minister. Although the results of its efforts were poor vis-à-vis Western countries, which ignored Hamas's approaches and stuck to the three-conditions principle, during the period when Hamas was at the head of the PA its ministers were received officially by many Arab and Asian countries. Only Egypt and Jordan, two countries that had signed peace treaties with Israel, were reluctant to receive Mahmoud al-Zahhar.

Official representation of Palestinians outside Palestine was also ensured by deputies elected in 2006 on the Hamas 'Change and Reform' list. Parliament President Aziz Dweik embodied this role.[14] According to sources within the Palestine Legislative Council of the PA, the aim of these visits was 'to extend relations at the regional and international levels, to strengthen relations with politicians and deputies from various states around the world, and to build real bridges to explain the Palestinian perspective to global political decision centers' (Palestinian Legislative Council 2010, 367). Contacts with Palestinian parliamentarians, and not government members, allowed certain European countries like Switzerland and Norway to receive Hamas, as they had not implemented the EU-imposed boycott. The Inter-Parliamentary Union Federation in Geneva thus welcomed Mouchir al Masri and Sayed Abu Musameh in January 2012, and their status as parliamentarians enabled them to interact with European diplomats. In January 2012, a high-ranking Hamas diplomat was received in Switzerland. He stated: 'I went to Switzerland as a member of Parliament, but it was really to represent the movement, since we did not even mention the Parliament's role. There were many officials and former ministers.'[15] In this way, some countries wished to open a dialogue with the leaders of Hamas, so that they could play the role of a mediator in the future. Hamas willingly mobilized its actors and mechanisms of representation, taking the preferences of external players into account, and it tried to use the visits of their MPs, which were all described as 'victories'. The movement used the formal status of the Palestine Legislative Council's members in order to forge diplomatic relations with Western states, with the goal of one day becoming recognized.

[13] *Ma'an News*, 28 November 2006.

[14] Hamas encountered various international representatives, such as Ohm Prakash, the Indian Ambassador to Palestine, a Belgian parliamentary delegation, Swedish Member of Parliament Gieri Mille, Chinese Ambassador Yang Wei and Egyptian Ambassador Ashraf Aqel.

[15] Interview with the author, Gaza city, March, 2013.

Whether it concerns the reception of Hamas parliamentarians abroad or the visits of European delegations to Gaza, these meetings have been sharply criticized by Israel,[16] which argued that they should not have been carried out in the name of the EU as it considers Hamas a terrorist organization. The meetings were also criticized by the PLO, which tried to prevent them, as was the case during the Non-Aligned Movement summit held in Kuala Lumpur in 2006. While Mahmoud al-Zahhar was on his second tour in Asia, Palestinian President Mahmoud Abbas imposed the presence of PLO representative Farouq al-Qaddoumi to represent Palestine. Mahmoud Al-Zahhar could, if he wished, be part of the delegation, but not lead it. Although he himself had pushed for the PA to have more power in diplomacy, Mahmoud Abbas backtracked after the 2006 election results and added a motion stipulating that for everything concerning Third World and European summits, it was indeed the PLO that should represent Palestinians. Thus, the example of diplomacy illustrates how Hamas's conducting of foreign affairs has generated intense tensions between Hamas and the PLO.

Implementing reforms in the Gaza Strip

After its electoral victory, Hamas assumed responsibility of the Palestinian government and the PA. The Prisoners' Document, signed by various Palestinian factions including Hamas in June 2006, demonstrated a form of respect for the PA's institutions. Indeed, the document calls for the protection and strengthening of the PA as the core of the future Palestinian state 'born of the struggle and sacrifices of the Palestinian people'.[17] After the division between Ramallah and the Gaza Strip in June 2007, Hamas decided to continue to embody the 'institutional' face of a government by duplicating the PA's institutions in Gaza. Isolated diplomatically, it sought legitimacy by conducting institutional reforms in the Gaza Strip. Some studies have highlighted how these reforms are part of a quest for legitimacy aimed at making Hamas an effective authority rather than an armed group (Berti 2015; Alijla 2021; Tuastad 2021). This is particularly true for the reform of the security sector. By separating the Al-Qassam Brigades, Hamas's armed branch, from the civilian police forces responsible for maintaining internal order, this reform aimed to convey the image of a responsible government accountable for its actions, following the model of the rule of law. Hamas cadre and inspector-general of the Ministry of the Interior and National Security in Gaza, Hasan Saifi, claimed to have established order not only through the construction of a new police force, departments, units, and procedures, but also by establishing an integrated criminal justice system: 'We have a security, political, and social vision of what we seek and how to attain it'.[18] In

[16] The members of the Palestinian Legislative Council also welcomed delegations in the Gaza Strip, including a British delegation from the European Parliament.

[17] National Conciliation Document of the Prisoners, 28 June 2006. Available online: https://www.un.org/unispal/document/auto-insert-208621/, accessed 18 February 2025.

[18] Hasan Saifi, quoted after Sayigh (2011, 119).

Hamas's self-perception, the government can be seen as a means to realize this vision, and the security sector as a means to ensure the public order required for the realization of this vision. Moreover, the resistance factions should 'absolutely not ... intervene in internal affairs—we have courts and police and government institutions for that'.[19]

Yezid Sayigh attests to the success of this reform, which reportedly managed to reduce crime rates and end the security anarchy (*falatan amni*) that had prevailed in the Gaza Strip since 2004 and for which both Fatah and the PA were responsible due to their failure to provide essential public goods (Sayigh 2011). Despite difficulties arising not only from the blockade but also from a decree of the PA ordering all civil servants not to work under the Hamas administration, the Haniyeh government managed to impose 'virtually uncontested, and uninterrupted, control over the entirety of its territory and population—enabling it to attain a level of policy coherence, administrative cohesion, and political mastery that its predecessors had never enjoyed, even at the height of Palestinian–Israeli cooperation in 1994–99' (Sayigh 2011, 7).

Yet, where Hamas gained domestic legitimacy in its conduct of internal affairs, it lost popularity on the external front with its struggle against Israel. Indeed, conducting reforms required maintaining a truce with Israel.[20] This truce, concluded in 2008, led Hamas to play the role of border police, preventing other factions from firing rockets at Israel. Hamas's image then significantly deteriorated, with the movement being accused of following the same path as the PLO or even being likened to neighbouring Arab regimes, proclaiming resistance while accommodating Israel to defend their own interests.

In sum, if the post-June 2007 period closed any possibility for the Gaza government to pursue an official role in diplomacy, this period did not prevent other modes of international recognition. The Israeli offensive in the Gaza Strip in December 2008 and January 2009 enabled parliamentarians to resume diplomatic tours, while the assault on the Turkish *Viva Palestina* flotilla in May 2010 set up similar dynamics with the visit of the Arab League General Secretary Amr Moussa's trip to Gaza in June 2010. This was the first visit to the Gaza Strip made by an official political representative at a ministerial level in four years. Israeli hostile acts allowed Hamas to mobilize international public opinion and to participate in organizing humanitarian convoys in the framework of European campaigns aimed at lifting the blockade. However, the Palestinian division had the effect of drastically limiting Hamas's capacity for diplomatic action. With the exception of a symbolic form of recognition, Hamas continued to face isolation until the upheavals of the Arab Spring.

[19] Hasan Saifi, quoted after Sayigh (2011, 118).

[20] While Israel unilaterally disengaged from the Gaza Strip in 2005, the UN and a number of human rights organizations continue to consider Israel the occupying power of the Gaza Strip due to its blockade of the territory; Israel rejects this characterization.

Recognition thanks to new external patrons?

The Arab uprisings of late 2010 and early 2011, which led to the fall of the Tunisian and Egyptian presidents, brought Muslim Brotherhood parties to power. This entirely new regional context was thus favourable to the recognition of movements associated with the Muslim Brotherhood. Hamas's leaders from Gaza were then able to travel again and bypass the blockade. Thus, they first tried to regain the official status they enjoyed in January 2006 but, driven by the influence of their new patrons Egypt and Qatar, the movement quickly moved closer to Fatah.

In less than a year, Hamas signed two agreements with Fatah: the Cairo Agreement in May 2011, and the Doha Agreement in February 2012. The first reiterated the central elements of previous reconciliations, namely the transfer of negotiations with Israel to the PLO. The second stipulated that Mahmoud Abbas, in addition to being the president of the PA, would be the prime minister of a government intended to prepare for the next legislative elections. The agreement also raised hopes for a reactivation of the PLO to include Hamas and Islamic Jihad. These agreements, like previous ones, were a means to end the quarantine imposed on the Hamas government and attempts to gain international recognition though the formation of a national unity government. However, unlike earlier agreements, these reconciliations stemmed less from a Hamas initiative than from pressure from external actors. For Hamas, conceding to the demands of Egypt and Qatar was supposed to facilitate the opening of a dialogue with Western countries engaged in discussions with the region's Islamist parties. In return for the Doha Agreement, Qatar had indeed committed to helping Hamas in establishing official relations with the United States and Europe after some European countries had developed behind-the-scenes relations with Hamas. A former adviser to Prime Minister Haniyeh said about the Cairo Agreement of 2011: 'Hamas wanted to test the international community to see if it would accept a technocrat government so that, if so, it could be recognized and receive funds.'[21] For the analyst Nathan Brown, Hamas leader Khaled Mesh'al saw regional opportunities opening for Hamas and he attempted to adopt a 'sanitized' version of political Islam, which would allow Hamas to integrate into a Western process of recognizing the Muslim Brotherhood. 'In short, those Hamas leaders who have led the movement into reconciliation efforts have embarked on a process that would allow Hamas to be incorporated as a diplomatic actor and subjected to authoritative Palestinian structures. In return, the movement would gain a voice in that diplomacy and a role in those structures' (Brown 2012).

The victory of Islamist parties in Egypt and Tunisia indeed led the movement to believe that Western states might soften their stance and start a dialogue with Hamas especially if the group emphasized Palestinian reconciliation. This hope was reinforced by the fact that long-standing underground contacts existed, which could have taken on an official dimension in line with the political change in the region. As a

[21] Personal conversation with the author, Ankara, June 2011.

Hamas spokesman put it: 'With the Arab Spring, we can now seize the opportunity to expand our relations with European countries that had recognized the national unity government in 2007 following the Mecca agreement. We believe today that several of these countries are ready to establish direct relations with Hamas, and this is not an analysis I am making, but it comes from authentic sources: there are several states, but we cannot reveal their identity. It is up to them to do so if they wish, but we cannot publicly identify them because of American pressures on them, and pressures from the Zionist lobby.'[22]

Ready to gain international recognition, Hamas renewed its discourse and made flexible declarations. At the Cairo meeting, Khaled Mesh'al reiterated his recognition of the 1967 borders and adopted the notion of 'peaceful resistance'. Although not new, in the post-revolutionary context of the Arab spring, one could interpret it as an attempt by Hamas to consolidate Egyptian support, as the Muslim Brotherhood's opposition had a high chance of winning the majority of votes in the upcoming legislative and presidential elections in Egypt. After the victory of the Muslim Brotherhood in the elections in January and June 2012, Mohammed Morsi attempted to transform Hamas into a movement that resembled other Muslim Brotherhood movements not listed as terrorist organizations. The adoption of 'peaceful resistance' was made easier as it corresponded to the line of the Egyptian Brotherhood, which criticized armed struggle and thus allowed Hamas to emphasize its ideational affinity with an organization which was about to normalize its relations with the United States and the European Union. However, these pursuits by Mesh'al were criticized by a significant part of the Hamas leadership in Gaza. The criticism focused both on the procedure, i.e., the unilateralism of the Political Bureau's president, and on the loss of decision-making autonomy due to the influence of Qatar, Egypt, and the Egyptian association that interfered in Hamas's internal electoral process to support the re-election of Khaled Mesh'al as the head of the political bureau. This phase quickly closed with the overthrow of Morsi in July 2013. Even by the end of that same month, General Al-Sissi's rise to power in Egypt opened a period of crisis that significantly impacted Hamas, both in its relations with the PA and in how it presented itself to the world and the Palestinians.

Embodying national unity through armed struggle

Isolated on the regional scene after the coup in Egypt in July 2013, Hamas made several strategic changes. Whereas it had relied on PA institutions to gain legitimacy both internally and externally, Hamas no longer seemed concerned by its representation within the government. Indeed, the various reconciliation agreements then left it to Fatah alone to form a government of national consensus. While the need to present itself as capable of maintaining order and security still constituted a major legitimating tool vis-à-vis its own constituency as well as Egypt, Hamas began to

[22] Personal conversation with the author, Cairo, February 2012.

mobilize armed struggle as a central resource, presenting itself as the embodiment of resistance in order to claim leadership of the Palestinian national movement.

Gradual detachment from the executive structures of the PA

Aware of the difficulties in gaining recognition through a national unity government with Fatah, in 2014 Hamas decided to leave the responsibility of forming the PA government to Fatah. Thus, the movement chose to withdraw from official representation. Hamas had no ministers in the consensus government led by Rami Hamdallah, a crucial factor that allowed its recognition by the European Union and the United States. This recognition led Israel to take major steps to prevent any unification process between Ramallah and Gaza, from economic sanctions to the PA to a declaration of war against Gaza. Three years later, in October 2017, Hamas reiterated its position of detachment from the PA, signing another agreement with Fatah in Cairo in which it recalled its wish to withdraw from official representation in government and hand over Gaza to the president of the PA.[23]

This changing position towards the PA was gradually reflected in official documents signed between Hamas and other Palestinian factions. Whereas the Prisoners' Document from 2006 called for the protection of the PA as the future Palestinian state, eleven years later the focus had shifted with the Document of General Principles and Policies from 2017.[24] Article 27 of the Document of General Principles implied that the PA had nothing to do with the state born from liberation (Legrain 2020). It stipulated that the Oslo Accords contradicted the rules of international law in force because they entailed commitments that violated the inalienable rights of the Palestinian people (Article 21).

Even in September 2017, Hamas had already announced the dissolution of its administrative committee, a structure that was created in 2014 to manage the daily affairs of the Gaza Strip and was considered by Fatah to be a separate government. The rise to power of Yahya Sinwar, known for his opposition to the establishment of an independent government in the Gaza Strip, had allowed the signing of the Cairo agreement of 2017. Thanks to this agreement, which was made possible by Egypt, Hamas obtained several concessions from Cairo. This included the opening of the Rafah crossing and the allowance of cement and tanker trucks filled with millions of litres of fuel. The Egyptian goodwill could be interpreted as a recognition of Hamas. It should also be seen with regards to the Document of General Principles which does not mention Hamas's affiliation with the Muslim Brotherhood.[25] If this silence primarily reflected Hamas's diplomatic interest in warming its relations

[23] The agreement involves the control of the Rafah border crossing by Mahmoud Abbas's Presidential Guard, the absorption of the Gaza police into the security forces of the Ramallah Authority, and the payment by the PA of Hamas civil servants.

[24] Available online: https://web.archive.org/web/20170510123932/http://hamas.ps/en/post/678/

[25] While the link between Hamas and the Egyptian Muslim Brotherhood was explicit in Article 2 of the Charter, this mention disappeared from its General Principles Document.

with General al-Sisi (Hroub 2017), it also represented a means of claiming the movement's decision-making independence at a time when Hamas was perceived as being extremely permeable to external influences (Legrain 2020).

From Hamas's point of view, the structures of the PA no longer represented a source of legitimacy. Eleven years after seizing power, Hamas did not see the benefit of continuing to conduct governmental affairs in a context of blockade. More importantly, the Document of General Principles no longer mentioned the status of the PLO as the 'sole legitimate representative of the Palestinian people', as it did in the June 2006 Prisoners' Document.[26] Article 29 of the Document of General Principles stated that the PLO was only to be a 'national framework for the Palestinian people inside and outside Palestine'.[27] This absence reflected a desire to capture all those disillusioned with nationalism as embodied by the PLO and the PA, offering them armed struggle as the only path to liberation, and thus an approach to the conflict that was free of religious references but just as rich in references to the PLO as it was before everything that led to Oslo (Legrain 2020). In its Document of General Principles, Hamas also emphasized armed resistance as a strategic choice aimed at protecting principles and restoring the rights of the Palestinian people.[28]

From popular resistance to military means: uniting the Palestinian ranks though armed struggle

In February 2017 Hamas elected Yahya Sinwar as head of the Gazan leadership. Sinwar was among the 1,023 Palestinian prisoners released in 2011 in exchange for Gilad Shalit, a French-Israeli solider captured in 2006 and detained since then in the Gaza strip. Sinwar's rise to power marked a significant shift in the movement's strategy. Convinced of the key role that armed struggle could play when synchronized with negotiations, he sought to mobilize the military tool in a different way by breaking away from the tempo set by Israel during previous operations against Gaza (2009, 2012, 2014), which forced Hamas to react to Israeli attacks while being perceived as the aggressor. Since Hamas was consistently portrayed as the aggressor, it chose to take the offensive, attempting to overturn an unfavourable balance of power by no longer merely responding to Israeli attacks but by initiating confrontations.

The 'Great March of Return' demonstrations that began in March 2018, initially served as a platform for peaceful mobilizations to demand the end of Israel's blockade and the right of return for refugees. Although it was only intended to last six weeks, these weekly demonstrations at various border areas with Israel continued for more than a year. Part of the old, non-violent forms of resistance (*sumud*), the march

[26] National Conciliation Document of the Prisoners, 28 June 2006. Available online: https://www.un.org/unispal/document/auto-insert-208621/, accessed 18 February 2025.

[27] Available online: https://web.archive.org/web/20170510123932/http://hamas.ps/en/post/678/

[28] The source of inspiration is clearly the revised 1968 PLO Charter, especially regarding the legal character of armed struggle, presented as a legitimate activity of self-defence and the expression of people's natural right to self-determination.

militarized when different political factions entered and replaced the march's high coordination committee with a new one, the Great March of Return and Breaking the Siege (Mohammad Saleh 2019). From a dynamic of popular protest, the march was overseen by Hamas, which sought to attract international attention and the support of global public opinion (Hamed 2023). Moreover, this popular resistance was punctuated by rocket firings, part of a strategy oscillating between appeasement and confrontation, aimed at forcing Israel to negotiate and make concessions (Abou Rekbe 2021).

In November 2018, negotiations between Hamas and Israel in Cairo led to an initial agreement concluded under the auspices of the United Nations, Egypt, and Qatar. Hamas obtained the opening of border crossing points and Qatari aid for civil servants. Other points discussed included the construction of a gas pipeline to supply electricity to Gaza, combating unemployment and solutions for making the Gaza Strip viable. From then on, the challenge for Hamas was to ensure that this agreement was implemented in practice. When talks proceeded well, Hamas maintained calm at the borders; when the talks stalled, it used military instruments to force Israel back to the negotiating table. Yahya Sinwar thus chose to reactivate the march during the run-up to the Israeli legislative elections in April 2019. Aware that Prime Minister Netanyahu could not afford to launch a large-scale military operation in the midst of an election campaign and would have no choice but to negotiate, he decided to send a series of rockets towards Tel Aviv on 14 and 25 March 2019. Conversely, on 30 March 2019, during the first anniversary of the march, Hamas chose to keep the peace in the midst of negotiations in Cairo. Its men in orange suits were deployed to secure the security barrier separating Gaza from Israel and prevent Gazan youth from heading towards confrontation. This appeasement dynamic allowed the conclusion of an agreement on 1 April under the auspices of Egypt, supported by the UN and Qatar: Hamas received an increase in Qatari aid from 15 to 40 million dollars, without interrupting the weekly march.[29] While Gazans welcomed the agreement, they also criticized Hamas for using rockets only for its own interests, far from the initial demands of the march.[30] As for the rest of the Palestinian population, they criticized Hamas for using armed struggle only for partisan ends or for the limited purpose of easing the blockade in the Gaza Strip. These accusations were overcome during a subsequent operation.

In May 2021, Hamas launched an 11-day major military operation against Israel, the 'Sword of Jerusalem', which allowed the movement to confront Israel, to appear as the defender of Al-Aqsa, and to unite the different Palestinian localities in a common struggle. This confrontation needs to be viewed in the context of the mobilizations against the expulsion of Palestinians from Sheikh Jarrah in East Jerusalem and the

[29] http://eng.alzaytouna.net/2019/05/13/political-analysis-gaza-understandings-snatching-rightsor-mined-gains/

[30] According to surveys conducted in Gaza and published by the PCRPS (Palestinian Center for Policy and Survey Research), Hamas's popularity increased from December onwards thanks to the conclusion of the agreement. Three-quarters of the respondents welcomed Qatar's involvement and criticized the policy of sanctions imposed by President Abbas against Gaza.

Israeli repression in the Al-Aqsa Mosque. In response to these attacks the Al-Qassam Brigades, along with other armed groups in Gaza, fired rockets at Tel Aviv, Jerusalem, Ashkelon, and Ashdod. According to Zaher Jabarin, a member of Hamas's Political Bureau, the fact that Hamas fired first is a key element of this strategic change (Jabarin 2022). Comments also emphasize the unprecedented nature of a military confrontation decided by the Palestinians, the first since 1973 (Al-Hajj Hassan 2021). This operation tarnished Israel's image, as the rockets bypassed the Iron Dome, hitting targets over 200km from Gaza, and Hamas demonstrated its ability to disrupt Israeli military plans, as evidenced by the 'Hamas metro' operation, a fiasco for Israeli intelligence (Abou Amer 2022). Armed struggle allowed Hamas to present itself as the best actor to defend Jerusalemites and unite the Palestinian people around a common cause: solidarity with and defence of Jerusalem (Abou Amer 2021). Indeed, without any direct connection to Hamas's action, unity was expressed in the streets through spontaneous uprisings in all cities of the West Bank, pushing their marchers towards Al-Aqsa, and, for the first time, protests were carried out by Palestinian citizens of Israel in cities like Lod, Acre, and Wadi Ara, who clashed with the police in solidarity with Jerusalem. By displaying a force of resistance, Hamas managed to position itself at the centre of this unprecedented configuration. Every evening at the Al-Aqsa Mosque, Al-Qassam flags were waved and slogans such as 'Ya Qassam, ya habib ("oh Qassam, oh beloved") strike strike Tel Aviv' were chanted (Abou Amer 2021). Abu Ubayda, spokesman for the Al-Qassam Brigades, announced that 'Gaza, Jerusalem, the West Bank, and the Palestinians of 1948 are one body, one homeland, sharing one destiny and one resistance'.[31]

Armed struggle has long been considered to be less of a strategy to fight occupation and more of an instrument to unite Palestinians dispersed in fragmented societies. The heroic imagery of the *fedayeen* replaced the figure of the stateless refugee with that of a revolutionary people actively determining their destiny (Sayigh 1999). This connection between Gaza, the West Bank, and more broadly the rest of Palestine occurred in a context where, a few months earlier, the United Arab Emirates, Bahrain, and Morocco had concluded the Abraham Accords, showing they were ready to sign strategic partnerships with Israel despite threats that it would annex the West Bank. While many Arab countries were not ready to defend Palestine, an enclosed Gaza came to the rescue of the Occupied Territories and Jerusalem, both a national symbol of the Palestinians and the third holiest site in Islam.

While at the beginning of 2021 Fatah enjoyed popularity among Palestinians, surveys following the 'Sword of Jerusalem' battle reveal a significant reversal. The battle produced a consensus: Hamas was the winner of the confrontation, and its rocket fire was motivated by the desire to defend Al-Aqsa and Sheikh Jarrah. Support for armed struggle increased, and the majority of respondents claimed that Hamas, not Fatah, deserved to represent the Palestinian people. Analyses emphasize that while Hamas

[31] https://felesteen.news/post/108004/خطابات-أبو-عبيدة-كلمات-ساهمت-في-تعزيز-حتمية-الانتصار

would certainly not liberate Palestinian land, armed struggle undoubtedly strengthened the Palestinian position in negotiations. The prospect may not be liberation in the short or medium term, but Hamas certainly imposed an alternative model of resistance (Hassan 2021).

Since 2018, Hamas has adopted a strategy of uniting the Palestinian ranks through armed struggle. From the start of the 'Great March of Return', Hamas began to gather all Palestinian groups under a common banner, either in the 'National High Committee for the Great Return March and Ending the Blockade' or the 'Joint Operations Room', which includes armed groups from all Palestinian political organizations present in Gaza. On the Al-Qassam Brigades website, the Palestinian flag has replaced that of Hamas. In speeches, Yahya Sinwar took care to avoid any partisan identification. The resistance was carried out in the name of the Palestinian nation seeking a new path towards the formation of a national army that should include Fatah. On the anniversary of the 'Sword of Jerusalem' battle in May 2022, Sinwar delivered a speech with a background photo of an Israeli soldier attacking Palestinians praying, an image as Sinwar himself stated indicates a warning for the future. In this speech, he addressed the Palestinians of 1948, also focusing on the protection of Al-Aqsa, whose fate is intertwined with that of Gaza: 'We defended Al-Aqsa, and it is Al-Aqsa that defended us.'[32] National unity was also present in another speech given on the anniversary of the 'Sword of Jerusalem' battle, this time by Ismail Haniyeh, who asserted that the battle 'erased all geographical barriers of historic Palestine, unified the land, the people, and the Palestinian cause, and touched the entire world.'[33]

The surprise attacks of 7 October 2023, also known as 'Al-Aqsa Flood', followed the same logic as the 'Sword of Jerusalem' battle: the goal was to bring Gaza out of its isolation and reconnect this territory with the rest of Palestine. While Israeli forces and settlers had assassinated more than 400 Palestinians in the West Bank since early 2022, and attacks on Al-Aqsa continued,[34] Hamas once again opened hostilities, displaying solidarity with Jerusalem. It could also count on the armed actions of militants in Jenin and Nablus as a second front capable of mobilizing a part of the Israeli forces. As in 2021 or April 2023,[35] Hamas received support from Palestinians in the West Bank in its strategy of unifying the fronts (*wahdat al sahat*). 'Al-Aqsa Flood' also called upon Arab countries in the 'second circle', pushing other actors like Hezbollah to follow an impulse initiated by the Palestinians in a strategy of multiplying fronts (*jabahat muta'adida*). The capacity of the Al-Qassam Brigades to hold against one of the world's most powerful armies also encouraged other political organizations to confront Israel. This included Shia brigades in Iraq and the Houthis in Yemen, who attacked international trade in the Red Sea to stop aggression against Gaza. Thus, 'Al-Aqsa Flood' first activated a front of resistance that many thought

[32] https://www.youtube.com/watch?v=nkXP6mzMZfI
[33] https://www.youtube.com/watch?v=L0b0uHPrXug
[34] https://www.amnesty.org/en/location/middle-east-and-north-africa/middle-east/israel-and-occupied-palestinian-territories/report-israel-and-occupied-palestinian-territories/
[35] In April 2023, in reaction to the Israeli assault against the Palestinian in the West Bank, Hamas opened a new front from Lebanon, firing hundreds of rockets towards Israel.

was merely rhetorical. However, this regional cooperation was quickly undermined by Israel and its US ally, who together set about to dismantle these fronts. Hamas was thus ultimately confronted with the failure of its strategy.

Conclusion

Looking at the ongoing extermination war in the Gaza strip, Hamas's strategy of uniting the Palestinian ranks though armed struggle has not produced the anticipated results. The opening of multiple fronts did not yield the group's expected outcomes, neither externally—where Hamas had anticipated that the involvement of other actors, notably Hezbollah, would exhaust the Israeli army—nor internally, as mobilizations by Palestinians in Israel and those in the West Bank also fell short of the movement's expectations. However, Hamas was successful in obtaining a ceasefire that allowed the population of Gaza some relief. In January 2025, after months of negotiations through Egyptian and Qatari mediators, the Israeli government agreed to sign a ceasefire agreement which stipulated an exchange of prisoners and hostages and a progressive retreat of the Israel army from the Gaza strip. This agreement, identical to the Biden plan accepted by Hamas as early as spring 2024, can be considered as a backlash for the government of Benjamin Netanyahu who failed to achieve its declared war objectives: rescuing hostages by force and destroying Hamas. While pressures from the US government of Donald Trump in the conclusion of this accord remain debated, it is clear that the ability of the al-Qassam Brigades to hold their ground and inflict substantial losses on the Israeli army was a decisive factor.

In March 2025, Israel broke the ceasefire and embarked on a new phase of the war aimed at starving out the population by blocking aid. It also halted its ground incursions to focus on aerial bombardments. However, this strategy did not prevent Palestinian resistance forces from carrying out daily ambushes, inflicting heavy losses on Israeli soldiers.

Although Hamas was significantly impacted by the losses in both its political and military leadership, the movement has shown a remarkable resilience. On the ground, it continues to dominate the coastal strip from north to south. Despite two years of intensive military attacks by Israeli forces which led to the death of thousands of civilians, the complete destruction of the Gaza strip, and the forceful displacement of nearly all of the population, Hamas is still legitimate in the eyes of Palestinians both inside and outside Palestine. Post-war plans concerning Gaza's future governance unanimously exclude Hamas's presence. While it is too early to determine Hamas's future, it is clear that any American, Israeli, or Arab attempt to destroy or neutralize the movement may face significant challenges on the ground, as Hamas is likely to remain in Gaza.

References

Abou Amer, Adnan. 2017. 'Hamas Relations with the Arab World'. In *Islamic Resistance Movement (HAMAS): Studies of Thought and Experience*, edited by Dr Mohsen Mohammad Saleh, pp. 282–324. Beirut: Al-Zaytouna Center for Studies and Consultations.

Abou Amer, Adnan. 2021. 'The Joint Chamber for Resistance Operations in Gaza: The Seed of the "Liberation Army"' (in Arabic), 28 May, *Arabi21*, accessed 26 March 2025: https://www.aljazeera.net/politics/2021/5/28/غرفة-العمليات-المشتركة-للمقاومة-في.

Abou Amer, Adnan. 2022. 'The Occupation Army Acknowledges its Failure in the Operation Against "Metro Gaza"' (in Arabic), *Arabi21*, 13 April, accessed 26 March 2025: https://arabi21.com/story/1431213/اعترافات-لجيش-الاحتلال-بفشل-عملية-تدمير-مترو-حماس.

Abou Rekbe, Mohamed Mansour. 2021. 'The Marches of Return and Breaking the Blockage: From a Thought to an Open Confrontation 2018–2019' (in Arabic), Gaza: Dar al Kalima.

Al-Hajj Hassan, Saleh. 2021. 'Policy Paper: Sayf al Qods Battle in the Balance'. Al-Zaytouna Center for Studies and Consultations, May 2021, accessed 26 March 2025: https://www.alzaytouna.net/2022/04/29/ورقة-علمية-معركة-سيف-القدس-في-الميزان/#.ZCrdLHZBxD8.

Alijla, Abdalhadi, 2021. 'The (Semi) State's Fragility: Hamas, Clannism, and Legitimacy'. *Social Sciences* 10(437): pp. 1–18.

Alijla, Abdalhadi. 2022. 'Thorny Identity?: Non-state Actors, Service Provision, Identities, and Hamas in Gaza'. *Small Wars & Insurgencies* 34(1): pp. 195–220.

Awad, Hani. 2021. 'Understanding Hamas: Remarks on Three Different and Interrelated Theoretical Approaches.' *Al-Muntaqa* 4(2): pp. 42–62.

Bao, Hsiu-Ping. 2016. *Hamas' Political Transformation and Engagement (2003–2013).* PhD diss., University of Exeter.

Berti, Benedetta. 2015. 'Non-State Actors as Providers of Governance: The Hamas Government in Gaza between Effective Sovereignty, Centralized Authority, and Resistance'. *Middle East Journal* 69(1): pp. 9–31.

Brenner, Björn. 2017. *Gaza Under Hamas: From Islamic Democracy to Islamist Governance.* London: I.B. Tauris.

Brown, Nathan. 2012. 'Is Hamas Mellowing?'. Carnegie Endowment for International Peace, https://carnegieendowment.org/research/2012/01/is-hamas-mellowing?lang=en (accessed 21 February 2025).

Gunning, Jeroen. 2007. *Hamas in Politics: Democracy, Religion, Violence.* London: Hurst.

Hamed, Qossay. 2023. *Hamas in Power: The Question of Transformation.* Hershey, PA: IGI Global.

Hassan, Mohammad Shabrawi. 2021. Can Hamas and other Palestinian Factions Liberate Palestine?, in: *Al-Jazeera*, 13 June, accessed 26 March 2025: https://mubasher.aljazeera.net/opinions/2021/6/13/هل-تحرر-حماس-وفصائل-المقاومة-فلسطين؟ .

Hroub, Khaled. 2000. *Hamas: Political Thought and Practice.* Washington, DC: Institute for Palestine Studies.

Hroub, Khaled. 2017. 'A Newer Hamas?: The Revised Charter'. *Journal of Palestine Studies* 46(4): pp. 100–111.

Jabarin, Zaher. 2022. 'Sayf al Qods Battle … Strategic Plan for Liberation Battle' (in Arabic). *Mayadeen*, 16 May, accessed 26 March 2025: https://www.almayadeen.net/Blog/معركة-سيف-القدس-التخطيط-الاستراتيجي-لمعركة-التحرير.

Klein, Menachem. 1996. 'Competing Brothers: The Web of Hamas–PLO Relations'. *Terrorism and Political Violence* 8(2): pp. 111–132.

Legrain, Jean-François. 2020. 'Le "Document" de Hamas (2017) ou l'ouverture comme garante des invariants'. *Les Carnets de l'IREMAM*, 10 September, https://iremam.hypotheses.org/7122 (accessed 22 February 2025).

Milton-Edwards, Beverley and Stephen Farrell. 2010. *Hamas: The Islamic Resistance Movement*. Cambridge: Polity Press.

Mishal, Shaul and Avraham Sela. 2000. *The Palestinian Hamas: Vision, Violence, and Coexistence*. Updated edn. 2006. New York: Columbia University Press.

Mohammad Saleh, Mohsen. 2019. 'The Marches of Return After One Year'. May. Al-Zaytouna Center for Studies and Consultations, accessed 26 March 2021: https://www.alzaytouna.net/2019/04/04/مقال-مسيرات-العودة-بعد-عام-وقفة-مراجعة/#.ZCvxw3ZBxD8.

Palestinian Legislative Council. 2010. Majlis al-tashri'î al-falastînî, 2010: Arbâ sanawât min al-'ita raghma l-hisâr, 2006–2010 (Four years of sacrifice despite the blockade), Publication of the Palestinian Legislative Council.

Sayigh, Yezid. 1999. *Armed Struggle and the Search for State: The Palestinian National Movement 1949–1993*. Oxford: Oxford University Press.

Sayigh, Yezid. 2011. *'We Serve the People': Hamas Policing in Gaza*. Crown Center for Middle East Studies, Brandeis University, Crown Paper, 5 April 2011.

Seurat, Leila. 2022. *The Foreign Policy of Hamas: Ideology, Decision Making and Political Supremacy*. SOAS Palestine Studies. London: I.B. Tauris.

Signoles, Aude. 2006. *Le Hamas au pouvoir: et après?*. Toulouse: Milan Éditions.

Signoles, Aude. 2010. 'Le Hamas, organisation de résistance ou organisation terroriste?'. In *Résistances, insurrections, guérillas*, edited by Corentin Selin, pp. 55–76. Rennes: Presses Universitaires de Rennes.

Tuastad, Dag. 2021. 'Hamas and the Clans: From Islamisation of Tribalism to Tribalization of Islamism?'. *Third World Thematics: A Third World Quarterly Journal* 6(1–3): pp. 88–104.

10

Recognition on their Own Terms?

The Paradox of the Taliban's Quest for International Legitimacy

Ahmed Elsayed

Introduction

The Taliban first seized power in Afghanistan in September 1996 amid the Mujahideen civil war, following the collapse of the communist regime in 1992.[1] Their government, 'the Islamic Emirate of Afghanistan', characterized by draconian domestic rule (Rubin, 2020, 106), was overthrown by the 2001 US-led invasion on the heels of the 11 September attacks. After two decades of insurgency, the group recaptured Kabul on 15 August 2021, marking a startling turning point in international politics. As the current rulers of Afghanistan, the Taliban actively seek legitimacy in the international system. More precisely they want to be recognized as the legitimate government of the Afghan state. Nonetheless, many of their policies and edicts—especially those related to human rights, inclusive governance, and foreign militants on Afghan soil—create considerable barriers to this recognition. What explains this paradox on the Taliban's side: seeking international recognition while simultaneously undermining the likelihood of its attainment?

This chapter offers an answer to this question. It argues that this apparent contradiction is resolved by acknowledging that the Taliban do not seek international recognition as a top strategic objective. Instead, they seek recognition only *on their own terms* and not at any cost. Put differently, the Taliban are cognisant of the perks of external legitimacy and desire them, but not at the expense of their doctrine—that is, their distinctive religious and identity beliefs. Therefore, they are likely to shun requirements or preconditions for upgrading their international status if these are deemed detrimental to their worldview or ideology. This is not to suggest that the Taliban are not calculating actors. Rather, it is to emphasize that the group's practical reckoning and policy priorities are often defined by the sanctity they accord to their idiosyncratic value system. In other words, drawing on Max Weber's concepts (1978, 24–25), the Taliban's behaviour on the international stage suggests that 'value

[1] Research for this chapter has been made possible by a grant from the German Research Foundation (Deutsche Forschungsgemeinschaft, DFG), project no. 424426045.

Ahmed Elsayed, *Recognition on their Own Terms?*. In: *Armed Groups and the Politics of International Legitimation*. Edited by: Stephan Hensell and Klaus Schlichte, Oxford University Press. © Ahmed Elsayed (2025).
DOI: 10.1093/9780198974529.003.0010

rationality' (*Wertrationalität*) has often taken precedence over 'instrumental rationality' (*Zweckrationalität*) as their guiding compass (Elsayed 2024a, 3–4).[2] As such, a deeper insight into the Taliban's politics of legitimation is possible when their belief and value commitments are factored in.

This chapter analyses the legitimation politics of the Taliban: how they legitimized themselves, de-legitimized their rivals, and how external interventions in Afghanistan contributed to or were shaped by these dynamics. The analysis focuses on the group's ideation-based (doctrine) and performance-based (behaviour and conduct) claims to legitimacy. Using this volume's terminology, this chapter examines how the Taliban leveraged *fighting*, *voicing*, and *mimicking* as practices to claim legitimacy and explores the extent to which *acknowledging*, *supporting*, and *officializing* by certain regional and international actors have directly or indirectly served to validate these claims.[3]

A key observation of this chapter is that, unlike most other armed groups governing regional or national territories, which struggle to adapt their legitimacy claims when addressing disparate domestic and international audiences,[4] the Taliban seem to have relieved themselves of this burden. Since their inception, they have anchored their legitimating ideas and practices in what they perceive as true Islam and Afghan identity. Jihadi and Islamic legitimacy, among their staff, foot soldiers, and supportive local constituencies, has consistently remained their top priority. According to Anas Haqqani, a senior Taliban leader, the essence of legitimacy in Afghanistan is 'religious credibility'; therefore, as he sees it, the Taliban have 'the strongest ... religious and popular legitimacy' (Jalal Foundation 2023, 26). In what follows, the group's legitimacy claims and doctrine—examined in more detail later in the chapter—are briefly sketched and contextualized.

Afghanistan has long been described as a 'religious nation' (Long 2022, 61; see Al-Afghani 1879, Chapter 3). Wars with imperial invaders, such as the British and the Russians during the 19th and 20th centuries, helped instil in many Afghans a scepticism towards foreign cultures, secular modernity, and Western hegemony (Kakar 2022).[5] For many years, the presence of the Afghan government in the countryside, where most of the population resides, was largely symbolic, allowing regional and tribal customs and affiliations to remain undiluted (Rubin 2020). Elders and

[2] According to Weber, instrumentally rational behaviour is utilitarian, focused on the 'means' and the 'ends', and guided by the maximization of benefits. In contrast, value-rational behaviour is devotional, driven by fidelity to intrinsic beliefs. This distinction is purely analytical; in practice, these rationalizations are *inseparable*. It is important to note that the issue is not whether the 'value' in question is moral or immoral, commendable or condemnable, or deserving of advocacy or resistance. Rather, the emphasis lies in highlighting its ideational basis, as opposed to the material foundation characteristic of the other form of reasoning.

[3] For a detailed discussion of this conceptual vocabulary, see the Introduction to this volume.

[4] On this problematic see the chapters by Christian Olsson on Fatah and Leila Seurat on Hamas in this volume.

[5] The pursuit of modernist reforms, particularly those aimed at Afghan women, cost King Amanullah Khan his throne in 1929 and later galvanized widespread armed rebellion against the Afghan communist regime.

Mullahs (clerics), who were revered and respected, guided the people's everyday life and arbitrated their disputes (see Johnson et al. 2017, 25; Thabet 1934, 228).

The Taliban's doctrine originated in this socio-cultural milieu. During the Mujahideen civil war, the Taliban (religion students) emerged in 1994 as a rural Pashtun vigilante movement adhering to a strict form of religious orthodoxy and conservatism. This orthodoxy has been shaped by two interrelated epistemological streams. The first is the 'Deobandi' teachings—a variant of the Hanafi school of Sunni Islam—which the Taliban received in *madrasas* (religious schools) located in refugee camps and rural areas of Pakistan and Afghanistan.[6] The second is the tribal mores of Pashtuns, known as 'Pashtunwali'. Egyptian Islamist writer Fahmy Howeidy (2001, 105–6) described these madrasas as places focused on imitating ancestors, rejecting modern sciences, fixation on individual behaviour and outward appearance, and fostering strong loyalty to religious leaders. In this environment, the Taliban, according to Pakistani journalist Ahmed Rashid (2010, 33), were taught 'primitive Islamic injunctions'. The confluence of certain Deobandi/Hanafi interpretations and Pashtun principles intensified particular traits of the Taliban, including their stance on women's participation in public life.[7] In both traditions, the ideal place for women—unless there is a compelling reason to leave—is within the home. Still, the value system of the Taliban represents neither Pashtuns nor Hanafis in their entirety. Instead, to this day, it reflects a complex amalgam that both intersects with and diverges from Sufism, Salafism, political Islamism, and jihadism. This hybridity has been influenced by the lived experiences of the group's members, marked by continuous warfare, limited access to formal education, hardships and poverty in remote and arid rural areas, and, later, interactions with foreign Islamist currents and jihadi militants.

Overall, from the outset, the Taliban have claimed that their doctrine represents the pathway to establish what they perceive as an ideal Afghan and Islamic society and government, wherein all facets of life are regulated by Sharia as they interpret it.[8] Anchored in this conviction are the group's ideational and performance-based claims to legitimacy as guardians of Afghanistan's religion, culture, and honour, fighting against what they perceived as the immorality and injustice of foreign-imposed rule. These claims have remained remarkably consistent over the years and were

[6] Deobandism originated in the 1860s as a pedagogical movement in colonised India. It sought to revitalize and preserve Islamic values and identity against British and foreign influences. Over time, however, later Deobandi madrasas diverged significantly from the intellectual vitality and reformist agenda that defined their early predecessors.

[7] Limitations on women's rights in Afghanistan extend beyond Pashtuns (see Emadi 2002; Malikyar 1997, 396).

[8] Yusuf al-Qaradawi (2020 [2001], [25:23–40:40]), a renowned Islamic scholar who debated the Taliban Ulama in Kandahar in 2001 and is regarded by many Muslims as moderate yet often seen in the West as extremist, opined that the Taliban 'would like to do everything according to the *fatwas* [legal opinions or edicts] of the Sharia', but 'their problem', is that 'they live in the past more than in the present'. According to *Fiqh al-Waqi'* (the jurisprudence of current affairs and circumstances), fatwas must consider present contexts to remain relevant and accurate. In this regard, scholars like al-Qaradawi argue that the Taliban's detachment from contemporary realities undermines the validity of their fatwas.

particularly central to the domestic success of the Taliban's insurgency (2001–2021). For many insurgents, the ideological antagonism to their Western enemies turned the battlefield into a sacred space where the fault lines between good and evil were clearly demarcated. These fighters harboured little doubt about the righteousness of their resistance. Consequently, the war continued unabated until they had outgoverned the Western-backed Kabul regime, outlasted the US military occupation, and regained control of Afghanistan in 2021.

However, ideological commitments that served the group well on the battlefield were and still are a source of contention in the international corridors of power. Here arises the contradiction which is at the heart of the Taliban's government: While the Taliban view international recognition as their lawful entitlement, based on their domestic power and effective control of Afghan territory, their normative agenda—as manifested in their policies and proclamations—neither aligns with the rights and aspirations of diverse Afghan societies nor with prevailing material and moral considerations of the 'international community'. For example, despite pledging to prevent Afghanistan from serving as a base for threats against other nations and to form an inclusive government, the Taliban continue to shelter foreign militant groups and maintain exclusive rule. This is because, in their view, expelling fellow jihadis and sharing power with others are incompatible with their doctrinal purity and sovereignty. Also, contrary to the basic principles of modern governance, the Taliban appear to believe that the objectives of their version of Sharia are best achieved through a system in which ordinary citizens are denied the right to choose their leaders or hold them accountable, while women are subjected to systematic discrimination.[9] Despite international condemnations, the Taliban have maintained their unyielding stance. On multiple occasions, the group's current *amir* (supreme leader), Hibatullah Akhundzada, warned the outsiders, 'I will not move even one step with you or interact with you ... at the cost of this Sharia' (Gul 2023). He further assured his followers, 'We pledged to Allah ... [to] refrain from taking any action that endangers, undermines Islam' (Tolo News 2023).

Having very briefly outlined the Taliban's doctrine, it is important to consider some methodological points. First, this study is concerned not with the Taliban's genuine beliefs, but rather with what they *claim* to believe. Moreover, to circumvent orientalist and Western-centric readings of Afghanistan and the Taliban (see Manchanda, 2020), this enquiry focuses primarily on the group's own statements rather than the others' portrayals of them. Accordingly, the collected data include primary sources of the Taliban's discourse, such as public pronouncements and the online Arabic magazine *al-Somood* (perseverance or steadfastness), as well as non-Western critical analyses of the group, such as works by Mawlawi Hafizullah Haqqani (1997) and Ahmed Rashid (2010). The research is further informed by ten in-depth interviews conducted between October 2022 and February 2023.[10]

[9] UN experts opine that the comprehensive restrictions on women in Afghanistan amount to 'gender apartheid' (UN Press 2023).

[10] These interviews and three informal discussions were conducted with twenty participants in English and Arabic, both in-person, in Germany, and online via Zoom, and through WhatsApp calls and audio

In addition, it draws on governmental statements, secondary sources, and Western scholarly and media coverage of the Afghan conflict. Second, it is increasingly evident that the Taliban's doctrinal interpretations remain largely unchanged from the 1990s (Ibrahim 2021). However, this does not mean that the group's ideology and politics are entirely immutable (Gopal and van Linschoten 2017, 44). Instead, it suggests that most changes in the second Emirate (2021–present), compared to the first (1996–2001), are likely to be more cosmetic, rhetorical, technical, or tactical than structural or strategic. That is, while the Taliban's core beliefs appear consolidated, their peripheral perspectives and practices might remain open to incremental revisions under conducive social conditions, including popular dissatisfaction and contentious internal debates. For instance, the second Taliban Emirate has prioritized economic development and adopted a relatively less crude foreign policy in contrast to the first.[11] Although this reflects the group's capacity to adapt based on previous mistakes, it also highlights what kind of insights they have drawn from their past. Fundamental questions—such as those concerning political participation, dissent, and individual rights and freedoms—have yet to register, if ever, as 'lessons learned'. Furthermore, there is some validity to the assertion by Taliban co-founder Mullah Abdul Salam Zaeef's (2010, 153) that dividing the group into moderates and hardliners is not productive. Nonetheless, the Taliban, like other political groups, comprises leaders and members whose positions reflect varying shades of dogmatism and politicization. Although internal diversity typically remains behind closed doors—occasionally surfacing in public statements—its contribution to policy-making is not guaranteed, as ultimate authority and decision-making power rest firmly with the amir.[12] While some analyses rightly address divisions and disagreements within the Taliban, they often overlook the pivotal role of the *Bay'a* (oath of allegiance and obedience) to the amir in averting full-scale schisms or rebellions.

The remainder of this chapter is structured as follows. It starts with a review of the rise and fall of the Taliban's first Emirate, exploring how the group prioritized their identity and religious beliefs over the survival of their government. Second, it examines how the insurgent Taliban legitimized themselves and de-legitimized local and foreign rivals. Third, the group's precarious international standing is scrutinized. This section presents key external validations of the Taliban's status over time and regional reactions to the revival of their 'jihadi nationalist' government in 2021.[13]

messages. Participants included former and current Afghan government officials, Afghan migrants and experts, as well as Western scholars and analysts.

[11] The Taliban have retained most of their unconventional social restrictions from the past while introducing new ones of a similar nature (see, for instance, Oppenheim 2025).

[12] See Abubakar Siddique's (2024) report on the Taliban's internal discord over the ban on women's education.

[13] The Taliban describe their regime and mission as 'jihadi' (see Yaqeen Network 2024a: [02:30]). The Taliban's chief justice, Abdul Hakim Haqqani (2022, 177), views the purpose of the Emirate's military as serving 'a jihadi function'. The term 'nationalist' highlights the Taliban's focus on Afghanistan, in contrast to other transnational jihadi groups, such as al-Qaeda Central and the Islamic State, whose operations and purposes are not defined by national borders. This contrast is not new within the Salafi jihadi milieu, where the debate on whether to prioritize targeting the 'near enemy' (e.g. domestic ruling elites) or the 'far

It also discusses why widespread official recognition of this government remains elusive. Finally, the chapter concludes by briefly contrasting the Taliban's quest for international legitimacy with that of other jihadi and nationalist armed groups and providing a glimpse into their government's future trajectory.

Rise and demise of the first Emirate

The Taliban established themselves in 1994 with the initial aim of restoring order and peace in Kandahar that had been ravaged by lawlessness during the Mujahideen civil war. The group's religious nature and ability to suppress criminal activities and reopen transportation routes for people's movement and trade garnered significant support from Afghans and neighbouring Pakistan, which sought commercial opportunities in Afghanistan and Central Asia. This backing fuelled the Taliban's victories across Afghanistan, culminating in the conquest of Kabul and the establishment in September 1996 of the regime that would later be known as the Islamic Emirate of Afghanistan.

Two months earlier, at an assembly of *Ulama* (religious scholars) in Kandahar, Mullah Mohammed Omar, the Taliban's founder, proclaimed the group's ultimate goal to be the implementation of the rules of Islam. In response, the enthusiastic crowd bestowed upon him the title of *Amir ul-Mo'mineen* (Commander of the Faithful), a historic designation previously held by caliphs, who wielded supreme spiritual and temporal authority over the Muslim community. Perceiving themselves as the interpreters and enforcers of divine authority, the Taliban instituted exceptionally severe decrees. The shocking practices of the group's morality police, called *Hisba* or *al-Amr bil Ma'roof* (propagation of virtue and prevention of vice) attracted significant negative international press coverage. This department carried out public amputations, stonings, and lashings of ordinary people for various, often peculiar, offences. Women in particular were subjected to inhumane treatment and oppressive restrictions. The Hisba practices were so harsh that they drew condemnation not only from human rights advocates but also, at times, from allied Arab militants. For instance, jihadi strategist and Taliban supporter Abu Musab Al-Suri (1998, 31–2) criticized the 'repulsive roughness' and 'un-Islamic' behaviour of some Hisba squads, attributing them to the 'ignorance or harshness ... of [some of] the Pashtuns of Central and Southern Afghanistan'.

On their path to power, the Taliban showed competence in navigating the complex domestic Afghan landscape and deploying political strategies to further their military goals against rival Afghan factions. They mobilized co-ethnics, occasionally

enemy' (e.g. Western powers) has persisted for decades. As the occupation of Afghanistan (2001–2021) brought the near and far enemies together, the reciprocal influence between Salafi jihadism and the Taliban became most evident. Finally, one should bear in mind that the term 'nationalist' does not negate the presence of global aspirations or hopes. Rather, the apparent disregard for such aspirations arises from necessity rather than choice, as the jihadi ethos is intrinsically tied to the *Ummah* (the global community of Muslims), regarded as the ultimate nation, to be ruled by one system—the *Khilafah* (the Caliphate).

co-opted former communists, incentivized defections and demobilizations of enemy commanders, and rewarded loyal local power brokers (Sinno 2009, 59–89). Nonetheless, the group's grasp of foreign relations and state politics was limited (Maley 2006, 10; Cristol 2019, 26–32; Harpviken 2021, 2). Al-Suri (1998, 29) observed that the group demonstrated a 'general ignorance of world affairs'. Indeed, the Taliban, at least in their earliest years, did not seem to concern themselves with the world beyond the Islamic countries. A summary of a rare introductory pamphlet of the group—penned for them by Pakistani scholar Sher Ali Shah—highlighted the Taliban's intention to 'deepen' and 'improve' relations with 'all Islamic countries and organizations' against the background that 'arbitration in all political and international issues would be based on the Book [Quran] and the Sunnah [the Prophet's words and deeds]' (Haqqani, 1997, 108–9). No mention was made of other countries. This would change after 1996 as the newly ruling Taliban sought to establish ties with international organizations, oil corporations, and foreign states. Still, the group's insular approach and mistrust of external agendas would lead to clashes with actors in their region and beyond. One issue is that the Taliban appeared to treat Afghanistan's borders as a demarcation of their obligations and prerogatives. They disavowed responsibility for external matters and asserted their exclusive authority over internal affairs, even though these two domains often overlapped (see Zaeef 2010, 119). Put differently, the Taliban did not seem adequately considerate or apprehensive of the potential challenges arising from their adherence to certain values, policies, or goals within Afghanistan which could pose security threats or other anxieties to outside nations.

A salient example in this regard is the sheltering of Osama bin Laden and other foreign fighters, many of whom took part in the 1980s anti-Soviet resistance. From the early days of their relationship, Taliban officials provided assurances of religious solidarity and fraternal support to bin Laden and the Arabs around him (Al-Suri 1998, 34). While it is true that the Taliban's connections to al-Qaeda and other non-Afghan fighters can be viewed as driven by expediency, this perspective fails to fully capture the Taliban's reasoning behind hosting and protecting these militants.[14] After all, despite the enormous costs—including the loss of their regime and lives in 2001 due to their association with al-Qaeda—the Taliban remained steadfastly aligned with the transnational group.

With the 11 September attacks, the fate of the Emirate hung precariously in the balance. Shortly thereafter, George W. Bush (2001) issued an ultimatum to the Taliban: 'hand over the terrorists or … share in their fate'. Beyond Afghanistan, Bush declared, 'Every nation … now has a decision to make. Either you are with us, or you are with the terrorists' (AP Archive 2015: [02:16]). In stark contrast to countries like Pakistan, whose instrumental reckoning led them to embrace the so-called 'War on Terror' to placate the US and reap potential rewards (see Musharraf 2006, 201), the Taliban's doctrinal calculus guided them along a divergent path. The US threat invoked the

[14] Foreign jihadis fought alongside the Taliban, and al-Qaeda assassinated the group's domestic arch-rival Ahmed Shah Massoud (CBC 2021).

group's perennial dilemma: choosing between the survival of their government and the preservation of their ideological beliefs.[15] 'Osama issue is our faith issue, and we are not going to change our faith for anyone', declared Mullah Zaeef, then the Taliban ambassador in Islamabad, as the US invasion to uproot al-Qaeda and the Taliban was underway (AP Archive 2001: [01:30]). Mullah Omar based his stance on the fatwa from the Taliban's Ulama that bin Laden could not be expelled from Afghanistan: 'Islam and the honour of the Afghans do not allow us to meet [the Americans'] demands. [...] We understand that they can badly damage us but their demands are more devastating for our religion and faith' (Mutma'in 2022, 229–30). Under Mullah Omar's leadership, the Taliban made the ultimate sacrifice for their beliefs, resulting in the swift collapse of their Emirate after the US invaded Afghanistan on 7 October 2001.

On home turf: the insurgents' legitimacy claims in battle and governance

In late 2001, the Taliban were thought to be finished. However, they emerged from adversity, staged a protracted insurgency, and ultimately forced the US to come to terms with them. This culminated in the 2020 Doha Agreement, which marked the conclusion of negotiations on the US withdrawal from Afghanistan. The Taliban earned this transformative legitimation after they had proved their capabilities in both battle and governance. This section examines the discourses and practices that underpinned the Taliban's legitimacy claims—as expressed through armed actions and local governance—which re-established the group as the foremost Afghan political power.

Although external conditions, such as Pakistani sanctuaries and détentes with Iran and Russia, aided the Taliban insurgency, they alone cannot explain how the Taliban recovered from a shattering defeat to reconquer Kabul in 2021. In fact, the advent of some of these exogenous factors was contingent on the Taliban 'demonstrat[ing] some capacity to fight' (Giustozzi 2019, 31) and establishing themselves as a dependable military force in Afghanistan. Fighting, as elucidated in this volume's Introduction, was the Taliban's principal practice to promote and legitimize what they called 'defensive jihad' against the US. Moreover, the significance of these external factors only grew over time. To paraphrase, these factors held limited explanatory power for the early stages of the Taliban insurgency, when the stakes for the rebellion were exceedingly high, and failure seemed almost certain. This raises two key questions: what motivated the surviving Taliban to resist the US military, knowing that their death was highly likely,[16] and why did many Afghans welcome the Taliban back into their regions?

[15] Although some Taliban leaders were reportedly more inclined to distance the group from al-Qaeda (Loidolt 2022, 138–142), others, including the amir, held a different view.

[16] The Taliban were largely a volunteer army.

Part of the answer to these questions lies in the Taliban's legitimating ideas and actions, which tapped into Afghanistan's history, culture, and religion. Among these values were resistance to occupation, defence of the Afghan way of life, martyrdom, and jihad. These ideals were seemingly precious goals for the Taliban, and the group pursued them relentlessly, undeterred by the looming costs. The insurgents' performance on the battlefield and in governance reified these values, demonstrating to Afghans that resistance is the way for those unwilling to live a 'reduced mode of being' (Varshney 2003, 93) and endure humiliation under foreign authority. Another part of the answer involves the popular perceptions of the internationally-recognized Afghan Republic's moral standing and functionality. This is important because understanding rebels' legitimacy necessitates incorporating their opponent regime in the picture. The legitimation of the self and the de-legitimation of the enemy are two sides of the same coin.[17] The relevant point here is how the Taliban discredited Kabul and its foreign sponsors. Although the US, Kabul, and the Taliban committed numerous atrocities against civilians,[18] the group's legitimacy claims and narratives seemed to resonate more strongly with the values and interests of a significant local constituency (Johnson et al. 2017, 2–14). While the Taliban claimed to be fighting to liberate Afghanistan and restore a native Islamic government, the US claimed to be assisting the Afghans in building a modern liberal democracy. Over time, the Taliban's words and deeds appeared notably coherent, whereas those of the US did not.

The Taliban's self-legitimation: fighting, voicing, and state mimicking

The Taliban continued to mount a fierce insurgency despite sustaining heavy causalities. This won them the respect of many countrymen and even some enemies. '[T]his is not about money or jobs. The Taliban are fighting for something larger', stated US diplomat Michael McKinley in October 2014 (Malkasian 2021, 5). Later, as he sat across the negotiation table from the Taliban in Doha, US General Austin S. Miller conveyed that 'he respected them as fighters' (Mashal 2019). Yet, why the Taliban's fighting spirit seemed exceptionally unrelenting has been largely overlooked in most Western analyses of the Afghan conflict. The failure to understand this crucial dimension of the insurgency stemmed from a broader negligence of acknowledging the legitimating power and role of identity and religion in Afghanistan. Carter

[17] See the author's theoretical discussion on this topic (Elsayed 2024b, 20, 27, 44–45) and a related empirical case study examining the Syrian Kurdish YPG/PYD and Turkey (Elsayed 2023, 72).

[18] Distinctions between those whom the US considered terrorists and their sympathizers were blurred, enabling large-scale human rights violations and war crimes against Afghan civilians. The torture of insurgents by Americans and allied Afghans became routine in detention centres like Pul-e Charkhi, Bagram, and Guantanamo (see Human Rights Watch 2004). On the other side, the Taliban assassinated and intimidated civilian officials, religious scholars, and individuals who openly opposed their ideology and political agenda. The Taliban viewed opponents as collaborators with the occupiers, labelling them as apostates and deeming their killing permissible.

Malkasian's (2021) *The American War in Afghanistan* offered a valuable contribution to this particular debate. Malkasian (2021, 6) found it perplexing that 'the possibility that Islam and resistance to occupation played a role in America's Afghan War has gone oddly unnoticed, almost shunned—in a country where ... religion defines daily life'. However, the assumption here is not that all the Taliban were true believers or that their ideational claims were universally sincere. None of these is a must for ideas and beliefs to impact the conflict and its parties in significant ways (Maynard 2019). Therefore, bearing in mind George C. Marshall's words that it is '*morale* that wins the victory' (Vaughn 1983, 28), the exploration of the Taliban's stated beliefs merits attention.

The group declared that its war against the US is a continuation of the Afghan nation's historical confrontations with the British and Soviet empires (see Mawlawi Shamsuddin in Abu Abed 2014, 15; Baradar 2023). By portraying the conflict as a struggle for national independence, the Taliban sought to voice the legitimacy and nobility of their cause to the public. They presented themselves as the vanguard against the US attempt 'to obliterate the Afghan identity' (al-Sharky 2006, 24) and its 'injustice and marginalization' towards the Pashtun community (Al-Somood 2006, 1). During the early insurgency years, the Taliban self-identified as 'Mujahideen' and 'the Islamic Movement of the Taliban'. However, in October 2007, they announced the 'restoration of the Islamic Emirate's government' on 44% of the country's territory (Maiwandi 2007b). The Taliban have since adopted the primary self-designation of 'the Islamic Emirate', signifying a resurrection of the previous order, overturned by foreign armies and domestic rivals in 2001. Subsequent events would prove how the group deemed this name, along with the white flag inscribed with the *Shahada* (Islamic testimony of faith) as crucial legitimating tools. In 2013, in their quest for mimicking the state in order to be perceived as one, the Taliban refused to relinquish their regalia of statehood—the name and the flag—resulting in the closure of their short-lived diplomatic office in Doha. Similarly, despite clear objections from the UN Security Council (2020) and the Troika (China, Russia, and the US), the Taliban disregarded the calls to forgo 'the restoration of the Islamic Emirate' upon taking control of Kabul (State Department 2021).

Furthermore, the Taliban's assaults on Afghan and Western troops, characterized by high risks and minimal chances of survival for the attackers, alongside their utilization of suicide operations, served as evidence that the group acts in accordance with its professed convictions. Longing for martyrdom—a 'pure form of value-rational behaviour' (Varshney 2003, 94)—has been a major theme in the Taliban's discourse. The Amir Akhundzada (2016, 3) asserted to 'the American occupiers' in July 2016 that the Afghans 'consider martyrdom in confrontation with you a cherished goal of their life'. On the battlefield, the Taliban's death-embracing attitude sharply contrasted the focus of Western forces on having no casualties.

Besides the insurgents' fighting morale, the Taliban leadership communicated in words and deeds its cohesion and unity. Although the group consisted of multiple nodes of command and control, the amir remained the final arbiter of power. Within the Taliban, the amir is considered to be the successor of Prophet Mohammed.

Consequently, under normal circumstances, nothing can absolve the members from obeying him as long as he remains committed to the rules of Islam as they understand them. Put differently, they must follow the amir's command even if they disagree with it.[19] Sirajuddin Haqqani, then the Taliban's deputy amir, stated that his father, the powerful commander Jalaluddin Haqqani, believed that 'unity was the reason why the whole world failed against Amir ul-Mo'mineen' (Al-Somood 2016, 6). Gradually, the Taliban's parallel state consolidated on the ground. Their relatively efficient and accessible policing and Sharia courts notably solidified their territorial control, boosted their local legitimacy, and contributed to their eventual success (see Baczko 2021). Nevertheless, the insurgents were exhausted. By late 2000s, the Taliban's more strategist leaders were empowered to launch diplomatic and media charm offensives and explore troop withdrawal talks with the US. The group's approach to negotiations was later described by Mullah Akhtar Mansour, the group's second amir, as follows: 'We—besides the armed struggle—engage in politics ... according to the conditions defined by the Islamic Sharia, ... we will not come to the negotiation table under external pressures ... those who want to reconcile with us must submit to our demands' (*Al-Somood* 2015, 4).

The de-legitimation of the opponent: the image crisis of Kabul

The initial optimism surrounding the US 'democratic' project in Afghanistan quickly waned. The legitimacy of that endeavour was questioned as the US aligned itself with 'the worst of the worst' of corrupt local warlords and strongmen (Coll 2018, 268), turning a blind eye to their crimes against the Afghans.[20] Many people saw the Afghan police as 'the main criminals' (Rubin 2007, 67) and the court system as a marketplace where justice was up for sale. As former US State Secretary Mike Pompeo (2023, Chapter 16) put it, in the Afghan Republic, 'corruption was "a feature, not a bug", as it was all that held the government together'.

The Taliban's de-legitimating discourse aimed at the Afghan Republic had a corrosive and damaging effect. The group attempted to expose Kabul's malperformance and subservience to the Americans at every possible occasion. For the Taliban, this government was the 'stooge' or 'corrupt' Kabul 'administration', and the Afghan presidents Hamid Karzai and Ashraf Ghani were just 'puppets' or 'agents' of foreign masters. Both leaders were frequently ridiculed for their lack of real authority and incapacity to stand up to the transgressions of the US military (Maiwandi 2007a, 23; Mumand 2018, 19). Kabul failed to refute the Taliban's portrayals of it. It could not

[19] Quranic directives, such as 'Obey Allah and obey the Messenger, and those in authority among you' (Ayah 59, Surat al-Nisa'), combined with lessons from the Mujahideen civil war, seem to have instilled obedience, submission, and loyalty within the Taliban as key to victory. The group's leaders often cite this Ayah to remind followers of their obligation to obey their superiors (see Yaqeen Network 2024b: [05:00]).

[20] These crimes include multiple documented cases of child sexual abuse and torture (see Gossman 2019).

protect small farmers from the economic harm of flawed American anti-narcotics programmes or prevent US military night raids and airstrikes that killed thousands of civilians. Indiscriminate US counter-terrorism military operations undermined Kabul, alienated allied Afghans, including President Karzai, and contributed to the legitimation of the Taliban's insurgency. In addition, the electoral process in Afghanistan was riddled with fraud, where the presidency's winner was determined by external actors not Afghan voters (see UChicago, 2023: [32:00]; SIGAR 2022, 36). Foreign funds that poured into Afghanistan favoured a few select beneficiaries, while the average Afghan continued to struggle with insecurity and poverty. Widespread discontent with Kabul was compounded by crimes committed by US soldiers, such as the 2012 Quran burning, which bolstered the Taliban's narratives about the occupation and the war on Islam. In this context, more Afghans supported the insurgency, seeing the Taliban as protectors who punished predators, avenged the grievances of vulnerable peasants, and defended Islamic and national values.

Internationally, the Taliban's most impactful strategy to discredit the Republic was their refusal to acknowledge Kabul as even a party to the conflict. The Taliban's logic was that the principal concerned parties were exclusively themselves and the US, and not 'the traitor ... administration, which is but the new-born of the Crusader occupation' (Mukhtar 2007, 24). The Taliban firmly rejected engaging in a dialogue with the Republic in order not to 'confer legitimacy on [its] puppet government' (Ghaznawi 2009, 14). During peace negotiations with the Americans, the Taliban maintained the exclusion of Kabul from the talks as a non-negotiable condition.

The international realm: the Taliban's legitimacy gains and losses

Since 2001, the Taliban have made considerable strides in their quest for international legitimacy. They rebounded from a defunct authority to a de facto government with multiple international diplomatic ties and connections. Nonetheless, the Emirate still lacks definitive official recognition as a legitimate state actor. This section presents instances of external validations that bolstered the Taliban's claims to be a rightful resistance (as an insurgency) and a legitimate sovereign (as a government) in Afghanistan. Potential factors contributing to the group's lack of an explicit formal status in international politics are then examined.

External validations of the Taliban's legitimacy claims

Between 2001 and 2024, the Taliban garnered varying degrees of external support, acknowledgement, and officialization. Each instance of validation—regardless of its motivation—contributed to advance the group's domestic and international standing. One of the earliest and most consequential earnings of the insurgent Taliban was the revival of Pakistani support by 2003.[21] After a brief time of repudiating the defeated Taliban, the Pakistanis rekindled their support for the insurgency to counter India's politics in Afghanistan. Over time, Islamabad has been the Taliban's most reliable and sometimes sole ally. Pakistani assistance took multiple forms and came from different constituencies within the country,[22] but most importantly, the Taliban received sanctuary, logistics, and military aid (Giustozzi 2019, 31). It is conceivable that the absence of Pakistan's help would likely have hindered the restoration of the Emirate. Upon the return of the Taliban to power in 2021, Islamabad urged the world to 'be realistic' and 'engage' with them (Al Jazeera 2021). Similarly, around 2005, Iran started providing training and funding while leveraging local networks in Western Afghanistan to help the Taliban. Over the next two years, the Taliban's growing strength prompted a clandestine dialogue between the group and German, Norwegian, and UK officials (see Harpviken 2021).

Moreover, the insurgents' opposition to the presence of both the US and the 'Islamic State—Khorasan Province' (ISKP) in Afghanistan expanded their regional approval and transformed former adversaries, such as Russia, into a friendly audience (Mujahid 2016, 6; Reuters Staff 2015). ISKP endangered regional security and challenged the Taliban's religious and jihadi credentials. Initially, in June 2015, the Taliban made a fraternal appeal to Abu Bakr al-Baghdadi, the IS supreme leader, not to threaten 'the unity of the jihadi rank and its cohesion against global *kofr* [apostasy]' in Afghanistan (Mansour 2015, 2). The failure of this appeal propelled the Taliban to turn against ISKP with full force. This development aligned the insurgents paradoxically with Kabul, Washington, Tehran, and Moscow. The alignment further fuelled the post-2001 regional trend in Central and South Asia towards a gradual acceptance of the Taliban as a reality in Afghan politics.[23] Although the Taliban's foreign militant allies continued to evoke apprehension among neighbouring countries, the threat from ISKP overshadowed that from others. The insurgents seized this opening to boost their legitimacy by undertaking serious anti-ISKP operations while simultaneously doubling down on their positive self-representation as a nationalist organization that has no aspirations beyond the Afghan borders. The Taliban utilized new communication tools and contemporary political discourse of sovereignty, neutrality, regional stability, economic cooperation, and non-interference in others' affairs, to voice the legitimacy of their agenda. This pragmatic and politically nuanced language marked a significant departure from the group's earlier, purely dogmatic and dismissive stance during its first tenure in power towards its neighbours' cross-border concerns and demands. Nonetheless, whereas most regional

[21] Pakistan was the first country to officially recognize the Taliban's first Emirate in 1997, preceding Saudi Arabia and the United Arab Emirates. However, the Taliban's alliance with al-Qaeda and the imminent US attack led all three countries to withdraw their recognition.

[22] The relationship between Pakistan's official authorities and the Taliban was not immune to double-crossings and back-stabbings.

[23] In contrast, during the 1990s, several regional countries—including Russia, Iran, India, Tajikistan, and Uzbekistan—allied themselves with anti-Taliban Afghans before joining the US-led coalition against the group. This regional stance was a response to the Taliban's avowed sheltering of foreign jihadis who aimed to destabilize or overthrow ruling regimes across the region.

capitals—with the exception of Islamabad—did not favour the prospect of returning to a Taliban-ruled Afghanistan, the contacts that these capitals established with the insurgent group seemed to prepare them for this looming reality.

Furthermore, the failure of the 2009 US troop surge to quell the insurgency led President Barack Obama to recognize diplomatic talks with the Taliban as the viable path forward. That is, the group's perseverance on the battlefield compelled the US to shift from 'we will never negotiate with terrorists to we must negotiate with terrorists' (PBS NewsHour 2021: [12:50]). Preliminary talks between the US and the Taliban commenced on 28 November 2010, in Munich, Germany, paving the way for the group's first major legitimacy breakthrough—the establishment of their US-approved representation office in Doha on 18 June 2013. As noted previously, the office was swiftly closed after the Taliban refused to surrender their symbols of sovereignty and statehood. Still, this episode was a turning point that signified the group's political weight and broke their international isolation.

Amid the Taliban's military resilience and Washington's changing administrations and priorities, negotiations were resumed, culminating in the signing of the Doha Agreement on 29 February 2020. This accord was a watershed moment, in which the US officialized the Taliban as a key party to the conflict and peace in Afghanistan. It validated the group's claim to be a legitimate political power in the country and shattered the image of the Kabul government. The Agreement catalysed the Taliban's subsequent diplomatic ascendancy that was furthered with the announcement of the second Emirate in August 2021. Overall, the Taliban secured this recognition in Doha without compromising their distinct normative beliefs, including their associations with groups like al-Qaeda and their vision for the future government in Afghanistan. The Doha Agreement thus represented a recognition achieved largely on the Taliban's own terms.

Return of the jihadi nationalist government: regional reactions

The US holds a significant importance for the Taliban, with few other countries possessing a comparable capacity to impact them and Afghanistan.[24] However, the US and the West have been distant and hesitant in their interactions with the Emirate. Differently, regional powers and countries, including China and Russia, embraced the return of the Taliban. They kept their embassies operational in Kabul and hosted Taliban-designated diplomats on their soil. Beijing, for instance, appointed an ambassador to Afghanistan in September 2023 and accredited a Taliban ambassador in January 2024.[25] To Anas Haqqani, these facts constitute a 'silent' international recognition of the Emirate (Jalal Foundation 2023, 26).

[24] The meaning of 'jihadi nationalist' is elaborated in footnote 13.

[25] China has refrained from explicitly characterizing these diplomatic developments as a formal recognition of the Emirate (Al Jazeera 2023; MFA China 2024).

Indeed, neighbouring countries to Afghanistan have incrementally been integrating the Taliban authority into the regional landscape, and in the process acknowledging its governmental status. They often refer to the Taliban as 'the Afghan interim government' (MFA China 2023) or 'the interim Afghan authorities' (MFA Uzbekistan 2023). The neighbours also engage in routine bilateral meetings with Taliban officials, who are frequently invited to regional forums like the Tashkent International Conference. High-profile receptions and visits have become customary regional gestures towards these officials. For instance, on 24 March 2022, China's foreign minister Wang Yi met with the 'Acting Foreign Minister of the Afghan Interim Government', Amir Khan Muttaqi, in Kabul (MFA China 2022). Also, on 6 July 2022, the Amiri Diwan (2022), the seat of the Qatari government, announced the Qatari amir's meeting with 'HE Defense Minister in the Afghan caretaker government', Mohammad Yaqoob. In addition, on 4 June 2024, the Emirati official news agency publicized the meeting between President Mohammed bin Zayed and Sirajuddin Haqqani with the headline, 'UAE President receives Afghan Interior Minister' (WAM 2024). Unfettered by the ideological contradictions between the Taliban and the West, regional states eye the potential benefits of Afghanistan's stability and inclusion into regional development and economic projects. Predictably, the ruling Taliban's primary legitimacy claims tap into these states' expectations. According to Muttaqi (2023), the Emirate is a 'powerful central government' under which a 'stable and secure Afghanistan has the potential of playing a pivotal role' in the region. He warned that an alternative to Taliban rule could lead to 'regional instability, narcotics boom, arms proliferation, illegal migration and other challenges'.

Corroborating these claims, the UK chargé d'affaires to Afghanistan, Robert Dickson, stated that engaging with the Taliban became the 'only' way for other states to pursue their interests in the country (Amu TV 2023: [01:36]). Indeed, as highlighted by Dickson, there seems to be an implicit international consensus acknowledging the Taliban's control and the current absence of a viable alternative to their rule. At least for the time being, the perception among foreign actors that replacing the Taliban could exacerbate Afghanistan's humanitarian crises is a substantial, albeit inconclusive, legitimacy gain for the group. While seizing control of Kabul was insufficient to secure official international recognition, it remains a sine qua none for the Taliban's claim to be the rightful representative of the state.

The Taliban's internal barriers to external recognition

The Taliban, as of the time of writing, are still on the US list of Specially Designated Global Terrorists (since 2002) and the UN sanctions list (since 1999).[26] The persistence of these forms of institutional de-recognition of the group largely stems from the strategic and normative conundrums associated with its domestic and foreign

[26] In October 2024, Russia announced its intention to remove the Taliban from its list of terrorist organizations (Politico 2024).

policies. These conundrums perturb the region and Western states. Various countries engage with the Emirate partly because of the Taliban's fierce clampdown on ISKP. However, the Taliban also shelter other armed groups, including the anti-Pakistan 'Tehreek-e Taliban Pakistan' (TTP), anti-China 'Eastern Turkistan Islamic Movement', and al-Qaeda (MFA China 2023).[27] Pakistan, for instance, has experienced a surge in TTP cross-border attacks since the Taliban regained power.[28] It is assumed that many foreign militants in Afghanistan have sworn allegiance to the Taliban and fought alongside them during the past 20 years (see Mir 2021). Thus, the prospect of the group forcefully turning against its jihadi allies seems doubtful. Second, the Taliban dismissed the international demand—put forward by China, Russia, the US, the EU, and regional countries—for the formation of an inclusive government. These actors argue that lasting peace and stability in Afghanistan, a home for diverse communities and traditions, cannot be achieved under one-party rule. Despite the group's promise during the Doha negotiations to establish a broad-based system (Haqqani 2020), it continues to reject pluralism, viewing it as a means for corrupt and secular former Afghan elites to infiltrate the Emirate, tarnish its Islamist purity, and subvert its governance. Third, the Taliban similarly reneged on their pledge to allow women's education and work (Afghanistan News 2022), which antagonized the UN, Western countries, and the Islamic world (Dawi 2022).[29] Fulfilling this promise could have improved the Taliban's besmirched image, but the group's ideologues, led by the amir, chose not to do so. They marginalized the Taliban's more politicized figures and settled internal disagreements over girls' education to the detriment of both the Afghan people and the Taliban themselves.

In April 2022, the Taliban's chief justice Sheikh Abdul Hakim Haqqani, a leading doctrinaire, published a book entitled *The Islamic Emirate and its System*. This unprecedented treatise represents a 'manifesto', something the Taliban had never produced before. The endorsement of the book by the group's amir and Ulama underscores its authoritative status and significance as a window into the beliefs and principles underlying the Taliban's governance philosophy and practice. In *The Islamic Emirate*, Haqqani (2022, 151–52, 251) contends that 'the so-called education, culture, and women's rights' are 'a sedition [against] the divine system' and that 'fixing women [lies] in [them staying] at home'.[30] The book indicates that the material welfare and prosperity of the Afghan people are not the Taliban's utmost

[27] The US killed al-Qaeda's Ayman al-Zawahiri in Kabul in August 2022 (SIGAR 2023).

[28] Although the Taliban repeatedly deny the presence of the TTP in Afghanistan, their acting minister of information reportedly stated—following Pakistani strikes in December 2024 on alleged TTP hideouts in eastern Afghanistan—that 'the TTP will be allowed to stay in Afghanistan' (see Gannon in Al Jazeera English, 2024: [20:08]).

[29] In April 2023, the group banned Afghan women from working for the UN. Conceivably, the Taliban's policies have influenced the UN's deferral of the decision on whether to hand over Afghanistan's UN seat to them.

[30] Haqqani's argument reflects a prevalent view within most jihadi and ultra-conservative Salafi circles worldwide: 'the most evident contradiction between Western civilization and Islamic civilization is [their stance on] women' (al-Shatti in al-Menayyes 2021: [21:40]). Accordingly, the question of women's place in society is viewed as the domain where true believers must unequivocally distinguish themselves from infidels. In these circles, women's rights in the West are seen as disrupting established gender roles, undermining traditional family structures, and corrupting society.

concern. Instead, the central priorities are guiding people towards God, enhancing their morals, and ultimately 'guarding the religion' (Haqqani 2022, 20–21, 44).

Beyond Haqqani, it seems that many ideologues within the Taliban do not value international engagement and integration. They reportedly believe that limiting ties with the external world safeguards the purity of their creed and way of life.[31] However, they would still welcome external legitimacy for their government, as long as it aligns with their doctrine and ideology. For the US, this represents a 'faulty illusion' on the Taliban's side: attempting to deprive women and minorities of their rights while seeking improved international relations (Price 2023). Similarly, the Taliban's doctrinaires oppose moderate policies, even though such policies are widely accepted in Muslim-majority countries. This is conceivably because these countries are not considered valid references by the Taliban, who believe that their Emirate is the world's *only* true Islamic government (Afghani 2023; Haqqani 2022: 21, 151). Apparently, these purists harbour an intense apprehension, fearing that even the slightest political compromise could lead them down a precarious path—one that might erode their indigenous cultural and spiritual convictions, alienate their followers, and diminish their unity. Some in this clique even perceive the dissatisfaction of non-believers as a harbinger or sign of divine approval.[32]

Therefore, under current circumstances, it seems unlikely that the Taliban will reconsider their position on women's education and work. However, this remote possibility cannot be entirely dismissed if power dynamics within the group shift in favour of those who previously pledged in Doha to uphold these rights, and if advocacy by the Afghan people and Muslim interlocutors for human dignity and freedom in Afghanistan persists. Ultimately, since their return to power, the Taliban's rule has been marked by escalating repression and intransigence. Yet, they have made some notable progress in areas such as enhancing security, curbing opium cultivation, tackling corruption, and advancing infrastructure and development projects. That said, the group's current political dispensation—where the amir remains secluded, and the top command in Kandahar steers or marginalizes the government in Kabul—curtails their capacity to build on these achievements and expand their diplomatic ties. Under too rigid supreme leadership, the ideologues' subjective cultural and religious opinions often attain near-sacred status, impeding dialogue, negotiations, and

[31] In a conversation likely held in the late 1990s, Mullah Omar explained his refusal to leave Kandahar for Kabul, stating, 'I am staying in Kandahar so that I do not see [anyone] except Muslims' (al-Nafisi in Taqi 2019: [04:38]). He expressed discomfort at the prospect of meeting foreigners in Kabul, 'with whom I do not accept to shake hands', citing Ayah 28 of Surat al-Tawbah, which he interpreted as deeming non-Muslims physically 'impure' (ibid.). While some interpreters shared this understanding, other prominent Quranic exegetes argued that the Ayah refers to spiritual rather than physical impurity. The Taliban's current amir has likewise refused to leave Kandahar.

[32] Interview by the author with an Afghan analyst, 10 January 2023, via WhatsApp call; the Taliban believe that Western opposition to their Emirate stems not from a genuine 'defence of women, but rather [from] deep-seated hostility towards Islam, ... [The Westerners] will not cease their campaigns, nor will they be pleased with us unless we become their followers and imitators' (Howeidy 2001, 71). Reflecting this, a Taliban loyalist criticized Switzerland's burqa ban, which took effect in January 2025, stating that the West demands women's freedom in Afghanistan yet prohibits them from wearing what they choose: 'You will not find from us what pleases you, God willing!' (Alafghani 2025).

compromise with both domestic and foreign interlocutors, even within the widely accepted confines of Islam. A real conundrum for the Taliban is that a reformed political setup might lead them towards greater international engagement, but it could equally bring them to the brink of erosion.

Conclusion

The Taliban's seizure of Kabul in August 2021 marked the emergence of the sole jihadi nationalist government on the world stage, with which some major powers interact or have considerable diplomatic relations. Such an accomplishment places the Taliban in a unique position among jihadi organizations in particular and armed movements in general. No other jihadi group has managed to conquer a capital city and install its government by the force of arms, nor has any sought normal diplomatic ties with other countries or admission to the UN.[33] Unlike groups such as IS and al-Qaeda, which outright reject the secular international order and aim to overthrow or subvert it, the Taliban strive for inclusion within this very system. They also set themselves apart from other rebels that prioritize achieving international legitimacy as a top strategic goal. For this reason, the Taliban navigate uncharted territory among armed groups, pursuing recognition and its advantages, but strictly on their own terms: that is, not at the expense of their core doctrine and worldview. This explains the apparent paradox on the Taliban's part—seeking recognition while implementing policies that hinder its realization.

As this chapter has shown, a more nuanced understanding of the group's behaviour and precarious international standing can be achieved when their ideological and identity beliefs are integrated into the analysis of their policy priorities and preferences. Such beliefs have proved to be a double-edged sword in their pursuit of an international status. On one side, the insurgent Taliban's legitimating ideas of jihad, independence, an Islamic way of life, and Afghan honour resonated with various local communities and endowed the group with a sense of invincibility on the battlefield. This facilitated the Doha Agreement, which effectively recognized the Taliban as the principal Afghan political power and paved their way back to Kabul. As Afghanistan's undisputed authority since August 2021, engaging with the Taliban became imperative for countries pursuing their interests there, marking a key milestone in the group's path to legitimacy. On the other side, the international de facto acknowledgement of Taliban control has not yielded broad de jure recognition of their government. This is partly because the Taliban's policies, rooted in their idiosyncratic religious and cultural interpretations, raise security and normative concerns for some neighbouring and global actors. The group's refusal to sever

[33] It is unclear whether the current Syrian interim authority is yet another exception. Syrian opposition forces, led by Hayat Tahrir al-Sham (HTS) entered Damascus on 8 December 2024, bringing an end to Bashar al-Assad's rule. HTS, previously known as Jabhat al-Nusra, severed ties with al-Qaeda in 2016 and has since clashed with jihadi groups in north-western Syria. While the long-term trajectory of the HTS-led transitional administration in Syria remains uncertain, it currently espouses open, inclusive, and conciliatory rhetoric and is actively engaging diplomatically with Arab and Western countries.

ties with foreign militants, establish an inclusive government, and uphold women's and minority rights makes it difficult for other countries, especially in the West, to work with them. Therefore, the Taliban's aspiration of achieving widespread official recognition on their terms remains elusive.

Still, the global order is no longer as unipolar as it was in 2001 when the US invaded Afghanistan. Almost two decades into the unwinnable war against the Taliban, the hegemony of the US deteriorated, and its top national security interest shifted from counter-terrorism to containing China. The rise of Beijing signals the possible emergence of new paradigms of international politics, and as this global transition continues to unfold, the Taliban returned to power, declaring their desire to join the China caravan. The group's foreign minister predicted that 'the pace towards multipolarity has accelerated. [...] and our region particularly, will play a key role in this new world order' (Muttaqi 2023).

Both the People's Republic of China and the Emirate—alongside powers such as Russia—have a shared preference for a post-Western international order that prioritizes state sovereignty and non-interference in domestic affairs over democratic governance, human rights, and individual freedoms. Certainly, the growing Chinese influence will not render these rights and freedoms irrelevant in international politics. Instead, it may foster new dynamics that offer legitimacy-seeking governments alternative pathways—such as regional alliances or plurilateral state organizations—to secure external validation and support beyond the US and the West. It is clear that China's interests in Afghanistan are primarily security and investment. Yet, the relevance of the landlocked country may transcend these interests. Afghanistan could provide China with a new front to set its global leadership apart from that of the US. Consistent with its approach elsewhere, China's foreign policy in Afghanistan is centred on respect for local cultures, diplomacy, trade, and 'win-win co-operation'.[34] Should this policy contribute to stabilizing Afghanistan and improving the well-being of its people, it would not only undermine Washington's isolation of the Taliban, but also bolster China's regional influence and its self-portrayal as a benevolent superpower. This contrasts with the US approach to Afghanistan, which has been marked by military interventions, economic sanctions, and a disregard for domestic traditions. China openly expressed *schadenfreude* over the chaotic US withdrawal from Afghanistan, viewing the 20-year occupation of the country not only as a military fiasco for the US but also as a blow to its credibility and international politics (Observer Network 2021).

To conclude, evolving configurations of regional and world politics might offer the Taliban an avenue for legitimacy outside the Western realm. However, this opportunity could slip away unless the group addresses the concerns of regional capitals, shows flexibility, and empowers its politically astute officials, while also responding to the aspirations of the Afghan people. Should the Taliban's characteristic political dispensation and approach to domestic governance and foreign affairs remain unchanged, rather than moving towards more normalization, they risk becoming entangled in a recurring cycle of adversarial relations and international conflicts.

[34] A common Chinese characterization of their approach to foreign relations, aid, and trade.

References

Abu, Abed. 2014. '*Al-Somood* Interviews a Member of the Islamic Emirate's Committee for Dawah and Guidance'. *Al-Somood*, October.

Afghani, Anas. 2023. 'Afghanistan: The Only Country in the World Where the Islamic System Rules'. *X*. 2 August. https://x.com/Anas_Afghani1/status/1686818827021991952 (accessed 5 August 2023).

Afghanistan News. 2022. 'Taliban's Promises in Qatar'. *X*. 28 December. https://x.com/IAfghanNEWS/status/1608107906263822337 (accessed 1 June 2023).

Akhundzada, Hibatullah. 2016. 'Statement of Amir ul-Mo'mineen'. *Al-Somood*, July.

Al-Afghani, Jamal ad-Din. 1879. *Sequel to the Statement on the History of the Afghans*. Arabic E-Book: Hindawy Foundation.

Alafghani, Rihan. 2025. 'The Niqab has been Banned in Switzerland'. *X*. 1 January. https://x.com/RihanAlafghani/status/1874393297638302095 (accessed 2 January 2025).

Al Jazeera. 2021. 'Pakistan's Message to World on Taliban: "Be Realistic ... Engage"', 23 September. https://www.aljazeera.com/news/2021/9/23/pakistan-taliban-afghanistan-qureshi-interview (accessed 5 September 2023).

Al Jazeera. 2023. 'Taliban gives a Warm Welcome to China's New Ambassador to Afghanistan', 13 September. https://www.aljazeera.com/news/2023/9/13/taliban-gives-a-warm-welcome-to-chinas-new-ambassador-to-afghanistan (accessed 14 September 2023).

Al Jazeera English. 2024. 'Why are Relations between Pakistan and Afghanistan so Tense? | Inside Story'. *YouTube*. 29 December. https://youtu.be/OXQhQ91RrEk?t=1208 (accessed 3 January 2025).

Al-Menayyes, Muhammad. 2021. Taliban Towards Recognition or the Caliphate? | Eng. Abdul Latif al-Shatti. Juzoor. *Apple Podcasts*. 6 September. Arabic. https://podcasts.apple.com/de/podcast/%D8%AC%D8%B0%D9%88%D8%B1/id1554766478?i=1000534460727 (accessed 2 January 2025).

Al-Qaradawi, Yusuf. 2020. Visit of the Delegation of Scholars to Afghanistan | Sheikh Yusuf al-Qaradawi | Sharia and Life. *YouTube*. 18 August. https://www.youtube.com/watch?v=pfUd8YfhbQE (accessed 4 October 2024).

Al-Sharky, Abu Khaled. 2006. 'The US Seeks to Obliterate the Afghan Identity'. *Al-Somood*, May.

Al-Somood. 2006. 'On the Fifth Anniversary of the Occupation of Afghanistan', September.

Al-Somood. 2015. 'Translating and transcribing the speech of Amir ul-Mo'mineen', December.

Al-Somood. 2016. 'Deputy Amir Ul-Mo'mineen Sirajuddin Haqqani in an Open Dialogue', December.

Al-Suri, Abu Musab. 1998. *Afghanistan, Taliban and Islam's Battle Today*. Kabul: al-Ghuraba Centre.

Amiri Diwan. 2022. 'HH The Amir Meets Defense Minister in Afghan Caretaker Government'. 6 July. https://www.diwan.gov.qa/briefing-room/news/receiving-guests/2022/july/6/hh-the-amir-receives-defense-minister-of-the-afghan-caretaker-government?sc_lang=en (accessed 6 July 2023).

Amu TV. 2023. 'Interview with Robert Dickson'. *YouTube*. 21 July. https://youtu.be/pvVw_KpapN4?t=96 (accessed 1 September 2023).

AP Archive. 2001. 'Briefing by Taliban Ambassador in Islamabad'. *YouTube*. 19 October. https://youtu.be/RGqUZ1KY4hk?list=PLZETXkNzwH8vdenq7zMem45Uys9N0_Ghu&t=90 (accessed 4 September 2023).

AP Archive. 2015. 'Presidential Address to Congress on Terrorism'. *YouTube*. 23 July. https://youtu.be/zB145D3XjzE?t=136 (accessed 5 September 2023).

Baczko, Adam. 2021. 'Winning War Through Law in Afghanistan?'. *Sciences Po*. https://www.sciencespo.fr/ceri/en/content/winning-war-through-law-afghanistan-interview-adam-baczko (accessed 5 September 2023).

Baradar, Abdul Ghani. 2023. 'Highlights from the Speech of Mullah Abdul Ghani Baradar'. *X*. https://x.com/FDPM_AFG/status/1630930415157407747 (accessed 5 September 2023).

Bush, George W. 2001. 'Address to a Joint Session of Congress'. 20 September. https://georgewbush-whitehouse.archives.gov/news/releases/2001/09/20010920-8.html (accessed 30 August 2023).

CBC. 2021. 'He Tried to Warn the World about Al-Qaeda'. 10 September. https://www.cbc.ca/radio/asithappens/as-it-happens-friday-edition-1.6171560/he-tried-to-warn-the-world-about-al-qaeda-then-he-was-assassinated-2-days-before-9-11-1.6171563 (accessed 4 September 2023).

MFA China. 2022. 'Wang Yi Holds Talks with Acting Foreign Minister of the Afghan Interim Government'. 25 March. https://www.fmprc.gov.cn/eng/zxxx_662805/202203/t20220325_10655541.html (accessed 5 June 2023).

MFA China. 2023. 'Joint Statement of the Second Informal Meeting on Afghanistan'. 14 April. https://www.fmprc.gov.cn/mfa_eng/zxxx_662805/202304/t20230414_11059063.html (accessed 5 September 2023).

Coll, Steve. 2018. *Directorate S: The CIA and America's Secret Wars in Afghanistan and Pakistan*. Ebook: Penguin Press.

Cristol, Jonathan. 2019. *The United States and the Taliban Before and After 9/11*. Cham: Palgrave.

Dawi, Akmal. 2022. 'Muslim Countries Blast Taliban for University Ban for Afghan Women'. *VOA News*, 22 December. https://www.voanews.com/a/muslim-countries-blast-taliban-for-university-ban-for-afghan-women/6887860.html (accessed 20 May 2023).

Elsayed, Ahmed. 2023. From the Mountains to the Élysée: The Precarious International Legitimacy of the Syrian Kurdish YPG and PYD. *The Middle East Journal*, 77(1): pp. 53–77. https://doi.org/10.3751/77.1.13

Elsayed, Ahmed. 2024a. *Beliefs and International Legitimacy: The Janus Face of the Taliban's Political Worldview and Value Rationality*. InIIS Working Paper Nr. 45/2024. Bremen: InIIS.

Elsayed, Ahmed. 2024b. *The Legitimacy Battlefield: How Armed Groups Leverage Violence and Values in the Quest for International Recognition*. Dr. rer. pol. dissertation. University of Bremen. https://doi.org/10.26092/elib/3297

Emadi, Hafizullah. 2002. *Repression, Resistance, and Women in Afghanistan*. Westport, CT: Praeger Publishers.

Jalal Foundation. 2023. 'Tolo Channel's Interview with Anas Haaqani', 19 July. https://jalal.af/wp-content/uploads/2023/07/Anas-Magazine-tolo-TV-Interview.pdf (accessed 5 August 2023).

Ghaznawi, Shihabuddin. 2009. 'The Rumour of Negotiations'. *Al-Somood*, January.

Giustozzi, Antonio. 2019. *The Taliban at War: 2001–2018*. Ebook: Oxford University Press.

Gopal, Anand and Alex Strick van Linschoten. 2017. 'Ideology in the Afghan Taliban'. *Afghanistan Analysts Network*. https://www.afghanistan-analysts.org/wp-content/uploads/2017/06/201705-AGopal-ASvLinschoten-TB-Ideology.pdf.

Gossman, Patricia. 2019. 'Afghanistan Papers Detail US Role in Abuse'. *Human Rights Watch*, 11 December. https://www.hrw.org/news/2019/12/11/afghanistan-papers-detail-us-role-abuse (accessed 21 January 2025).

Gul, Ayaz. 2023. 'Taliban Under Fire.' *VOA News*, 30 March. https://www.voanews.com/a/taliban-under-fire-for-alleged-afghan-civil-society-crackdown-/7028856.html.

Haqqani, Abdul Hakim. 2022. *The Islamic Emirate and its System*. Afghanistan: Maktabat Darul Uloom al-Shar'iya.

Haqqani, Hafizullah. 1997. *Taliban: From the Mullah's Dream to the Emirate of the Believers*. Islamabad: Institute of Policy Studies.

Haqqani, Sirajuddin. 2020. 'What We, the Taliban, Want'. *The New York Times*, 20 February. https://www.nytimes.com/2020/02/20/opinion/taliban-afghanistan-war-haqqani.html (accessed 15 July 2023).

Harpviken, Kristian Berg. 2021. 'The Foreign Policy of the Afghan Taliban'. *Peace Research Institute Oslo Policy Brief*, 2. Oslo: PRIO.

Howeidy, Fahmy. 2001. *Taliban: God's Soldiers in the Wrong Battle*. 2nd edn. Cairo: Dar El Shorouk.

Human Rights Watch. 2004. '"Enduring Freedom": Abuses by US Forces in Afghanistan'. https://www.hrw.org/sites/default/files/reports/afghanistan0304.pdf (accessed 4 October 2023).

Ibrahim, Arwa. 2021. 'Explainer: The Taliban and Islamic law in Afghanistan'. Al Jazeera, 23 August. https://www.aljazeera.com/news/2021/8/23/hold-the-taliban-and-sharia-law-in-afghanistan.

Johnson, Thomas, Matthew DuPee, and Wali Shaaker. 2017. *Taliban Narratives: The Use and Power of Stories in the Afghanistan Conflict*. Oxford: Oxford University Press.

Kakar, Ahmed-Waleed. 2022. 'Afghanistan's Century of Politicised Education'. *The Afghan Eye*, 21 December. https://afghaneye.org/2022/12/21/afghanistan-girls-education/.

Loidolt, Bryce. 2022. 'How the Al-Qaeda–Taliban Alliance Survived'. *Survival* 64(3): pp. 133–152. https://doi.org/10.1080/00396338.2022.2078052

Long, Ding. 2022. 'Ideological Transition of the Afghan Taliban'. *China Academic Journal* 23(2), March/April: pp. 39–61. https://mideast.shisu.edu.cn/_upload/article/files/0a/b4/0c8b82df4b6882d3bc654c97852e/17372d78-2f78-457c-96d5-a1f4f2856c19.pdf

Maiwandi, Ekram. 2007a. 'The Whining of "NATO": Between the Bitterness of Despair and the Savagery of Madness'. *Al-Somood*, June.

Maiwandi, Ekram. 2007b. 'The Holy War Has Begun to Bear Fruit'. *Al-Somood*, October.

Maley, William. 2006. *Rescuing Afghanistan*. Sydney: UNSW Press.

Malikyar, Helena. 1997. 'Development of Family Law in Afghanistan: The Roles of the Hanafi Madhhab, Customary Practices and Power Politics'. *Central Asian Survey* 16(3): 389–399. doi:10.1080/02634939708400998.

Malkasian, Carter. 2021. *The American War in Afghanistan: A History*. Oxford: OUP.

Manchanda, Nivi. 2020. *Imagining Afghanistan: The History and Politics of Imperial Knowledge*. Cambridge: Cambridge University Press.

Mansour, Akhtar. 2015. 'Message to Sheikh Abu Bakr Al-Baghdadi and His Mujahideen Brothers'. *Al-Somood*, June/July.

Mashal, Mujib. 2019. 'Once Jailed in Guantánamo, 5 Taliban Now Face US at Peace Talks'. *The New York Times*, 26 March, section: World. https://www.nytimes.com/2019/03/26/world/asia/taliban-guantanamo-afghanistan-peace-talks.html (accessed 2 September 2023).

Maynard, Jonathan Leader. 2019. 'Ideology and Armed Conflict'. *Journal of Peace Research* 56(5): pp. 635–649. https://www.jstor.org/stable/48596219.

MFA China. 2024. 'Foreign Ministry Spokesperson Wang Wenbin's Regular Press Conference'. 31 January. https://www.mfa.gov.cn/eng/xw/fyrbt/lxjzh/202405/t20240530_11347691.html (accessed 8 October 2024).

MFA Uzbekistan. 2023. 'European Union and Central Asian Special Representatives for Afghanistan Met in Ashgabat'. https://mfa.uz/en/press/news/2023/european-union-and-central-asian-special-representatives-for-afghanistan-met-in-ashgabat—34017 (accessed 5 September 2023).

Mir, Asfandyar. 2021. 'Twenty Years After 9/11: The Terror Threat from Afghanistan Post the Taliban Takeover'. *CTC Sentinel* 14(7): pp. 29–42. https://ctc.westpoint.edu/wp-content/uploads/2021/09/CTC-SENTINEL-072021.pdf

Mujahid, Zabihullah. 2016. 'Interview with Al-Sharq al-Awsat Newspaper'. *Al-Somood*, June.

Mukhtar, Ahmed. 2007. 'Truth of the Peace Negotiations between the Taliban and the Stooge Government'. *Al-Somood*, October.

Mumand, Salahuddin. 2018. 'The Emirate Seeks Peace, But'. *Al-Somood*, March.

Musharraf, Pervez. 2006. *In the Line of Fire: A Memoir*. New York: Simon and Schuster.

Mutma'in, Abul Hai. 2022. *Mullah Mohammed Omar, Taliban and Afghanistan*. Syria: Khattabi Centre for Studies.

Muttaqi, Amir Khan. 2023. 'Speech by Afghan Foreign Minister Mawlawi Amir Khan Muttaqi at 4th Foreign Ministers Meeting among the Neighboring Countries of Afghanistan'. 13 April. https://justpaste.it/aswpz (accessed 4 September 2023).

Observer Network (观察者网). 2021. 'Exclusive Interview with YUE Xiaoyong, China's Special Envoy on the Afghan Issue'. *YouTube.* 29 August. https://www.youtube.com/watch?v=t94SlfsXmLA (accessed 5 September 2023).

Oppenheim, Maya. 2025. 'Taliban Bans Windows to Stop Women from being Seen at Home', *The Independent*, 1 January. https://www.independent.co.uk/asia/south-asia/taliban-afghanistan-ban-windows-women-b2672332.html (accessed 1 January 2025).

PBS NewsHour. 2021. Watch Live: Sec. Austin, Gen. Milley and Gen. McKenzie Testify on Afghanistan Withdrawal. *YouTube.* https://www.youtube.com/watch?v=EpAEco8K0_4 (accessed 5 September 2023).

Politico. 2024. 'Russia set to Remove Taliban from Terrorist List, says Putin's Envoy'. 4 October. https://www.politico.eu/article/russia-remove-taliban-terrorist-list-vladimir-putin-envoy-zamir-kamulov/ (accessed 8 October 2024).

Pompeo, Mike. 2023. *Never Give an Inch: Fighting for the America I Love.* New York: Broadside Books.

Price, Ned. 2023. 'Department Press Briefing'. *US Department of State.* 4 January. https://www.state.gov/briefings/department-press-briefing-january-4-2023/ (accessed 3 September 2023).

Rashid, Ahmed. 2010. *Taliban: The Power of Militant Islam in Afghanistan and Beyond.* New York: I.B. Tauris.

Rubin, Barnett. 2007. 'Saving Afghanistan'. *Foreign Affairs* 68(1): pp. 57–74. http://www.jstor.org/stable/20032211.

Rubin, Barnett. 2020. *Afghanistan: What Everyone Needs to Know.* Oxford: Oxford University Press.

Reuters Staff. 2015. 'Russia's Interests Coincide with Taliban's in Fight against ISIS'. *Reuters*, 23 December. https://www.reuters.com/article/uk-mideast-crisis-russia-taliban-idUKKBN0U61EM20151223 (accessed 5 September 2023).

Security Council. 2020. 'Resolution 2513 (2020)'. 10 March. https://digitallibrary.un.org/record/3855892?ln=en (accessed 5 September 2023).

Siddique, Abubakar. 2024. 'The Azadi Briefing: Taliban Appears Split Over Women's Education Ban'. *RFE/RL*, 8 March. https://www.rferl.org/a/azadi-briefing-taliban-womens-education-ban-split-siddique/32854007.html.

SIGAR. 2022. 'Why the Afghan Government Collapsed'. Special Inspector General for Afghanistan Reconstruction. https://www.sigar.mil/pdf/evaluations/SIGAR-23-05-IP.pdf (accessed 28 August 2023).

SIGAR. 2023. 'Quarterly Reports'. Special Inspector General for Afghanistan Reconstruction. https://www.sigar.mil/pdf/quarterlyreports/2023-01-30qr-section3-security.pdf (accessed 5 September 2023).

Sinno, Abdulkader. 2009. 'Explaining the Taliban's Ability to Mobilize the Pashtuns'. In *The Taliban and the Crisis of Afghanistan*, edited by Robert D. Crews and Amin Tarzi, pp. 59–89. Cambridge, MA: Harvard University Press.

State Department. 2021. 'Joint Statement on Extended "Troika" on Peaceful Settlement in Afghanistan'. 18 March. https://www.state.gov/joint-statement-on-extended-troika-on-peaceful-settlement-in-afghanistan/ (accessed 5 September 2023).

Taqi, Ammar. 2019. 'Mullah Omar meets al-Nafisi for the first time'. *YouTube*. 31 October. https://www.youtube.com/watch?v=Bi0wckvvpuQ (accessed 8 October 2023).

Thabet, Mohammed. 1934. *A Tour of the Near East between Egypt and Afghanistan*. Arabic. Faggalah, Egypt: Saad Masr Printing House.

Tolo News. 2023. 'Leader: Non-Islamic Laws Will Never Be Implemented', 22 April. https://tolonews.com/afghanistan-183067 (accessed 5 September 2023).

UChicago. 2023. 'Life After Trump: Mike Pompeo in Conversation'. *YouTube*. 12 January. https://www.youtube.com/watch?v=cIcXAvv1oH8 (accessed 5 September 2023).

UN Press. 2023. 'Induce Taliban to End "Gender Apartheid" in Afghanistan through All Available Means', 26 September. https://press.un.org/en/2023/sc15421.doc.htm (accessed 3 October 2023).

Varshney, Ashutosh. 2003. 'Nationalism, Ethnic Conflict, and Rationality'. *Perspectives on Politics* 1(1): pp. 85–99. https://doi.org/10.1017/S1537592703000069.

Vaughn, Thomas B. 1983. 'Morale: The 10th Principle of War?' *Military Review* 63(5): pp. 23–39.

WAM (Emirates News Agency). 2024. 'UAE President Receives Afghan Interior Minister'. 4 June. https://www.wam.ae/en/article/b3hvdz0-uae-president-receives-afghan-interior-minister (accessed 8 October 2024).

Weber, Max. 1978. *Economy and Society: An Outline of Interpretive Sociology*. Berkeley, CA: University of California Press.

Yaqeen Network. 2024a. 'The Speech of Interior Minister'. *X*. 17 September. https://x.com/yaqeennetwork/status/1836092378710421595 (accessed 3 October 2024).

Yaqeen Network. 2024b. 'The Speech of Amir ul-Mo'mineen in Arabic'. *YouTube*. 27 September. https://www.youtube.com/watch?v=CRpgUSZLRWQ (accessed 1 October 2024).

Zaeef, Abdul Salam. 2010. *My Life with the Taliban*. Edited by Alex Strick van Linschoten and Felix Kuehn. New York: Columbia University Press.

11
The Astonishingly Easy Path to International Legitimacy of Khalifa Haftar's LAAF

Wolfram Lacher

Introduction

In May 2014, Khalifa Haftar rose from obscurity to prominence overnight when he assembled an alliance of armed groups he called the Libyan National Army (LNA) to attack rival factions in Benghazi, Libya's second-largest city. Haftar's offensive was central to the escalation into a civil war that ended the transitional process begun after the fall of the Qadhafi regime in 2011. Three months after Haftar launched his offensive, Libya's institutions split in two, and the east-based parliament threw its support behind Haftar's nascent forces. It took Haftar less than a year to be officially received as the commander of the Libyan Arab Armed Forces (LAAF)—the name his forces soon used instead of LNA—by Jordan's King Abdullah II, and just over three years for him to be hosted by France's President Emmanuel Macron for a summit meeting with Fayez al-Sarraj, the prime minister of the internationally recognized government in Tripoli. Following that meeting, public visits by Western foreign ministers and ambassadors to Haftar's headquarters outside Benghazi became routine, as did high-level meetings between Haftar and government officials in Rome, Paris, Berlin and Moscow. Haftar has been alone among the leaders of Libya's innumerable armed groups to enjoy such honours.

How Haftar could have gained such a degree of international legitimacy so quickly is puzzling for several reasons. First, since their foundation, his forces have stood out as Libya's leading perpetrators of war crimes. Moreover, together with local Islamic State affiliates and the Kaniyat militia in the town of Tarhuna, the LAAF have been the only armed group in post-2011 Libya to establish a highly repressive form of rule. Those violent practices, it could be assumed, should have stood in the way of Haftar getting the red-carpet treatment in Western capitals. Second, Haftar consistently proved to be the biggest obstacle to the success of Western governments' official Libya policies—namely, the attempt to reunify Libya under the Government of National Accord (GNA) in 2016–17 and the plan to hold elections following the formation of the Government of National Unity in 2021. And yet,

Wolfram Lacher, *The Astonishingly Easy Path to International Legitimacy of Khalifa Haftar's LAAF*. In: *Armed Groups and the Politics of International Legitimation*. Edited by: Stephan Hensell and Klaus Schlichte, Oxford University Press.
 DOI: 10.1093/9780198974529.003.0011

Haftar's function as a spoiler merely prompted tireless efforts by Western diplomats to bring him into the fold and extract concessions from his Tripoli-based adversaries. Third, in a context of worsening Western relations with Russia, Haftar became increasingly dependent on Russian support and even facilitated the establishment of an unofficial Russian military presence in areas under his control. But Haftar's close alliance with Russia did not cause Western governments to isolate him internationally.

This chapter argues that a particular set of domestic and international circumstances enabled Haftar's success in rapidly gaining a great deal of international legitimacy. First, Haftar benefited from the crisis of legitimacy that erupted along with the civil war Haftar himself helped to provoke. International legitimacy became an object of rivalry between two competing governments, and even the first UN-backed unity government proved so dysfunctional as to reduce its international recognition to a thin veneer.[1] For Haftar, this meant—at least initially—that he did not have to arduously chip away at the legitimacy of his adversaries. More broadly, in a context of unsettled international legitimacy, all actors—Libyan and foreign—were hypersensitive to the finest nuances in foreign states' engagement with Libyan protagonists. For the latter, international legitimacy became a question of degree: it could rise and decline along with shifting conflict dynamics, and Haftar never succeeded in gaining the outright recognition as Libya's ruler that he coveted—just as he never managed to take control of the entire country. Second, in an international context that was increasingly marked by multipolarity, the role of regional powers became far more important than during the unipolar moment that had preceded it. Haftar's regional sponsors proved crucial for his accreditation on the international stage and helped overcome Western governments' misgivings over his terrible record on human rights.

Among the practices of legitimation outlined in this volume, *mimicking*, *acknowledging*, and *officializing* played a prominent role in the case of Libya.[2] For Haftar and his foreign supporters, his expansion and consolidation of territorial control was crucial to his pursuit of international legitimation. Mimicking the state by establishing territorial control allowed Haftar to claim that he was an indispensable interlocutor for foreign governments, and possibly the future ruler of Libya. Foreign support both enabled the LAAF's territorial expansion and was premised on its continuation. That expansion, and particularly the gradual takeover of Libya's oil infrastructure, provided a powerful incentive for foreign states to establish stronger relations with Haftar, while increasingly relegating their support for the internationally recognized government in Tripoli to a matter of formality. The more Haftar's forces expanded, the more convincing another facet of mimicking as a practice of claim-making became: their imitation of state security forces, particularly through the steady stream

[1] For another case in which the crisis of a sitting government during a civil war pushed international actors to recognize an insurgent movement instead as government, see the chapter by Klaus Schlichte on the NRA in this volume.

[2] See the Introduction in this volume.

of imagery disseminated by Haftar's propagandists that portrayed the LAAF as a disciplined, regular army—in stark contrast to the actual makeup of these forces.

Foreign governments, in turn, displayed great virtuosity in choosing how far to go in validating Haftar's legitimacy claims by acknowledging and officializing him and the LAAF. They could do so through the names and language they used. To refer to Haftar's forces by their official name and to him in his official role was an act of legitimation—since doing so already distinguished the LAAF from armed groups and lent them official status. Over time, foreign governments and UN officials chose to use different names for Haftar's forces as their willingness to legitimize him waxed and waned. Foreign governments were also acutely aware that official meetings bestowed international legitimacy upon Haftar, and they could choose whether or not to publicize their representatives' meetings with him—as well as whether to allow Haftar's propagandists to shape the optics of those meetings. In this regard, the meeting between Haftar and then prime minister Sarraj hosted by Macron in July 2017 stands out as the single most important step in Haftar's international legitimation.

Following a brief summary of the LAAF's evolution, this chapter first shows that a confluence of specific international and domestic conditions was particularly conducive to Haftar's successes in gaining international legitimacy. Its then analyses the legitimation practices which Haftar and international actors engaged in, showing how they varied according to Haftar's military successes and failures.

A very short history of Haftar's LAAF

Haftar's forces emerged during the escalation that caused Libya's post-Qadhafi transitional process to collapse into civil war in 2014. They originated in a coalition of armed groups Haftar mobilized in May 2014 to fight against what he called 'terrorists' in Benghazi, Libya's second-largest city. Over the previous year, Benghazi had witnessed worsening tensions between the armed groups that had formed since the 2011 revolution, as well as a series of assassinations targeting primarily former members of Qadhafi's army and security services. Although the perpetrators were unknown, many suspected the jihadist group Ansar al-Sharia as well as several revolutionary armed groups, some of which had Islamist leanings, of being responsible (Fitzgerald 2023). Haftar, a former officer in Qadhafi's army who had spent decades in exile and returned during the 2011 revolution, had at this point been retired from military service by the authorities in Tripoli. Since February 2014, he had been subject to an arrest warrant after he had announced his intention to overturn the transitional institutions (Kirkpatrick 2014). Haftar exploited the tensions in Benghazi to craft an alliance against Ansar al-Sharia and the revolutionary armed groups, putting himself at the head of a renegade army leadership. Simultaneously, tensions between rival militias escalated in western Libya, where Haftar allied with armed groups from Zintan, thereby playing a leading role in the eruption of civil war.

In many respects, Haftar was a classical warlord who drew on his military legitimacy, charisma, and skill in organizing a warring coalition to carve out a growing sphere of influence for himself amid the chaos that followed Qadhafi's demise.[3] Contrary to Haftar's claims that he headed a resurrected Libyan army, his forces were in fact largely composed by civilian fighters, as well as military officers who had formed militias. These forces included groups pursuing autonomy for eastern Libya; militias drawn from particular tribal constituencies, such as the Awagir from Benghazi's outskirts; and Salafist fighters. The militias that joined Haftar's coalition had their own leadership and weaponry, and therefore acted largely autonomously; Haftar, by contrast, initially had few forces under his direct, effective command. Haftar and his associates claimed that they were fighting 'terrorists', and their enemies did in fact include Ansar al-Sharia as well as, from late 2014 onwards, a local affiliate of the Islamic State. But the reality of the war was far messier: commanders in Haftar's forces publicly called for the expulsion of all families of Misratan origin from Benghazi, and they often seized the properties of such families, enriching themselves through the war. Haftar himself openly used his ostensible counter-terrorist operation to challenge the transitional government and attempt to seize overall power—and many of his opponents resisted him for his transparent dictatorial ambitions (Lacher 2020b; Wehrey 2023).

With support from the United Arab Emirates (UAE), Egypt, Jordan, and France, Haftar gradually gained the upper hand over his enemies in Benghazi during years of urban warfare. At the same time, he used his monopoly over foreign military assistance to progressively assert his authority over the forces in his alliance. He also built a counterweight to the autonomous armed groups in his coalition by inviting former Qadhafi regime loyalists to return from exile and by hiring mercenary fighters from Sudan and Chad. Over time, he strengthened his control by building new units under the direct control of his sons and close relatives while progressively having his former allies disarmed, imprisoned, exiled, or assassinated. In eastern Libya, the LAAF established a highly repressive form of rule, spreading fear with summary executions and enforced disappearances (Lacher 2020b, Eaton 2021).

From mid-2016 onwards, Haftar began gradually expanding his territorial control: first, over eastern Libya's oil export terminals; then, in mid-2017, over strategic airbases in central and southern Libya—in both cases largely without fighting. During 2018, his forces caused major destruction and displacement during their capture of Darna, the last eastern city to have resisted his authority. In early 2019, the LAAF expanded their control over much of southern Libya, mostly by co-opting local armed groups, with only few forces attempting to violently resist his advance. By spring 2019, then, Haftar controlled the bulk of Libya's territory, though not its most populous, north-western region, where most armed groups, many of them rooted in the 2011 revolution, were hostile towards Haftar and the return to dictatorship that he represented.

In April 2019, Haftar began a large-scale offensive to seize Tripoli, and overall state power. Despite major support from the UAE, which deployed combat drones and air defence systems, the offensive rapidly became stuck in Tripoli's southern outskirts

[3] For useful discussions of the loose concept of warlordism, see Giustozzi (2005), Marten (2006), and Driscoll (2015).

due to tenacious resistance. To tilt the balance, Haftar gained the backing of Russia's Wagner Group. Haftar's forces were responsible for the vast majority of civilian casualties during the 15-month offensive (UNSMIL 2020c, 2020e). The forces of the Tripoli government eventually obtained Turkish support and compelled the LAAF and Wagner to withdraw from western Libya in June 2020 (Wehrey 2020).

Haftar therefore came close, but ultimately failed to seize state power. Despite this failure, his Libyan Arab Armed Forces have continued to control eastern, central, and much of southern Libya. His sons have continued to centralize control over the various components of Haftar's forces, as well as over illicit activities such as the smuggling of drugs, fuel and migrants.

This brief overview suggests that Haftar should have faced substantial hurdles in gaining international legitimacy, owing to the nature and behaviour of his forces. And yet, Haftar was exceptionally successful in doing so compared to both other Libyan militia leaders and most armed groups in other conflicts. What, then, explains this puzzle?

Facilitating conditions of international legitimation

Haftar's path to international legitimacy was made easier by the combination of two specific circumstances: the fact that the legitimacy of Libya's executive and legislative authorities was unsettled to begin with, and the considerable influence Haftar's regional sponsors exercised in a situation of multipolarity.

Unsettled international legitimacy

Haftar's quest benefited from the fact that, early on in the formation of his forces, his adversaries lost international recognition, thereby levelling the playing field in the struggle for international legitimacy. Despite the formation of internationally recognized unity governments later on, the question of international legitimacy would remain unsettled throughout the following decade. This situation allowed foreign governments to send nuanced and often ambiguous signals to legitimize Haftar up to a degree, even as they engaged with the unity governments in Tripoli.

From mid-2014 onwards, as a civil war provoked a situation of split sovereignty, foreign governments adopted a position of constructive ambiguity towards Libya's rival authorities. When Haftar launched his May 2014 offensive against armed groups in Benghazi, he was in open defiance of authorities in Tripoli, including of the majority dominating the General National Congress (GNC) elected in 2012. At the same moment, mounting tensions between armed groups in Tripoli pushed Zintani-led militias opposed to the GNC majority into an alliance with Haftar. After the June 2014 elections to the new parliament, the House of Representatives (HoR), these tensions in Tripoli escalated into a major offensive against Zintani-led militias by armed groups from Misrata and other cities.

The HoR had been planned to convene in Benghazi in early August. But with war raging in both Tripoli and Benghazi, a number of pro-Haftar parliamentarians engineered its move to Tobruk, where army units had aligned with Haftar. Out of 188 parliamentarians, around thirty boycotted the Tobruk sessions. While Western governments and the UN recognized the Tobruk HoR as Libya's new legislative, they simultaneously emphasized the need for the body to overcome its internal rift, thereby suggesting that its legitimacy had flaws. The Tobruk HoR rapidly began siding with Haftar and his allies in the conflict. In late August, it called both Haftar's enemies in Benghazi and the Misratan-led militia alliance in Tripoli 'terrorists' who were 'a legitimate target for the Libyan army' (Al-Obeidi 2014). It also appointed a close associate of Haftar as chief of staff. Haftar himself still remained outside formal command structures. Meanwhile, a rump GNC reconvened in Tripoli in late August to contest the HoR's legitimacy and form a new government. This prompted the HoR, in turn, to task Abdallah Thinni, who had been acting prime minister since before the elections, with government formation. By September 2014, Libya therefore had two parliaments and governments vying for domestic legitimacy and international recognition (Wehrey 2014; Lacher 2020a).

The approach of Western governments and the UN towards this situation was to avoid granting unconditional recognition to either side, though they distinguished between the status of the eastern-based and Tripoli-based authorities. The Tripoli government—which opposed Haftar and his allies—was unable to secure any meetings with Western diplomats and UN officials. HoR chairman Agila Saleh, by contrast, spoke at the UN General Assembly in September 2014, and Thinni's foreign minister was able to get a limited number of meetings with foreign officials, including one with US Deputy Secretary of State Antony Blinken in February 2015 (UN News 2014; Alamy 2015). But as the crisis wore on, the position of the UN and Western governments towards the east-based authorities became increasingly ambiguous. In November 2014, Libya's Supreme Court ruled that the basis on which the HoR elections were held was unconstitutional. The UN merely responded by saying that it was 'studying' the ruling, and henceforth neither the UN nor Western governments would state clearly whether they still considered the HoR to be Libya's legitimate parliament (UNSMIL 2014). The UN then began convening talks between HoR and GNC representatives, which would ultimately lead to the conclusion of the Libyan Political Agreement in December 2015 and the formation of the Government of National Accord (GNA) a month later.

But throughout 2015, the question of which among Libya's competing authorities was internationally recognized remained unresolved. It was in this context that the HoR on 2 March 2015 adopted a law establishing the post of a 'General Commander of the Armed Forces', a position superior to that of the chief of staff and including the competencies of the defence minister. The same day, HoR chairman Saleh appointed Haftar to the post, thereby finally giving him a formal and leading position (Lacher 2020b). As would become increasingly common in the HoR, it remained unclear whether the law was adopted with the necessary quorum. But Haftar's foreign supporters, including foreign ministers Le Drian of France and Lavrov of Russia, would later point to this episode as justifying their cooperation with Haftar, since he had been appointed by the internationally recognized parliament (Gubert et al. 2019). One month after his appointment, Haftar had his first publicized meeting with a foreign head of state, Jordan's King Abdullah II (Al Jazeera 2015).

Even though Haftar had started out in open rebellion against what was then Libya's sole government, he had always styled himself as leading 'the army' rather than a rebel group. The fact that international recognition then dissolved in Libya's legitimacy crisis became a structural advantage for Haftar in his quest for international legitimacy: from March 2015, he could claim an official position as the head of Libya's army, and until early 2016, he did not have to laboriously undermine the international legitimacy of his adversaries—whom he nevertheless denounced as terrorists and criminals.

Paradoxically, however, Haftar made giant strides towards international accreditation *after* foreign governments once again had an official interlocutor in Tripoli, in the form of the GNA—even though Haftar was by far the GNA's biggest opponent. Formal international recognition of the GNA did not settle the question of international legitimacy, which remained there for the taking. The reason was that the GNA was unable to back up its international recognition with de facto authority. In fact, it became increasingly weak and dysfunctional, while Haftar grew ever stronger due to the foreign support he enjoyed. In this context, Libyan actors remained hypersensitive to the signals foreign governments were sending through the optics and semantics of their engagement with the conflicting parties, Haftar first and foremost among them.

The role of regional powers amid multipolarity

Along with Syria, Libya from 2014 onwards became the first warzone in which patterns of foreign intervention clearly exhibited the international system's evolution towards multipolarity: a decline in the relative power of the US and European states vis-à-vis rising middle and regional powers.[4] Alarmed by the Iran nuclear agreement and Western support for democratic transitions from 2011 onwards, the Gulf Arab states and—after the 2013 military coup—Egypt, began pursuing increasingly independent and aggressive foreign policies. This development overlapped with Russia's pursuit of a revisionist foreign policy as its relations with Western states turned increasingly antagonistic after the 2014 annexation of Crimea. The shift towards multipolarity therefore coincided with a forceful push for counter-revolution by several authoritarian powers, most of which were considered important allies or partners by Western states (Kausch 2015; Kamrava 2018). In Libya, Haftar was the primary beneficiary of this trend, which allowed him to gain powerful regional backers who

[4] On these more contemporary developments in international politics and its effect on the legitimation of armed groups, see the section on the historical time of 'sovereigntism' in the Introduction to this volume.

were ready to validate his claims to legitimacy by supporting him militarily and politically.[5]

Much has been made of Haftar's Cold War relationship to the CIA.[6] However, when Haftar mobilized his militia coalition in 2014, he received no support from the US. The CIA later cooperated with Haftar on counterterrorism, but the overall US attitude towards him remained one of wariness, not least because of his growing reliance on Russian support.[7] Instead, Haftar's key backers were regional powers pursuing policies increasingly independently of the US. He obtained support from Egypt and the UAE in the form of weapons shipments very early on in his operation, in the summer of 2014 (UNSC, 2015). In August 2014, UAE fighter jets bombed the positions of forces opposed to Haftar's alliance in the Tripoli area—an unprecedented UAE intervention far from its immediate neighbourhood, and one that surprised the US government (Kirkpatrick and Schmitt 2014). At the time, he emphasized to Arab media outlets that he intended to eradicate the Muslim Brotherhood in Libya (Asharq al-Awsat 2014). This message was clearly designed to resonate with the Egyptian and Emirati leaderships: only a year before, the Egyptian army, backed by the UAE and Saudi Arabia, had toppled the elected president and Muslim Brother Muhammad Mursi. More broadly, Haftar fit the authoritarian model which the Emiratis were promoting across the region as a response to the 2011 uprisings.

Egyptian and Emirati support, in turn, was instrumental in securing French military assistance for Haftar, which began in early 2016. At the time, French defence minister Le Drian had developed close relations with the Emirati and Egyptian leaderships, at the core of which were a series of major arms deals. Another reason for France to covertly back Haftar were the terrorist attacks that hit Paris in 2015 and profoundly impacted French domestic politics and foreign policy (Lacher 2021). By then, an affiliate of the Islamic State had emerged in Benghazi, further facilitating Haftar's efforts to portray his campaign in the city as a war against terrorists. French special forces played a key role in helping Haftar break a long stalemate and make advances in Benghazi in early 2016. Although this assistance was covert, it quickly became known (Guibert 2016). French backing for Haftar signalled to Libyan political actors—and to other states—that France's actual Libya policy contradicted its official position of supporting the internationally recognized GNA.

[5] On supporting as a practice of claim-validation, see the Introduction in this volume.

[6] Haftar's biography mirrors the evolution from the bipolar Cold War order to the dissolution of the US-dominated world order into multipolarity during the 2010s. In 1987, as a leading commander in Libya's war against Chad's Hissène Habre, he was made prisoner. While in captivity, he defected to the exiled Libyan opposition along with several other fellow prisoners-of-war. He became head of the military wing of the National Front for the Salvation of Libya, which received training by the CIA for operations against Qadhafi from bases in Chad. But before these forces could enter into action, Habre was toppled by his former ally Idris Déby with Libyan and French support, and the Libyan opposition fighters were hurriedly airlifted to the US. Haftar became a US citizen and continued to live in Virginia until 2011, but his group rapidly grew irrelevant and defunct, particularly after the US rapprochement with Qadhafi in the 2000s (Barfi 2014; Lacher 2020b).

[7] Author interviews with former US Ambassador to Libya Deborah Jones and former US Special Envoy to Libya Jonathan Winer, June 2022; author's informal conversation with senior US diplomat, October 2021.

Russia's resurgent role in the Middle East and North Africa also contributed to Haftar's rise. Foreign Minister Lavrov and Defence Minister Shoigu were the first top officials of a major power to publicly meet Haftar, in November 2016 (Tsvetkova 2016). This was followed in January 2017 by a well-publicized visit by Haftar to a Russian aircraft carrier in the Mediterranean off Libya's coast (Reuters 2017). The Russian rationale may well have been to use Haftar to undermine the GNA, which Western governments were supporting.

Why did Western governments not take a more robust approach towards Haftar in 2016, given that he remained the biggest holdout against the GNA and violently opposed the GNA's attempts to exercise authority in eastern Libya, while also reaching out to Russia? Why did concerns about human rights not prove a bigger obstacle to Haftar's legitimation through public meetings with senior representatives of Western states in the years thereafter?

As with French military assistance, at least part of the answer must lie in the international weight of Emirati and Egyptian political backing for Haftar—in other words, in multipolarity. According to the US ambassador between 2013 and 2015, Deborah Jones, both Emirati Foreign Minister Abdallah ben Zayed and the ambassador in Washington, Yousef Otaiba, tried to influence senior US officials on Libya policy—presumably lobbying for Haftar as the only viable bulwark against perceived Islamist radicals, as Otaiba had with Jones.[8] After the election of Donald Trump in late 2016, the Emirati and Egyptian leaders persistently pitched Haftar to Trump himself, as well as to his close entourage. Both Egypt and the UAE, in turn, also cooperated with Russia in backing Haftar (Kirkpatrick 2020). The ambiguity and fluidity of international alliances amid multipolarity was also evident in the French attitude towards a resurgent Russia during the same period. After all, Macron repeatedly emphasized that he saw Russia as a partner in the fight against 'Islamist terrorism' (Vie publique 2017; Élysée 2018; Macron 2019). Russia's accreditation of Haftar on the international stage therefore did not stand in contradiction with Macron's own unprecedented efforts in that area.

In Libya, as elsewhere, Russia, Egypt, and the UAE promoted a model of the violent restoration of authoritarian rule that displayed utter disregard for ostensible Western concerns about human rights. Moreover, the proliferation of new intervening powers on opposing sides of conflicts also rendered Western states' template for ending civil wars through negotiated settlements increasingly impracticable (Howard and Stark 2018; Lacher 2022). Faced with the alternatives of continuing instability or stalemate, the return of strongman rule became more appealing again, including from the perspective of Western policymakers. At the same time, particularly for the US and France, engagement with Haftar was also anchored in continuities with the unipolar moment: the pragmatic cooperation on counterterrorism with unpalatable allies, from militias to autocrats, regardless of their human rights record (Byman 2006; De Lauri and Suhrke 2021; Chebli 2022).

[8] Interview with author, June 2022.

In sum, Haftar benefited from the new multipolar order in the Middle East and North Africa: the regional powers that backed him had seen their influence grow vis-à-vis that of the US and European states, with the latter increasingly aligning their policies with the regional powers, rather than the other way around. Despite the UAE's blatant violations of the UN arms embargo, US officials remained extremely reluctant to publicly mention the UAE's destabilizing role in Libya. This remained true even when Emirati backing for Haftar facilitated the deployment of the Wagner Group in 2019. Other aspects of the US relationship with the UAE were simply far more important than US interests in a stable Libya: arms sales, Emirati financial clout, mobilizing regional allies against threats such as Iran and the Islamic State; and normalization with Israel (Kirkpatrick 2019; Emmons and Cole 2020; Powers 2021). Egyptian and Emirati backing also helped explain why Macron and other French officials vociferously denounced the Turkish support for Haftar's enemies, but had only muted criticism for the Wagner Group's presence in Libya, and remained entirely silent on the Emirati role (Reuters 2020; Élysée 2020).

Legitimation practices

Through which practices did Haftar seek international legitimacy, and did his foreign backers seek to get him accredited on the international stage? Chief among these practices certainly was Haftar's claim to legitimacy by mimicking the state through establishing and demonstrating territorial control. This was not only central to his efforts to portray himself as the head of 'the army'. It was also the precondition both for him to continue receiving foreign military assistance—with which, in turn, he could keep consolidating control—and for foreign states to openly deal with him as a key player, thereby validating his claims. Another way in which Haftar's forces mimicked the state was by styling themselves as a regular army both through language and visual propaganda. Accepting Haftar's nomenclature was an important way in which foreign officials validated his claims, but more important were their public meetings with Haftar.

For Haftar, these practices served a dual purpose: in addition to helping him gain international legitimacy, they were also addressed to his domestic audience. Portraying his forces as 'the army' rather than just another militia was key to the LAAF's message that it was bringing stability and security. Showing that Haftar had international recognition, in turn, helped bolster his domestic legitimacy and counter his opponents' depiction of him as a warlord who had no legitimate official position. In critical phases such as the early weeks of the Tripoli offensive, an event such as the White House statement on Trump's call with Haftar was a boon for the LAAF, since it appeared to show not only that the US would not oppose Haftar's attempt to take power by force, but that it considered Haftar legitimate. Claims to domestic legitimacy and the validation of Haftar's international legitimacy therefore mutually reinforced each other, and both supported the impression that Haftar was winning.

Mimicking the state

Foreign support was critical to Haftar's gradual expansion, but this expansion was equally crucial for Haftar to retain foreign support and encourage foreign governments to legitimate him as a counterpart. This first became apparent after Haftar took over eastern Libya's oil ports in September 2016, thereby acquiring uncontested dominance over eastern Libya and ending any attempts by rival eastern figures to weaken his authority by allying themselves with the UN-backed GNA. Haftar gained substantial international goodwill by ensuring that the oil continued to flow, despite the fact that oil revenues kept accruing to the central bank in Tripoli. He further gained in stature after his forces expanded southwards in May and June 2017. By monopolizing control over entire regions, Haftar's forces set themselves apart from all other Libyan armed groups, most of which barely controlled entire neighbourhoods. The LAAF thereby gradually narrowed the gap between their claim of being a national army and the reality of their reliance on particular local and regional factions. Haftar's growing territorial control contrasted starkly with that of the GNA, which barely seemed able to exercise authority in Tripoli, let alone outside the capital.

Up until his defeat in the Tripoli war, Haftar was very successful in converting his territorial control into credibility among foreign interlocutors. In 2017–18, as Haftar continued to expand, it became a mantra among Western diplomats to state that there could be no solution without Haftar, and that he had to be engaged with because he was de facto in charge of vast parts of the country (Fernández-Molina 2023). Ostensibly, such engagement aimed at bringing Haftar into the fold—ideally under the umbrella of the GNA's civilian authority. The closer Haftar came to becoming the dominant force in Libya, however, the more lukewarm Western governments' support for the GNA as the internationally recognized government became. In February 2019, shortly after the LAAF had taken over much of southern Libya, a European diplomat said: 'we need to be realistic. Do we continue supporting the GNA if Haftar controls seventy or eighty percent of the country?'[9] By the time Haftar attacked Tripoli in April 2019, then, his territorial control and support from influential Middle Eastern powers had raised his international standing above that of the GNA. Western governments failed to come to the rescue of the government on whose internationally recognized status they had insisted for years.

From the outset, Haftar's forces displayed their ambition by mimicking the state through language and imagery. In this regard, Haftar's branding of his forces as 'the army' was a masterstroke. Since 2011, Libyan armed groups had been creative

[9] Interview with author, Tunis, February 2019.

in coining names that sounded both official and powerful—the 'Supreme Military Council of Cyrenaica' or the 'Special Deterrence Force', to take just two examples. But the 'Libyan National Army' was far more audacious—even more so as Haftar's forces were at first in open rebellion against Libya's government, and that they were composed primarily of civilians. At the same time, despite evoking the status of an official military, the names Haftar used—'Libyan National Army' at first, then increasingly 'Libyan Arab Armed Forces'—set his forces apart as his personal project, since Libyan law only knows the 'Libyan Army'. This remained the case even after Haftar was officially appointed as general commander of the armed forces.

From the beginning of his Benghazi offensive, Haftar benefited from supportive media coverage from TV channels and social media pages run by political allies. These media outlets popularized the designation of Haftar's forces as simply 'the army'. Adverse Libyan media coverage, however, tended to refer to the same forces as 'Haftar's militias' and avoid any use of the names they gave themselves. International media coverage, in turn, generally referred to the 'self-styled' or 'so-called' Libyan National Army, thereby making clear that the claim of Haftar's forces to official status was at a minimum contested. Likewise, sympathetic media coverage and the LAAF's own propaganda outlets systematically referred to Haftar by his official position and the rank he had been awarded by HoR chairman in September 2016: 'the General Commander of the Libyan Arab Armed Forces, Field Marshal Khalifa Belgasem Haftar'. Adverse media outlets, by contrast, called him 'war criminal Khalifa Haftar', while the international media typically referred to him as 'General Khalifa Haftar', thereby refusing to accept the grade and title Haftar claimed at face value.

As Haftar's enterprise developed, its own propaganda operations became increasingly sophisticated. Through its social media accounts and its unofficial satellite channel Libya al-Hadath TV, it disseminated professionally produced imagery aiming to portray the LAAF as a disciplined, regular army. Pictures and videos showed masses of uniformed soldiers operating sophisticated weaponry (LAAF Media Office 2016, 2021a, 2021b). Photos of recruits undergoing training and exams were a particularly popular theme (Libya al-Hadath 2020, 2021; LAAF Morale Guidance Division 2023), as were Haftar's speeches at military manoeuvres, in front of parading soldiers, or addressing large numbers of senior LAAF officers (Libya al-Hadath 2022a, 2022b, 2022c).

Such imagery contrasted starkly with the reality of the LAAF, whose members often shared videos on their private social media accounts that showed long-haired, bearded young men in motley clothing, at times randomly shooting in the air while under the influence of drugs, voicing threats against entire communities or committing vicious crimes such as desecrating corpses (Lacher 2019). And whereas the senior officers visible on LAAF imagery conveyed the idea of a military institution, actual power was wielded by Haftar's sons, in-laws, and field commanders who mostly had a civilian background. However, there is little doubt that foreign officials were exposed to LAAF propaganda far more regularly than to imagery that showed the real face of Haftar's forces.

Acknowledging and officializing

In a situation of unsettled international legitimacy, foreign officials did not approach the question of Haftar's international recognition or its absence as a binary choice. Instead, they engaged in a range of micro-practices that sent nuanced messages about their attitude towards Haftar and other Libyan actors (Fernández-Molina 2023). Two ways of finely calibrating these messages were the semantics by which they referred to Haftar, and the optics of their meetings with him. In both cases, international actors were keenly aware of the legitimacy they could bestow on Haftar with a particular act (Fernández-Molina and Casani 2023). Their willingness to do so fluctuated along with their assessments of Haftar's chances of taking power and whether they considered his behaviour to be cooperative.

For example, in the first half of 2016, as the UN Support Mission in Libya's (UNSMIL) backing for the GNA strained the relations of Special Representative of the Secretary-General (SRSG) Martin Kobler with Haftar, the UN secretary general's reports on Libya merely referred to the LAAF as 'forces loyal to General Khalifa Haftar' (UNSC 2016a, 2016b). Following the LAAF takeover of the oil ports in September 2016, however, these reports began referring to the 'Libyan National Army' (UNSC 2016c, 2017). In March 2019, after the LAAF's expansion in southern Libya but before its Tripoli offensive, SRSG Ghassan Salamé conferred the highest honours on the group by repeatedly referring to it simply as 'the army', and to Haftar as field marshal, in interviews with Libyan and regional TV channels (Libya al-Hadath 2019). Following Haftar's attack on Tripoli, which nullified all of Salamé's diplomatic efforts, he reverted to 'General Haftar's forces' while briefing the Security Council (UNSMIL 2019).

As Haftar's inability to capture Tripoli became clear and his forces conducted increasingly indiscriminate attacks on civilian areas, UNSMIL chose to put 'Libyan National Army' in quotation marks, thereby marking it as a name rather than an official designation (UNSMIL 2020b, 2020d). But when LAAF representatives signed a ceasefire with their GNA counterparts under UN auspices, the quotation marks disappeared (UNSMIL 2020f). Until March 2020, Italy and the US had respectfully referred to Haftar's forces as the 'Libyan National Army' or the 'LNA' (Italian Embassy in Libya 2020a; US Embassy in Libya 2020). But when in May, LAAF shelling hit close to the Italian embassy, Italy's foreign ministry condemned 'yet another attack by Haftarian forces' (Farnesina 2020), and when in July Haftar refused to lift his blockade on oil exports while also allowing the Russian Wagner Group to expand in southern Libya, the US embassy hit hard by referring to the 'so called Libyan National Army' (US Embassy in Libya 2020b).

Getting foreign governments to adopt Haftar's nomenclature was therefore synonymous with winning international recognition for his armed group as Libya's army. The more Haftar appeared to be winning, the more willing Western governments and the UN were to validate Haftar's legitimacy claims by adopting his preferred language—until they reversed course when Haftar lost the Tripoli war.

Libyan actors and observers—such as commentators on social media—registered even the finest nuances in foreign governments' choice of wording: while international legitimacy remained an unsettled question, they were hypersensitive to the signals foreign states were sending. Overall, however, foreign states' semantic accommodation of Haftar clearly supported his efforts to distinguish his forces from other Libyan armed groups, and thereby helped draw the emerging boundaries between state and non-state actors (Fernández-Molina and Casani 2023).

Foreign governments were equally aware that openly meeting Haftar bestowed legitimacy on him and lifted his status in the eyes of the Libyan public. For this precise reason, Western government officials were long reluctant to publicly meet Haftar. According to Deborah Jones, the US ambassador from 2013 to 2015, Haftar's associates reached out on several occasions to seek a meeting for him, but 'we didn't want to give him that kind of profile'.[10] Jonathan Winer, who was the US special envoy to Libya in 2015 and 2016, met Haftar twice, for lengthy meetings—but made sure that these meetings would not be publicized. According to Winer,

> we did not want to legitimize him. Our position was, if you give Haftar legitimacy, through public meetings, he will use that to strengthen himself at the expense of others, and we confer legitimacy upon him. That's a bad idea. No pictures with him, no meetings with him of a public nature.[11]

Winer's approach was made easier by the fact that he met Haftar in foreign capitals, rather than in eastern Libya, and was therefore able to control the setting. The UK ambassador Peter Millett tried to follow the same policy but was surprised when pictures of his meeting with Haftar at the latter's headquarters were leaked immediately afterwards.[12] The incident showed how eager the LAAF's propagandists were to use meetings with foreign officials to score points with their domestic audience.

Contrary to US diplomats, French officials apparently thought that bestowing international legitimacy upon Haftar was a very good idea. After all, that is what Macron's summit at La Celle-Saint-Cloud did in July 2017, in a wholly unprecedented way. In the preceding months, Haftar's regional state sponsors had elevated him from a covert proxy to their official Libyan counterpart. In April and again in July 2017, the Emirati de facto ruler Mohamed ben Zayed gave Haftar high profile receptions in Abu Dhabi (Sky News Arabia 2017; Emirates 71 2017). In May, Egyptian President al-Sisi publicly received Haftar, and in July, he hosted Haftar alongside ben Zayed, the crown prince of Bahrain and the Kuwaiti defence minister for the opening of a major military base in Egypt's western desert (Alwatan 2017; Alarabiya 2017). With the July 2017 summit meeting hosted by Macron, Haftar became fully accredited on the international stage (Semo 2017). The meeting effectively accorded Haftar the same status as Fayez al-Sarraj, the head of the internationally recognized government in Tripoli. Macron gained nothing in exchange other than vague commitments

[10] Interview with author, June 2022.
[11] Interview with author, June 2022.
[12] Interview with author, July 2022; see also Middle East Eye (2017).

by Haftar to hold elections. In retrospect, it appears likely that establishing Haftar on the international stage was a key objective of the meeting for the Élysée, quite regardless of the concrete outcomes it might yield.

After La Celle-Saint-Cloud, any inhibitions Western officials had in publicly meeting Haftar disappeared, as a rapid succession of meetings in the second half of 2017 showed. While previous meetings between US diplomats and Haftar had remained confidential, in August 2017, the US for the first time publicly announced that its ambassador had met Haftar, though no pictures of the meeting were published (US Embassy in Egypt 2017). The same month, Haftar met Lavrov and Shoigu in Moscow for a well-publicized visit, and UK Secretary Boris Johnson became the first senior Western official to visit Haftar at his headquarters outside Benghazi (Radio Free Europe 2017; Cusack 2017). Italian Interior Minister Marco Minniti and French Foreign Minister Le Drian followed hard on Johnson's heels, in early September (ANSA 2017; MEAE 2017). In mid-September, Tunisian President Béji Caid Essebsi received Haftar, and the month was crowned by the first of Haftar's two official visits to Rome that year, to meet with Defence Minister Roberta Pinotti (Nigro 2017; Business News 2017).

Haftar would continue to be received in Paris, Rome, Berlin, Moscow, and regional capitals even as his forces attacked Tripoli from April 2019 onwards. Egyptian and Emirati lobbying led Trump to hold a phone call with Haftar less than two weeks into his offensive (Kirkpatrick 2020). In the call, according to the White House readout, Trump did not demand a ceasefire; instead, he

> recognised Field Marshal Haftar's significant role in fighting terrorism and securing Libya's oil resources, and the two discussed a shared vision for Libya's transition to a stable, democratic political system.
>
> (Holland 2019, para 3)

Trump's stance was a clear endorsement of Haftar, and stood in direct contradiction with his administration's official position, which was to demand that Haftar stop his offensive. Trump's attitude would doubtlessly have presaged Haftar's recognition as Libya's ruler if he had been militarily successful. But when the Tripoli war raged on for months without Haftar making substantial progress, Trump reportedly expressed his disappointment to administration officials and lost interest in Libya.[13]

Once Haftar began receiving a steady stream of foreign visitors at his headquarters, his propagandists carefully orchestrated the imagery of those meetings. Pictures showed Haftar on a throne-like chair in the centre, against the background of a big golden eagle, the LAAF crest; his visitors were arranged in a row of ordinary chairs on the side, leaning towards the commander, showing them as supplicants (German Ambassador to Libya 2020; UNSMIL 2020a). The US ambassador, who was in a position of strength vis-à-vis Haftar, was not treated to such a hierarchical setup (Zaptia

[13] Author interviews with senior European and US diplomats, Tunis, Rome, and Berlin, September–December 2019.

2020). When standing for a handshake, Haftar was invariably on the left, enabling him to easily stretch out his right hand and look straight into the camera, while his visitors had to turn to him and twist their head towards the lens. Visitors who were oblivious to such *mises en scène*, such as UN Special Envoy Ján Kubiš, even agreed to having their handshake photographed against the background of a Quranic verse in calligraphic golden letters—'we have given you a clear conquest'—that had inspired the name for Haftar's Tripoli offensive, thereby giving the impression that they were endorsing Haftar's claim to power (LAAF General Command 2021b).

As with foreign governments' willingness to adopt Haftar's preferred nomenclature, their readiness to entertain his exploitation of meetings with foreign officials also fluctuated. As the Tripoli war dragged on and Haftar's inability to make further advances became clear, the stream of foreign visitors petered out. Just after Haftar's forces had been forced to withdraw from western Libya in June 2020, leaving behind mines and booby-traps across wide swathes of Tripoli's residential southern suburbs, a tone-deaf public visit by the German ambassador triggered not only an outcry on Libyan social media, but also incomprehension among other Western diplomats.[14] Egypt's President al-Sisi now pointedly received Haftar not as an actor in his own right, but together and on equal footing with HoR chairman Saleh, thereby relativizing Haftar's status (Asharq al-Awsat 2020; Ahram Online 2020). Italy's Prime Minister Giuseppe Conte and Foreign Minister Luigi di Maio were the only senior Western officials to visit Haftar in the year following his Tripoli defeat. In fact, according to di Maio, Italy's relations with Haftar had broken off for several months, but the government agreed to re-establish them with Conte's and di Maio's visit in exchange for Haftar releasing Italian fishermen he was holding captive (AGI 2020). Acting SRSG Stephanie Williams deliberately refrained from meetings with Haftar as she negotiated a ceasefire and a unity government in late 2020 and early 2021. Haftar had to wait for her successor Ján Kubiš before being honoured with a visit (LAAF General Command 2021a). When Macron held another Libya summit in November 2021, he invited two representatives of the new unity government, but not Haftar—in an obvious effort to rid himself of his reputation as having been one of Haftar's key foreign enablers (Élysée 2021).

Foreign officials had another way of shaping the message they were sending by publicly meeting Haftar: through their body language. In a context where Western ambassadors increasingly had their own social media accounts, Libyan reactions to the photos they published heightened foreign officials' sensitivity to how their posture would be perceived. The British ambassador's stony-faced performance at a February 2023 meeting with Haftar contrasted starkly with the French special envoy's cordial smiles a month later—with the former suggesting reluctant engagement with Haftar, the latter endorsement (LAAF General Command 2023a, 2023b).

Foreign governments, then, had multiple choices in deciding whether—and to what extent—to bestow international legitimacy upon Haftar with meetings: they could choose whether or not to meet him, whether or not to make these meetings

[14] Interviews with Western and UN diplomats, June 2020.

public, and whether or not to allow him to define the optics of these meetings. Their choices directly reflected their assessments of whether Haftar was on the path to ruling Libya—and as such they were carefully studied and debated by Libyan actors and observers.

As Haftar's sons began taking on increasingly important political and military roles from 2020 onwards, Western officials have made these same choices again. For several years, diplomats were reluctant to let themselves be seen with Haftar's sons, wary of the backlash this might provoke. But during the spring of 2024, public meetings with Saddam, Khaled, and Belgasem Haftar rapidly became routine. Western embassies' communications respectfully addressed them as 'lieutenant general' or 'engineer', and showed smiling diplomats.

Conclusion

With his reception by Macron at Madame de Pompadour's former chateau in July 2017, Haftar appeared to be approaching the peak of international legitimacy. He had done so within a mere three years of founding a coalition of unruly militias that included neighbourhood vigilantes, Salafist fighters, and foreign mercenaries, and which had rapidly become Libya's foremost perpetrators of war crimes. And he was being courted by Western states even as he gained a reputation for being Russia's man in Libya.

If Haftar's rise to international respectability appeared astounding, it owed much to the particular international moment in which it occurred. Indeed, Haftar's forces themselves were in many ways a product of the new age of multipolarity. The political and military support of newly assertive regional powers was crucial to the inexorable advance of the LAAF, and of the authoritarian model they represented. Wielding greater influence in Western capitals than ever before, these regional powers not only subverted the policies of the very Western states they claimed to be allying with, but increasingly encouraged states like France and the US to either emulate their Libya policy or refrain from opposing it too insistently. Moreover, the fluidity and ambiguity of the emerging multilateral order showed in the fact that Egypt, the UAE and France found themselves on the same side as Russia in backing Haftar.

Yet, regional powers' efforts at lobbying Western states to legitimate Haftar would have been far less persuasive if it had not been for Haftar's military successes—successes Haftar owed in no small measure to the foreign backing he received. In an otherwise hopelessly fragmented Libyan military landscape, Haftar's forces were the only ones to establish territorial control over entire regions. To sympathetic foreign observers, and at least for a while, it looked like Haftar could eventually take over the entire country—and in the meantime, he controlled much of Libya's oil production.

Haftar's mimicking of the state through territorial control enticed Western states to establish ties with him, and the possibility of his eventual victory made these states willing to do so in a way that validated Haftar's claims to legitimacy. Senior foreign officials met him publicly, often in the knowledge that his propagandists would exploit the optics of these meetings. Foreign governments indulged in Haftar's pretence that he was a field marshal and headed an army, or even 'the army'. Language and imagery were key tools both for Haftar's claims to legitimacy and for foreign governments to validate those claims.

The fact that Haftar initially did not have to compete against an internationally recognized government facilitated his quest for legitimacy. Thereafter, the dysfunctional nature of the unity government in Tripoli prevented it from settling the question of international legitimacy. In this context, Libyan actors and observers were hypersensitive to the nuances in the optics and semantics of Western governments' engagement with Haftar. Foreign governments constantly had to calibrate the *degree* of legitimacy they would confer upon Haftar. Western states' reactions to Haftar's eventual defeat in Tripoli suggested that legitimacy could be taken away even more quickly than it had been accorded. Just as the ultimate military victory, outright recognition has so far remained elusive for Haftar.

References

Agenzia Italia (AGI). 2020. 'Di Maio: "Per liberare i pescatori abbiamo ripreso i rapporti con Haftar"', 23 December. https://www.agi.it/estero/news/2020-12-23/libia-di-maio-pescatori-liberi-rapporti-haftar-10791309/

Ahram Online. 2020. 'Egypt's Sisi Receives Haftar, Aguila Saleh to Discuss Libyan Developments', 23 September. https://english.ahram.org.eg/NewsContent/1/64/383760/Egypt/Politics-/Egypt%E2%80%99s-Sisi-receives-Haftar,-Aguila-Saleh-to-disc.aspx

Alamy. 2015. 'Deputy Secretary of State Antony Blinken meets with Libyan Foreign Minister Mohamed Dayri at the U.S. Department of State in Washington, D.C, 19 February 2015', 19 February. https://www.alamy.com/stock-photo-deputy-secretary-of-state-antony-quottonyquot-blinken-meets-with-libyan-79303133.html

Alarabiya. 2017. 'Egypt's Sisi Opens Biggest Military Base in Middle East and Africa', 22 July. https://english.alarabiya.net/News/middle-east/2017/07/22/Egyptian-President-opens-biggest-military-base-in-the-Middle-East-and-Africa

Asharq al-Awsat. 2014. 'Haftar: Hadafna Tandhif Libya min al-Ikhwan al-Muslimin' [Haftar: Our Goal Is to Cleanse Libya of the Muslim Brotherhood], 20 May. https://aawsat.com/home/article/100236

Asharq al-Awsat. 2020. 'With Haftar's Support, Sisi Launches Cairo Declaration to Resolve Libya Conflict', 6 June. https://english.aawsat.com/home/article/2320866/haftar%E2%80%99s-support-sisi-launches-cairo-declaration-resolve-libya-conflict

Al Jazeera. 2015. 'Malik al-Urdun Yastaqbil Haftar w Yabhath Juhud Mukafahat al-Irhab' [King of Jordan Receives Haftar and Discusses Efforts to Combat Terrorism], 13 April. https://www.aljazeera.net/news/2015/4/13/ملك-الأردن-يستقبل-حفتر-ويبحث-جهود

Al-Obeidi, Salem. 2014. 'Majlis al-Nuwab: Ansar al-Shari'a w Fajr Libya Jama'at Irhabiyya' [Parliament: Ansar al-Sharia and Fajr Libya Are Terrorist Groups], *al-Wasat*, 24 August. https://alwasat.ly/news/libya/30200

Alwatan. 2017. 'Al-Sisi Yabhath ma' Haftar Raf' Hadhr Taslih al-Jaysh al-Libi' [Sisi Discusses with Haftar Lifting the Arms Embargo on the Libyan Army], 13 May. https://alwatannews.net/Arab/article/714512/السيسي-يبحث-مع-حفتر-رفع-حظر-تسليح-الجيش-الليبي

ANSA. 2017. 'Minniti met General Haftar in Benghazi', 6 September. https://www.ansa.it/english/news/2017/09/06/minniti-met-general-haftar-in-benghazi_a01b7586-4349-4927-9f89-4b27e8c78fdc.html

Barfi, Barak. 2014. *Khalifa Haftar: Rebuilding Libya from the Top Down*. Research Note. Washington, D.C.: Washington Institute for Near East Policy.

Business News. 2017. 'Béji Caïd Essebsi reçoit le maréchal Khalifa Haftar', 18 September. https://www.businessnews.com.tn/beji-caid-essebsi-recoit-le-mareal-khalifa-haftar,520,74892,3

Byman, Daniel. 2006. 'Friends Like These: Counterinsurgency and the War on Terrorism'. *International Security* 31(2): pp. 79–115.

Chebli, Denia. 2022. 'Économie transnationale de la violence et autonomisation des milices au Nord du Mali'. *Cultures et Conflits* 125: pp. 95–114.

Cusack, Robert. 2017. 'When Boris Met Khalifa: Libya's Haftar Receives UK Foreign Secretary in Benghazi'. *The New Arab*, 25 August. https://www.newarab.com/news/libyas-haftar-receives-uk-foreign-secretary-benghazi

Driscoll, Jesse. 2015. *Warlords and Coalition Politics in Post-Soviet States*. Cambridge: Cambridge University Press.

Eaton, Tim. 2021. 'The Libyan Arab Armed Forces: A Network Analysis of Haftar's Military Alliance'. Chatham House Research Paper, June.

Élysée. 2018. 'Transcription du discours du Président de la République à la conférence des Ambassadeurs'. 27 August. https://www.elysee.fr/emmanuel-macron/2018/08/27/discours-du-president-de-la-republique-a-la-conference-des-ambassadeurs

Élysée. 2020. 'Conférence de presse conjointe avec Angela Merkel'. 30 June. https://www.youtube.com/watch?v=6KDrwCPYxW4

Élysée. 2021. 'Conférence internationale pour la Libye à Paris'. 12 November. https://www.elysee.fr/emmanuel-macron/2021/11/12/conference-internationale-pour-la-libye-a-paris

Emirates 71. 2017. 'Muhammad bin Zayid Yastaqbil Haftar fi Abu Dhabi' [Mohammed bin Zayed Receives Haftar in Abu Dhabi], 9 July. https://emirates71.net/posts/45096

Emmons, Alex and Matthew Cole. 2020. 'Arms Sale to UAE Goes Forward Even as U.S. Probes Tie Between UAE and Russian Mercenaries'. *The Intercept*, 2 December. https://theintercept.com/2020/12/02/uae-arms-sale-wagner-group/

Farnesina. 2020. 'Italia condanna con fermezza ennesimo attacco di forze haftariane'. *Twitter*, 8 May. https://twitter.com/ItalyMFA/status/1258540485867483138

Fernández-Molina, Irene. 2023. 'The International Recognition of Governments in Practice(s): Creatures, Mirages, and Dilemmas in Post-2011 Libya'. *International Studies Review* 25(4), https://doi.org/10.1093/isr/viad050.

Fernández-Molina, Irene and Alfonso Casani. 2023. 'International Recognition Meets Areas of Limited Statehood: Practices and Effects on Hybrid Actors in Post-2011 Libya'. *Journal of Intervention and Statebuilding* 17(5): pp. 624–645.

Fitzgerald, Mary. 2023. '"Changed Utterly": How the 2014–18 War Transformed Benghazi's Social Fabric'. In *Violence and Social Transformation in Libya*, edited by Wolfram Lacher and Virginie Collombier, pp. 97–131. London: Hurst.

German Ambassador to Libya. 2020. 'In Today's Meeting with Gen Haftar I Expressed Concern'. *Twitter*, 10 June. https://twitter.com/GermanAmbLBY/status/1270725346657091590

Giustozzi, Antonio. 2005. 'The Debate on Warlordism: The Importance of Military Legitimacy'. LSE Crisis States Development Research Centre, Discussion Paper 13, October. https://eprints.lse.ac.uk/13316/1/dp13.pdf

Gubert, Romain, Jean Guisnel, and Sébastien Le Fol. 2019. 'Jean-Yves Le Drian, la grande confession d'un silencieux'. *Le Point*, 12 June. https://www.lepoint.fr/politique/jean-yves-le-drian-la-grande-confession-d-un-silencieux-12-06-2019-2318321_20.php

Guibert, Nathalie. 2016. 'La France mène des opérations secrètes en Libye' [France is conducting secret operations in Libya]. *Le Monde*, 24 February. https://www.lemonde.fr/international/article/2016/02/24/la-france-mene-des-operations-secretes-en-libye_4870605_3210.html.

Holland, Steve. 2019. 'White House says Trump Spoke to Libyan Commander Haftar on Monday'. *Reuters*, 19 April.

Howard, Lise and Alexandra Stark. 2018. 'How Civil Wars End: The International System, Norms, and the Role of External Actors'. *International Security* 42 (3): pp. 127–171.

Italian Embassy in Libya. 2020. '#Italy welcomes the humanitarian truce accepted by the Libyan National Accord Government and the Libyan National Army'. *Twitter*, 20 March. https://twitter.com/ItalyinLibya/status/1241684600612696065

Kamrava, Mehran. 2018. 'Multipolarity and Instability in the Middle East'. *Orbis* 62(4): pp. 598–616.

Kausch, Kristina. 2015. 'Competitive Multipolarity in the Middle East'. *The International Spectator* 50(3): pp. 1–15.

Kirkpatrick, David. 2014. 'In Libya, a Coup: Or Perhaps Not'. *New York Times*, 14 February.

Kirkpatrick, David. 2019. 'The Most Powerful Arab Ruler Isn't M.B.S. It's M.B.Z.'. *New York Times*, 2 June.

Kirkpatrick, David. 2020. 'The White House Blessed a War in Libya, but Russia Won It'. *New York Times*, 14 April.

Kirkpatrick, David and Eric Schmitt. 2014. 'Arab Nations Strike in Libya, Surprising U.S.'. *New York Times*, 25 August.

Lacher, Wolfram. 2019. 'Who Are the Forces Fighting Each Other in Tripoli?'. *Twitter*, 12 December. https://threadreaderapp.com/thread/1205112543901048833.html

Lacher, Wolfram. 2020a. *Libya's Fragmentation: Structure and Process in Violent Conflict*. London: I.B. Tauris.

Lacher, Wolfram. 2020b. 'A Most Irregular Army: The Rise of Khalifa Haftar's Libyan Arab Armed Forces'. SWP Working Paper, November.

Lacher, Wolfram. 2021. 'Unser schwieriger Partner: Deutschlands und Frankreichs erfolgloses Engagement in Libyen und Mali'. SWP Studie, 26 February.

Lacher, Wolfram. 2022. 'African Conflicts amid Multipolarity: Implications of a Changing Actor Landscape'. Megatrends Afrika Policy Brief, April. https://www.megatrends-afrika.de/assets/afrika/publications/policybrief/MTA_PB03_2022_Lacher_African_Conflicts_amid_Multipolarity.pdf

Libya al-Hadath. 2019. 'Liqa' Khass ma' al-Mab'uth al-Umami Ghassan Salama' [Exclusive Interview with UN Envoy Ghassan Salamé], 26 February. https://www.youtube.com/watch?v=lVIiFTm-k0Y

Libya al-Hadath. 2020. 'Idarat al-Tadrib Tastaqbil al-Duf'a al-Thalitha min al-Mujannidin al-Juddud bi-Madrasat Dhubbat al-Saff' [Training Department Receives Third Batch of New Recruits at Non-Commissioned Officers School], 4 September. https://www.youtube.com/watch?v=ah3B-tfiTLQ&lc=UgxqPqAu22qaDff6qRt4AaABAg

Libya al-Hadath. 2021. 'Al-Akadimiyya al-'Askariyya bi-Hay'at al-Saytara Tajri al-Imtihanat al-Niha'iyya li-Talabat al-'Ulum al-'Askariyya', [Military Academy Conducts Final Exams for Military Science Students], 3 March. https://www.youtube.com/watch?v=xJB6Rtn1nVI

Libya al-Hadath. 2022a. Coverage of Haftar Meeting with LAAF Officers on the Occasion of Eid al-Fitr, 7 May. https://www.facebook.com/LibyaAlhadathTV/posts/pfbid0v6xCgd4T3HwCFuXQB8Az4KdTAdwfDM9T1fSSmsHGaVNWZoCe2V57FweExPPTcuVgl

Libya al-Hadath. 2022b. Pictures of Military Recruits' Graduation from Training in the Presence of Haftar, 31 May. https://www.facebook.com/LibyaAlhadathTV/posts/pfbid0ppWVyAuwstTXcV3EEzcw7ZTho5DHwkbYagpqeujRTfphE1z65a7yiqnahxAbLm6sl

Libya al-Hadath. 2022c. 'Ru'asa' al-Arkan w 'Addad min al-Qiyadat al-'Askariyya w Dhubbat al-Quwwat al-Musallaha Yuhanni'un al-Mushir' [Chiefs of Staff and Several Military Leaders and Armed Forces Officers Congratulate the Field Marshal on Eid al-Adha], 13 July. https://libyaalhadath.net/?p=105590&fbclid=IwAR3lktWsC2vnU3t0AZrIuVs8IDdv5maDSb590PLn5YnTC9lNJdSNuc6Dckk

De Lauri, Antonio and Astri Suhrke. 2021. 'Armed Governance: The Case of the CIA-Supported Afghan Militias'. *Small Wars and Insurgencies* 32(3): pp. 490–508.

Libyan Arab Armed Forces (LAAF) General Command. 2021a. 'Coverage of Haftar's Meeting with UN Special Envoy Ján Kubiš'. 27 March. https://www.facebook.com/General.official.leadership/posts/pfbid0KNTvePRBXtPUXekJKj83KRkQfojPRXWp6cpzSNx4ER9x4gCN4su8tcTT3V4wtUSwl

Libyan Arab Armed Forces (LAAF) General Command. 2021b. 'Coverage of Haftar's Meeting with UN Special Envoy Ján Kubiš'. 10 July. https://www.facebook.

com/General.official.leadership/posts/pfbid0f9pneg71dLKmP9gsN5ySgwRRXDGxdBoSmxBYZxG4AQzMbpg3mrnncCNtUedi5AJFl

Libyan Arab Armed Forces (LAAF) General Command. 2023a. 'Coverage of Haftar's Meeting with UK Ambassador Caroline Hurndall'. 7 February. https://www.facebook.com/General.official.leadership/posts/pfbid0FajX1PFXpCEoHNk7j3iptEeFT6USZiC1aeXZfi5eaC2GpSZVDH2LkGbi55oMkrqCl

Libyan Arab Armed Forces (LAAF) General Command. 2023b. 'Coverage of Haftar's Meeting with French Special Envoy Paul Soler'. 14 March. https://www.facebook.com/General.official.leadership/posts/pfbid0MNE4gGstFPFCkN33zV1rBHhtvHG6i6HZyYbZm8JNwz4ztVx9FfHNcyynYtoEqtfSl

Libyan Arab Armed Forces (LAAF) Media Office. 2016. 'Nuharib, Nabni, Nu'assis w Nantasir' [We Fight, We Build, We Establish, and We Triumph]. 16 March. https://www.youtube.com/watch?v=0qLHAqo9vus

Libyan Arab Armed Forces (LAAF) Media Office. 2021a. 'Bi-Sawa'id Abna' al-Umma al-Libiyya w bi-Tadhhiyyat Sha'bna al-Abiy, Nuqaddim Arwahna Fida'an Lillah thumma al-Watan' [With the Hands of Libya's Sons and the Sacrifices of Our Noble People, We Offer Our Lives for God and the Homeland] 9 June. https://www.youtube.com/watch?v=Pdq8HUEEOZg

Libyan Arab Armed Forces (LAAF) Media Office. 2021b. 'Munawarat Ra'd 2021' [Thunder Manoeuvre 2021]. 28 October. https://www.youtube.com/watch?v=RVa1Ngsql_0

Libyan Arab Armed Forces (LAAF) Morale Guidance Division. 2023. 'Coverage of Officers' Exams, 25 February. https://www.facebook.com/permalink.php?story_fbid=pfbid05yYRdfg5knfpYDBd4ogma2SKKLFkSMpe6Ma95ci33mCBGZtbaK3n37NWxSY36DkCl&id=477587285750110

Macron, Emmanuel. 2019. 'Qui est l'ennemi de l'OTAN?'. *Twitter*, 4 December. https://twitter.com/EmmanuelMacron/status/1202288236820926464

Marten, Kimberly. 2006. 'Warlordism in Comparative Perspective'. *International Security* 31(3): pp. 41–76.

Middle East Eye. 2017. 'British Ambassador Meets Haftar as Clashes Rage in Libya's Oil Crescent', 21 March. https://www.middleeasteye.net/news/british-ambassador-meets-haftar-clashes-rage-libyas-oil-crescent

Ministère de l'Europe et des Affaires Étrangères (MEAE). 2017. 'Libyen—Reise von Jean-Yves Le Drian (04.09.17)', 4 September. https://www.diplomatie.gouv.fr/de/neuigkeiten/article/libyen-reise-von-jean-yves-le-drian-04-09-17

Nigro, Vincenzo. 2017. 'Libia, Haftar incontra Minniti e Pinotti [Libya, Haftar Meets Minniti and Pinotti]'. *La Repubblica*, 26 September. https://www.repubblica.it/esteri/2017/09/26/news/libia_haftar_italia-176595587/

Powers, Colin. 2021. 'Dollars and Decadence: Making Sense of the US–UAE Relationship'. NORIA Research, April. https://noria-research.com/wp-content/uploads/2021/03/Noria-Research_Dollars-and-Decadence_2021_ENG.pdf

Radio Free Europe. 2017. 'Libyan Strongman Haftar Meets Lavrov, Shoigu in Moscow', 14 August. https://www.rferl.org/a/russia-libya-haftar-meeting-moscow-lavrov-shoigu/28676446.html

Reuters. 2017. 'East Libya Strongman Visits Russian Aircraft Carrier in Mediterranean: RIA', 11 January.

Reuters. 2020. 'France's Macron Accuses Turkey's Erdogan of Breaking Libya Promises'. 30 January. https://www.reuters.com/article/world/frances-macron-accuses-turkeys-erdogan-of-breaking-libya-promises-idUSKBN1ZS29G/

Semo, Marc. 2017. 'Emmanuel Macron parraine un accord en Libye, sans garantie de résultat. *Le Monde*, 27 July: p. 3.

Sky News Arabia. 2017. 'Muhammad bin Zayid: Nad'am al-Jaysh al-Libi bi-Harbih dhidd al-Irhab' [Mohammed bin Zayed: We Support the Libyan Army in Its War Against Terrorism], 10 April. https://www.skynewsarabia.com/middle-east/939379-محمد-بن-زايد-ندعم-الجيش-الليبي-بحربه-ضد-الإ رهاب

Tsvetkova, Maria. 2016. 'Libyan Commander Visits Russia to Ask for Help Fighting Islamists'. *Reuters*, 29 November. https://www.reuters.com/article/world/libyan-commander-visits-russia-to-ask-for-help-fighting-islamists-idUSKBN13O1XS/

US Embassy in Egypt. 2017. 'Statement by US Embassy Libya'. 11 August. https://eg.usembassy.gov/statement-u-s-embassy-libya/

US Embassy in Libya. 2020a. 'Ambassador Norland Commends the GNA and LNA'. 20 February. https://www.facebook.com/usembassytripoli/photos/a.241731185889960/2968777793185272/?type=3&theater

US Embassy in Libya. 2020b. 'US Embassy Regrets Foreign Interference Against Libya's Economy'. 12 July. https://www.facebook.com/USEmbassyLibya/photos/a.241731185889960/3324009520995429/?type=3&_rdr

UN News. 2014. 'At UN Assembly, Senior Libyan Official Calls for Sanctions against Political Spoilers', 27 September. https://news.un.org/en/story/2014/09/479432

UN Security Council (UNSC). 2015. 'Final Report of the Panel of Experts Established Pursuant to Resolution 1973 (2011)'. S/2015/128, 23 February.

UN Security Council (UNSC). 2016a. 'Report of the Secretary-General on the United Nations Support Mission in Libya'. S/2016/182, 25 February.

UN Security Council (UNSC). 2016b. 'Report of the Secretary-General on the United Nations Support Mission in Libya'. S/2016/452, 16 May.

UN Security Council (UNSC). 2016c. 'Report of the Secretary-General on the United Nations Support Mission in Libya'. S/2016/1011, 1 December.

UN Security Council (UNSC). 2017. 'Report of the Secretary-General on the United Nations Support Mission in Libya'. S/2017/726, 22 August.

UN Support Mission in Libya (UNSMIL). 2014. 'UNSMIL Studying Supreme Court Ruling, Emphasises Urgent Need for Political Consensus'. 6 November. https://unsmil.unmissions.org/unsmil-studying-supreme-court-ruling-emphasises-urgent-need-political-consensus

UN Support Mission in Libya (UNSMIL). 2019. 'Remarks of SRSG Ghassan Salamé to the United Nations Security Council on the situation in Libya'. 21 May. https://unsmil.unmissions.org/sites/default/files/srsg-salame-briefing-unsc-libya-21-may-2019-english_0.pdf

UN Support Mission in Libya (UNSMIL). 2020a. 'SRSG @GhassanSalame & DSRSG-Political Stephanie Williams met today in Rajma with Field Marshal Khalifa Haftar'. *Twitter*, 1 February. https://twitter.com/unsmilibya/status/1223687954318659584

UN Support Mission in Libya (UNSMIL). 2020b. 'One Year of Destructive War in Libya, UNSMIL Renews Calls for Immediate Cessation of Hostilities and Unity to Combat COVID-19'. 4 April. https://unsmil.unmissions.org/one-year-destructive-war-libya-unsmil-renews-calls-immediate-cessation-hostilities-and-unity-combat

UN Support Mission in Libya (UNSMIL). 2020c. 'Civilian Casualties Report 1 January–31 March 2020'. 30 April. https://unsmil.unmissions.org/civilian-casualties-report-1-january-31-march-2020-1

UN Support Mission in Libya (UNSMIL). 2020d. 'UNSMIL Welcomes Acceptance by the Libyan Parties of Resumption of the Joint Military Committee (5+5) Talks'. 1 June. https://unsmil.unmissions.org/unsmil-welcomes-acceptance-libyan-parties-resumption-joint-military-committee-55-talks

UN Support Mission in Libya (UNSMIL). 2020e. 'Civilian Casualties Report 1 April–30 June 2020'. 29 July. https://unsmil.unmissions.org/civilian-casualties-report-1-april-30-june-2020

UN Support Mission in Libya (UNSMIL). 2020f. 'UNSMIL Welcomes Agreement between Libyan parties on permanent country-wide ceasefire agreement with immediate effect'. 23 October. https://unsmil.unmissions.org/unsmil-welcomes-agreement-between-libyan-parties-permanent-country-wide-ceasefire-agreement

Vie publique. 2017. 'Conférence de presse conjointe de MM. Emmanuel Macron, Président de la République, et Vladimir Poutine, Président de la Fédération de Russie, sur les relations franco-russes et sur les conflits en Syrie et en Ukraine, à Versailles le 29 mai 2017'. 29 May. https://www.vie-publique.fr/discours/203301-conference-de-presse-conjointe-de-mm-emmanuel-macron-president-de-la-r

Wehrey, Frederic. 2014. 'Ending Libya's Civil War: Reconciling Politics, Rebuilding Security'. *Carnegie Endowment for International Peace*, 24 September.

Wehrey, Frederic. 2020. '"This War is Out of Our Hands": The Internationalization of Libya's Post-2011 Conflicts from Proxies to Boots on the Ground'. *New America*, 14 September. https://www.newamerica.org/international-security/reports/this-war-is-out-of-our-hands/

Wehrey, Frederic. 2023. 'The War for Benghazi: Violence, Paramilitarism and Social Rupturing, 2014–18'. In *Violence and Social Transformation in Libya*, edited by Wolfram Lacher and Virginie Collombier, pp. 165–194. London: Hurst.

Zaptia, Sami. 2020. 'U.S. Ambassador Norland Pays First Visit to Libya—Meets Haftar First'. *Libya Herald*, 19 February. https://www.libyaherald.com/2020/02/u-s-ambassador-norland-pays-first-visit-to-libya-meets-haftar-first/

Conclusions

Stephan Hensell and Klaus Schlichte

In our Introduction, we presented a theoretical framework for studying the politics of legitimation of armed groups. This framework has two core elements: The first concerns the conceptualization of legitimation. We have argued for a process-oriented understanding that is based on practices and publics. Analysing claim-making and claim validation as well as the role of audiences and constituencies is central to this approach. The second element of our framework is the differentiation between four historical times which we understand as structural contexts of armed groups' legitimation. We have distinguished the time of decolonization, the Cold War, humanitarianism, and the new sovereigntism. International legitimacy is time bound, and we have argued that these historical times have set different incentives and constraints for armed groups' quest for legitimacy. Furthermore, we have argued that the politics of legitimation of armed groups are parts and parcel of these historical times. Based on these concepts, our eleven case studies have analysed the international legitimacy of armed groups in different world regions and historical times since 1945. The case studies explored the legitimation of rebel groups, guerrilla fighters, or liberation movements and how changing historical conditions affected the fates of these actors. Our cases unpacked various aspects of their struggle, such as strategies of internationalization, the role of legitimating norms and values, military and diplomatic tactics, as well as the impact of regional constellations and transcontinental support. Finally, we suggested with this book an explorative, and at the same time theory-guided, iterative-cyclical research process that we called 'reiterative case comparison'. We made use of this by applying the circle of empirical analysis, comparison, and theory-driven interpretation in repeated rounds. Thus, we have put the open comparative discussion of our cases centre stage in order to facilitate theory-building and explanation.

Based on our framework, our case studies, and our reiterative case comparison, we made a range of observations that we will discuss in this conclusion. We proceed in three steps: First, we point to some more general lessons for IR as a discipline and we argue that armed groups should be taken seriously as what they are: actors that co-shape and reproduce international politics. Second, we will present a number of empirical findings and insights from our case analyses and discuss their conceptual implications as well as open questions and lacunae. Finally, we give a short retrospect and prospect on armed groups and the social origins of armed rebellion.

Stephan Hensell and Klaus Schlichte, *Conclusions*. In: *Armed Groups and the Politics of International Legitimation*. Edited by: Stephan Hensell and Klaus Schlichte, Oxford University Press.
DOI: 10.1093/9780198974529.003.0012

Armed groups as actors of international politics

If we understand international politics as the set of cooperative and conflictual relations and interactions among political actors across boundaries, armed groups are very much at the heart of what international politics is about. The politics of legitimation exemplifies the manifold transnational and international ties armed groups maintain with a host of other actors, state and non-state alike involving, among others, politicians, diplomats, activists and civilians. Our cases have shown that armed groups maintain a multitude of relations with their supporters and allies. The role of the European leftist parties, solidarity movements, and governments for the legitimation of decolonization movements and social-revolutionary guerrillas in Latin America and Africa makes this apparent. The linkages between Islamist movements and insurgents in the Middle East are further cases in point. This is also true for the relations among armed groups themselves, be they liberation movements, Maoist rebels, or jihadi groups (Byrne 2016; Lovell 2019; Roy 2004). Transnational relations, based on solidarity, friendship, shared opponents, and support among actors fighting in the same historical time featured prominently in the age of decolonization and the Cold War and still matter today.

We assume that further study of these global political alliances would be another argument to revise the dominant institutionalist and North-Atlantic bias of IR and a contribution to decolonizing the history of internationalism (Lewis 2023). However, armed groups also maintain conflictual relations, first and foremost with their adversaries but also with competitors such as other armed groups. As early sociologists like Georg Simmel or Lewis Coser have stressed, enmity can become functional even for warring parties and lead to interdependent relationships rooted in conflict. Finally, the relations with third parties such as neighbouring states, regional powers, or international organizations often play a particular role in the politics of legitimation. We see the manifold entanglements in our cases as a powerful argument for thinking ahead in IR with approaches of relational sociology rather than along single variables or static units. As classical sociologists and contemporary works on relational sociology (Emirbayr 1997) but also scholars of global history and critical IR (Osterhammel 2016, Bhambra 2007) remind us, there is no such thing as a self-subsistent exclusive entity. Rather, actors are constituted and formed through dynamic processes, continuous interactions, entangled, and interdependent relationships.

As actors in their own right in international politics, armed groups also play a very active role in shaping these politics. Modifying an often-quoted title by the IR-scholar Alexander Wendt (1992), one could say international politics is also what armed groups make of it. Their self-legitimation practices are themselves 'constructive' work in the making of international politics. Be it the historical time of decolonization, the Cold War, humanitarianism, or the current period of new sovereigntism, armed groups actively try to steer how their fight is perceived and evaluated. And their practices of claim-making—giving speeches, inviting foreign media, lobbying in foreign capitals, or mobilizing transnational networks can over time not only alter power balances but ultimately also topple regimes and outmanoeuvre states. Armed groups thus have profound consequences in international politics. They produce major outcomes, however often in unintended and unforeseen ways. On the basis of our case material presented in this book, we see conservative and transformative effects of their fights in international politics.

One general insight concerns the 'structural conservatism' of armed groups. The state-centrism of armed groups has evidently stabilizing effects. Rather than embodying the opposite of the state system, armed groups reproduce the structure of international politics. In all cases considered in this volume armed groups aimed at getting into state offices. This, it seems, is true for the vast majority of armed groups after 1945 (Schlichte 2009, 178–202). Armed groups seek power, either by seizing positions within an existing state or by establishing a new one. Thus, the political structuration of international politics along the logic of states is not questioned by armed groups. While they seek to replace a government with violent means, they do not oppose the state as an institution. Armed groups have fully accepted the state as the dominant political form. The fighting of armed groups is thus rather a confirmation of the state system as a global legitimate order than a contribution to its dissolution. We therefore think that considering armed groups as alien to official institutionalized realms and juxtaposing legitimate states and illegitimate 'non-state' violent actors is not plausible. As this book has shown, armed groups can acquire full international legitimacy. Their 'rightfulness' is socially produced like that of any other actor, and it is a genuine outcome of social constructions in international politics.

The fact that armed groups tend to preserve stateness as a fundamental structure of international politics does not hinder them contributing massively to shifts in international politics. Our volume has considered cases where the international effect of armed groups' fights was quite considerable. During the historical time of decolonization this was most obvious. The resistance by armed groups, raising the costs of colonial domination exorbitantly, was a major causal force in the sea-change of international politics that meant the end of European empires. And that the Cold War became a global political structure was also partly due the activities of armed groups that dragged external powers into domestic conflicts or at least opened opportunities for such powers. Also, more recent cases show that armed groups can have considerable leverage on the course of things, be it by triggering large interventions, attracting great power involvement or prompting alliances with states that can even tip regional balances. Our cases of Indonesia and Vietnam, Western Sahara, Angola, El Salvador, Israel/Palestine, and Afghanistan are exemplary here. Often change is not a direct effect of the armed groups' action. Sometimes, it is rather the massive reaction to what armed groups do which causes shifts in international politics. The recent fate of Hamas is most telling in this regard.

Armed groups, thus, need to be understood as outright political actors. Creating and maintaining international relations are part of their fundamental repertoires and they need to be taken seriously as actors that co-shape and reproduce international politics.

The politics of legitimation of armed groups: comparative remarks

Why are some armed groups successful in attaining international recognition and becoming legitimate while others are not? From what became visible in this volume, there is no simple equation that would explain the variation we can observe here. To answer that question, a larger set of cases with consolidated data would be needed. Such a data-set, seemingly, does not yet exist. There are, however, also reasons to be sceptical about the chances for such an endeavour, if it is not combined with deep knowledge of cases. As our Introduction has argued, and as our case studies show in much more detail, the politics of legitimation of armed groups, as any social and historical phenomena, defies transhistorical law-like causal explanations. We thus view careful case descriptions as essential. What large-N comparisons or even qualitative case comparisons (QCA) cannot see seems to us of utmost importance: to relate social practices with historical contexts in order to understand the historical and socially embedded nature of the politics of legitimation.

This is not just a question of competition between theoretical and methodological schools of thought, but also of quite some importance for dealing with armed groups politically. The politics of legitimation of armed groups, so our more general finding, is characterized by a huge variety of historically structured domestic and international conditions. Thus, the question of why some armed groups become legitimate actors while others do not cannot be answered by hinting at one single factor. There is no unique trajectory for armed groups to become legitimate. However, on the basis of our case comparison, we can see some typical features, recurring dilemmas, and strategies in the politics of legitimation. A first set of our observations is related to the politics of legitimation as such and highlights its relational and multilevel character. A second set of observations relates to the strategies of armed groups, namely the use of military and diplomatic means as well as the role of coupling and timing in the politics of armed groups. A final set of observations discusses the role of legitimation in conflict resolution.

The relational character of legitimation

Legitimation is a deeply relational activity as numerous authors discussed in the Introduction of our book have argued before us. Our case studies confirm this. Armed groups make claims to legitimacy, but so do others. A wide range of activists and supporters take part in claim-making on behalf of armed groups, thus highlighting a web of relationships in the production of legitimacy claims. The relational character also becomes apparent by the significance of othering in the politics of legitimation. Armed groups base their legitimacy claims often on the perceived failures, injustices, or inadequacies of their opponents, rather than on their own qualities. This, too, shows the intricacy of relations in the making of claims to legitimacy. Claims are validated but also reiterated, disseminated, contested, or met with counter claims. All this involves a host of actors. Who exactly is participating in this process is oftentimes difficult to delineate as is the distinction between agents and addressees of claims. Clearly identifying the agency in the politics of legitimation and pinning down the entangled relations among all actors involved necessitates close attention to empirical detail. The legitimation of armed groups, as our case studies show, is truly transboundary work in which a host of other state and non-state actors participate, often in quite contradictory roles. Our case studies have broadly covered three processes here: the self-legitimation of armed groups, their legitimation by others, and the de-legitimation of the armed groups' opponents, which feeds back into the self-legitimation.

Armed groups are generally the ones to make the first move by engaging in self-legitimation. Our case studies contain enough evidence that armed groups are more often than not true believers in their cause. Outright conviction is an important motive for these actors. Be it the belief in the right to self-determination, international law, or true Islam, legitimating norms and values play an important role in the struggle of armed groups, that sometimes stretches over decades against an overwhelming enemy. The fate of long-standing groups such as the Taliban, who have successfully fought a vastly superior intervention force for two decades, or the Polisario, who have struggled to liberate Western Sahara for half a century, seemingly without any chance of success, make this very clear. This does not mean that pragmatic motives, mundane needs, and material interests do not matter for these groups. They surely do. However, it does mean that justifications and claims to legitimacy often figure prominently in the politics of armed groups, which quickly involve a multitude of international actors connected in a wide network of relationships.

The validation of armed groups' claims is highly interactive and relational. Solidarity movements, diasporas, or advocacy groups are cases in point as our studies on the Polisario and the FSLN show. As transnational constituencies of liberation movements, they often validate claims but also disseminate them. They are, thus, addressees and agents of legitimation at the same time. The varied claim validation by states, governments, or international organizations also exemplifies a complex web of relationships as well as a wide range of motives. It is sometimes a question of principle and genuine belief, but often also the result of strategic calculations, power struggles, media attention, or historical experiences. Actors grant recognition selectively, often in unforeseeable and erratic ways as the cases of Indonesia and Vietnam, El Salvador, Uganda, Angola, or Libya show. The politics of external legitimation seemingly defy a universal logic and we also see abundant hypocrisy in state practices towards legitimating some armed groups but not others. The pragmatic demands of realpolitik figure prominently in the reactions of governments to armed groups and their final recognition often seems to be contingent on a unique constellation of forces. Overall, the legitimacy of armed groups seems to be as much the result of intended as of unintended action.

The case studies in this volume also show that self-legitimation and de-legitimation are deeply intertwined processes. Our book has analysed the legitimation of actors that are often by default considered illegal or criminal. Thus, we have focused conceptually on how armed groups engage in self-legitimation and how others react to this. However, our case studies make clear that armed groups often as much engage in self-legitimation as in the de-legitimation of their opponents. Undermining the legitimacy of their enemy is part of the strategy of almost all armed groups and in some cases perhaps even the decisive strategic choice. A case in point are the Taliban who constantly sought to expose the Afghan government as a corrupt puppet of foreign masters and subservient to the US. Their most impactful strategy to discredit the Afghan Republic was their refusal to acknowledge Kabul as even a party to the conflict. Likewise, the NRA in Uganda took advantage of the de-legitimation of the Obote regime in the course of massive human rights violations which became publicly known to an international audience not least thanks to the NRA itself. Other groups rather oscillated between emphasizing their own legitimacy while de-legitimating the other. The FSLN and the FMLN both successfully denounced the human rights violations of their opponents and framed their struggle as a fight against an oppressive dictatorship while presenting themselves as moderate alternatives. The Polisario has always sought to de-legitimate Morocco as an occupying power in the Western Sahara while pointing to international law and UN resolutions in order to underscore the legitimacy of its liberation struggle. More generally, the crises and erosion of incumbent regimes, often as an indirect effect of warfare, seems to be an important causal element in the successful legitimation of armed groups as likely or acceptable credible alternatives as the cases of Indonesia, Uganda, Afghanistan, or Libya make clear. Legitimation and de-legitimation are two sides of the same coin—a fact that the further study of the politics of legitimation of armed groups could take into account more systematically.

The impact of domestic and regional legitimacy

While our book has focused conceptually on the international legitimacy of armed groups, our case studies have shown that the politics of legitimation are truly a multilevel phenomenon in which domestic and regional legitimacy matters too. There are in fact different levels of legitimacy that are entangled and reinforce each other and ultimately impact on the recognition of armed groups as legitimate actors.

Generally, armed groups direct their claims simultaneously towards domestic, regional and global publics and in doing so they address different audiences and constituencies with different messages. Often armed groups' domestic legitimacy feeds into their international legitimacy and vice versa, as the case of the decolonization movement in Indonesia has demonstrated. We can see similar dynamics also in contemporary cases such as the LAAF in Libya. Haftar showed that he had international recognition which helped him to bolster his domestic legitimacy and counter his opponents' depiction of him as a warlord who had no legitimate official position. Claims to domestic legitimacy and the validation of Haftar's international legitimacy therefore mutually reinforced each other, and both supported the message that Haftar's LAAF was bringing stability and security to Libya. However, local and international legitimacy requirements can also be at odds with each other when armed groups accommodate international concerns to the detriment of credibility in the eyes of their domestic constituencies.

That the relationship between domestic and international legitimacy can be outright contradictory becomes particularly evident in the cases of Fatah and Hamas. Fatah has been granted international recognition in the wake of the Oslo Accords. However, the group then became increasingly caught between, on the one hand, the responsibilities imposed by this recognition and the Oslo Accords and, on the other hand, demands from its Palestinian constituencies for taking decisive action in the face of obstacles on the way towards Palestinian independence. Fatah had attained international recognition but in the absence of concrete steps to end Israel's military occupation, it progressively lost domestic support. This presented Fatah with a stark choice: either resuming military struggle at the expense of international legitimacy or cracking down on dissent at the expense of domestic legitimacy. Hamas has faced similar constraints. Having won a landslide electoral victory in 2006 and administering the Palestinian Authority in the Gaza strip, Hamas initially also sought international legitimacy. Formal meetings with representatives from Arab and Asian countries indicated first tentative steps in that direction. However, Hamas has gradually relinquished the quest for recognition and favoured instead domestic legitimacy by presenting itself as the embodiment of Palestinian resistance. Winning the popular support of the Palestinians by showing itself capable of confronting Israel was more important for Hamas than an elusive international acceptance.

Regional legitimacy matters, too. Armed groups are often backed by neighbouring states whose practices of supporting can be seen as validating a claim to legitimacy. Also, states from the region often function as gatekeepers that help armed groups in getting access to other governments and international institutions. More often, however, the direct flow of funds and guns or the provision of safe haven is crucially important for armed groups. It has often been stated that the Cold War superpowers played a decisive role in backing liberation movements and rebel groups in the Third World. However, as our cases show, regional politics in fact often mattered more. In the case of the Polisario, Algeria has always provided an important backing for the liberation movement. The Polisario has strong regional legitimacy in Africa and Latin America but not much elsewhere. Likewise, the FMLN profited massively from Mexico's government which provided a safe haven for the group and officialized it through the Franco-Mexican declaration. Mexico, thus, played a critical role in the FMLN's bid to become a legitimate political actor in international politics. In the case of the NRA, the insertion of its leading personnel in the regional political class of post-colonial East Africa was a major factor in gaining informal support, acknowledgement, and finally recognition by other governments in the region. It preceded recognition by great powers who could not help following suit. A particularly evident case of regional legitimacy is the PLO, which was not only supported but even created by Arab states as a platform to further the regional legitimacy of the Palestinian cause that was inseparably linked to Arab unification and state-led Pan-Arabism. The regional dimension of the legitimation of armed groups is perhaps today even more pronounced than in past historical times. The new sovereigntism and with it the rise of assertive regional powers seems to make a difference. The political and military support of Haftar's sponsors from the United Arab Emirates, Egypt, and Russia was crucial to the advance of Haftar's LAAF. An exemplary case is also Afghanistan, where a wide range of regional neighbours including China, Russia, Iran, Pakistan, Qatar, and the United Arab Emirates embraced the return of the Taliban and engage in routine bi-lateral meetings with Taliban leaders.

A number of our case studies show that international legitimacy of armed groups is often mediated and prefigured on the regional level before it expands into broader international recognition. This contradicts the argument that great power support matters most in international recognition (Coggins 2014). It is not necessarily global constellations and global powers that explain the legitimation dynamic surrounding armed groups but rather the much less studied regional political constellations. In a comparative perspective, the question would merit some attention as to what role specific regional constellations have played here in suppressing, solving, or postponing the question of recognition of armed groups' claims. The intricate relationship between local, regional, and global levels of legitimation is a vast field for further investigation. This relationship 'shimmers'. Seemingly, it can take on very different forms, also changing over time. To know more about it is also essential for any effort to end violent conflicts with armed groups.

The use of military and diplomatic means

Armed groups are usually not only brave fighters but also professional diplomats. All groups investigated in this book used a mix of military and diplomatic means in order to improve their position in the conflict and with regard to their future political fate. But were both equally important? Seemingly not, but the variation we see does not show an unambiguous pattern. There is no zero-sum relation between the use of military or diplomatic means. Military action was a constituent part of all the groups' strategies as well as their self-understandings. They employed physical violence and used military action in order to force their opponents, gain attention, and control territory. Battlefield performance and military success seem to be a crucial factor in the success of armed groups, especially when combined with the control of territory and the establishment of alternative state-like institutions. Armed groups thus simply establish facts and if foreign governments believe that an armed group has the power to enforce politics in a large area, the propensity increases to acknowledge this group as a legitimate player or credible alternative to the incumbent regime. This underlines the importance of fighting and state mimicking as practices of claim-making for armed groups. The Viet Cong/NLF in Vietnam, the NRA in Uganda, the UNITA in Angola, the Taliban in Afghanistan, or Haftar's LAAF in Libya make this clear. In all these cases, powerful rebel movements challenged or toppled incumbent regimes by military means, established alternative governmental institutions and thus presented themselves as likely candidates for recognition.

However, there is apparently no necessity here that would support a realist argument about military might or 'hard power' as the decisive factor for becoming recognized as legitimate. Our contributions show evidence that contradict such reductionism. A case in point are the Taliban, who are still not formally recognized as the government of Afghanistan despite their coercive power and being in full control of the state. Moreover, diplomatic means likewise play an important role in the legitimation of armed groups. A paradigmatic historic case here is the FLN in Algeria, which succeed mainly on the diplomatic stage while they almost failed on the military front (Connelly 2002). The case studies in this volume show similar trajectories. The success of the FSLN in Nicaragua was much less based on military achievements and rather on the activities of 'grassroot-diplomats', cultural appeal, clever arguments, and convincing messages that allowed the Sandinistas to gain support from friendly governments. In the case of El Salvador, diplomatic initiatives by the Mexican and French governments seem to have been much more important than any battlefield performance of the group. And the UNITA in Washington invested huge efforts and resources in lobbying the US government with remarkable achievements while their military successes in the Angolan Civil War remained limited. In long-standing cases we can also observe shifts in the use of military and diplomatic means. The Polisario was considered in the 1980s one of the most effective guerrilla movements in Africa, but as its military fortunes later declined diplomatic manoeuvres moved centre stage. The fact that the liberation movement has not been very successful with its diplomacy until today also indicates the limits of a purely diplomatic strategy. Fatah long maintained a delicate balance between armed struggle against Israeli occupation and diplomatic initiatives that aimed at a broad-based international recognition. After the Cold War, however, as the group committed itself to the Oslo Accords, its successes in diplomacy increasingly contradicted the use of the armed action Fatah used to rely on.

The relative importance of military and diplomatic means differed in each case of this book but the pragmatic mix of the two in the end proved efficacious for almost all armed groups considered here. Military power does not seem to be the ultimate necessary condition for the success of armed groups, nor does it necessarily translate into international legitimacy. It is generally supplemented by diplomatic means which often yield remarkable results in terms of support, acknowledgement, and recognition. Yet, a minimum of military might seems indispensable in order to be taken seriously. Without sufficient battlefield performance and threat potential, armed groups risk remaining irrelevant and might ultimately fail to become legitimate actors.

The role of coupling and timing

As our case studies have shown, armed groups that seek international legitimacy have to develop their agendas, images, messages, and projects in order to convince others that they are rightful political actors. It would be misleading, however, to characterize armed groups' claim-making as a mere result of strategic choices. With their legitimating discourses and ideas, armed groups are part of globalized political figurations such as Marxism, Maoism, or Islamism. Hence, the self-legitimation of armed groups is also an expression of transnationally shared political attitudes. It is part of a shared space of experience and a horizon of expectation that define a particular historical time of which an armed group is part and parcel. We see coupling and timing as two modes which express this transnational social embeddedness and connectivity.

Coupling means establishing transnational linkages with other actors and agendas. Armed groups often insert themselves into like-minded transnational milieus and movements and, thus, become part of a transnational figuration. This enables them to connect with prevailing understandings of international legitimacy. Bringing forward claims that are normatively justifiable but that also resonate with the dominating legitimating ideas of the time and capture the imagination of an international public is paramount for armed groups. However, coupling also means to build ties with allies who are ready to support the struggle with practical means and resources. The 'Global Left' (Wallerstein 2022) as a historical transboundary political figuration is a case in point. Many armed groups fighting in the historical times of decolonization and the Cold War were an integral part of it with its various progressive political parties, Marxist circles, and Third World solidarity groups that shared an anti-imperialist outlook and leftist ideas. In connecting armed groups with supportive milieus elsewhere, intermediary actors such as intellectuals, priests, writers, poets, artists, students, or exiled activists played an important role. Our case studies show how closely different arenas can be coupled, while at times armed groups also purposefully seek to keep them apart because they are well aware of intermittent effects. Such "de-coupling" is reflected in a broad range of positions as well as careful references. While some groups were more openly communist or Marxist-Leninist such as the Viet Cong/NLF or the PLO, others tried to avoid being read along the lines of Cold War division and framed as Soviet-inspired communists. Sandinista diplomats played down the FSLN's connections to the Soviet Union, the Eastern bloc, and Cuba. The FSLN, they argued, was guided by political pluralism, nonalignment, and a mixed economy approach. Likewise, the open support of the Mexican government compelled the Salvadorian FMLN to tone down its Marxist rhetoric in order not to be perceived as a dogmatic organization reluctant to make any compromises. While a leftist tradition prevailed in the NRA's initial discourse and its Maoist conceptions of 'protracted warfare', socialist leanings remained contested to the extent that the NRA did not seek socialist countries' direct support. These groups sought to remain acceptable for moderate political milieus in Western Europe and the US, given widespread anti-communism in Europe and North America. Other armed groups, despite some Marxist leanings, followed an essentially nationalist course and remained ideologically non-aligned such as the Polisario and Fatah. Yet others were embedded in rather conservative milieus such as UNITA in Washington or were coupled with religious groupings such as the Catholic Church in the case of the FSLN or the Muslim Brotherhood in the case of Hamas.

Timing refers to the need to seize opportunities and the opening up of possibilities, and to sense the proper time for action. This is not only about when exactly to do things in order to receive public attention or influence agenda setting. It is also more fundamentally about the dependency of armed groups' claims on fitting into historical times, appealing to the prevailing zeitgeist and seizing unique historical opportunities with skill and foresight. In our case studies, we see lots of evidence that timing matters for armed groups. In the Dutch East Indies and French Indochina long-standing independence movements acted swiftly towards the end of the Second World War in order to bring forward their claims to legitimacy. By unilaterally declaring the independence of Indonesia and Vietnam, they immediately capitalized on the power vacuums opened by the sudden surrender of Japan's occupying forces in August 1945. Armed groups such as the FSLN and FMLN obviously profited from the favourable political conditions in the 1970s. In West European states, Latin America became the dream location of revolutionary change for a wide range of leftist actors who called for solidarity with and support of social-revolutionary movements in the Third World. The NRA can be seen as a particular case here as the group acted at the thresholds of historical times. Unconditional support for African allies in the Cold War was slowly superseded by a new discourse on human rights in the 1980s, and the NRA had a clear sense that denouncing its opponents' human rights violations was more rewarding diplomatically than joining the already weakened socialist camp. The successful de-legitimation strategy of the NRA, in that sense, heralded the new historical time of humanitarianism. With changing historical times, however, claims to legitimacy can also lose appeal and fail to get validation. The case of the Polisario is most telling here. Its claims revolving around decolonization were widely acknowledged in the late 1970s and 1980s but seemingly fail to capture the imagination of today's audiences and break the wall of disinterest around the Western Sahara conflict. UNITA in Angola suffered from a similar fate, being supported heavily by the US during the Cold War but dropped like a hot potato after its end. As US policymakers turned to new issues in a fundamentally altered international context, supporting UNITA was simply no longer a priority.

Studying coupling and timing brings up intriguing questions: why do some conflicts gain a lot of attention in international, particularly Western, media, while others do not? Were the fights of the Polisario and the NRA simply too remote to publics in Western Europe and the US or were their causes not coupled with larger agendas in Western metropoles? What is the role of timing of armed groups' agendas and actions in the successful internationalization of conflicts? Our case studies deliver a range of insights into practices and mechanisms here, but a lot of research is still needed to come to more general insights on how armed groups inscribe themselves into transnational spaces and figurations.

Towards conflict resolution? Legitimation as a double-edged sword

Does the legitimation of armed groups contribute to conflict resolution? Or does it have rather antagonizing effects? These politically relevant questions are debated in the literature (e.g., Geis et al. 2021). Our case studies show mixed results here. Apparently, international legitimacy is easier to achieve for armed groups that are seen as moderate. Already the early cases of decolonization show that the reactions of major powers varied depending on the perceived character of the respective insurgencies, as the case of Indonesia and Vietnam make clear. The US leaned towards the Indonesian independence movement which it saw as moderate nationalists but had deep misgivings about the communist Vietnamese national liberation movement that was oriented to China and the Soviet Union. Engaging armed groups diplomatically and validating their claims to legitimacy can contribute to peace negotiations. In Vietnam, to break the military stalemate on the ground, Hanoi, Moscow, and Beijing insisted that the Viet Cong/NLF had to be accepted as a legitimate entity and negotiating partner in any peace talks between the US and its South Vietnamese ally. Albeit with great reluctance, the US, increasingly desperate to bring an end to the divisive Vietnam War, accepted that demand as a condition for opening the Paris peace talks. The Sandinistas in Nicaragua, in their quest for international legitimacy, stressed the FSLN's moderate tendencies. The guerrillas portrayed themselves as a group fighting for democracy and human rights which greatly facilitated its recognition. Another case in point is the FMLN in El Salvador. When Mexico and France officialized the FMLN, the guerrillas were brought closer to a negotiated settlement. By coaxing the FMLN into a rather social democratic position, both governments made the guerrillas more acceptable for the US government that supported the Salvadorian regime.

To enter negotiations or even to establish contact openly is associated with considerable risks for all parties involved. This is especially true for armed groups themselves. The quest for international recognition and the negotiated settlements that come with it require restraint in violent action, but this concession can become costly for armed groups, who are often torn between their commitment to government and armed struggle.

The cases of Fatah and Hamas are exemplary here. The Oslo Accords in the 1990s and Fatah's renouncement of armed struggle in 1993 contributed to its international legitimacy and led more states to recognize the Palestinian claim to statehood. However, in the wake of the stalled peace process, Fatah became increasingly a guardian of the post-Oslo status quo while being deprived of its repertoire of political-military contention it used to have. This opened the political space for more radical groups like Hamas or Islamic Jihad and an inter-Palestinian conflict about who would embody the struggle for national liberation. Hamas faced similar difficulties. Although having assumed responsibility for administering the PA in the Gaza strip and profiting from a tentative officialization through diplomatic meetings and foreign patronage, Hamas ultimately favoured armed struggle over international recognition. Both groups face the dilemma that in an ongoing conflict too many concessions to the outside world can endanger domestic support as internal competitors claiming to represent the 'purity' of the cause are already in their starting blocks (see also King 1997; Duursma and Fliervoet, 2021).

Legitimizing armed groups might contribute to ending violent conflicts but it can also have unintended and adversarial effects. The Taliban are a case in point. The 2020 Doha Agreement was a watershed moment for this group. It not only marked the conclusion of negotiations on the US withdrawal from Afghanistan but also officialized the Taliban as a key party to the conflict in Afghanistan. It validated the group's claim to be a legitimate political power in the country and shattered the image of the internationally backed Kabul government. The Agreement also catalysed the Taliban's subsequent diplomatic normalization on the regional level after coming to power as the new rulers of Afghanistan in 2021. Overall, the Taliban secured their international recognition in Doha but without compromising their religious orthodoxy and their vision for the future government in Afghanistan. The Doha Agreement thus represented a recognition achieved largely on the Taliban's own terms.

Legitimation can also empower armed groups and thus prolong wars. A case in point is UNITA that secured covert military aid from the Reagan administration which prolonged the Angolan Civil War for several more years thanks to fresh infusions of US weapons and money. The same is true for the Polisario which has always been dependent on Algeria's patronage. The North African state supported the Polisario with arms, aid, safe haven, and diplomatic lobbying and has always justified these actions with the international legitimacy of the Polisario's struggle to liberate Western Sahara. Algeria provided an important backing for the liberation movement from the first days of the conflict and has guaranteed its political survival ever since. That international legitimation can also have adverse effects is exemplified by the case of Haftar and his LAAF in Libya. Haftar gained a great deal in international legitimacy in a civil war he himself helped to provoke. While receiving red-carpet treatment by Western powers this did not contribute to conflict resolution. Haftar's forces have established a highly repressive form of rule and stood out as Libya's leading perpetrators of war crimes. Moreover, Haftar consistently proved to be the biggest obstacle to the success of Western governments' policies, namely the attempt to set up a government of national unification. And yet, Haftar's function as a spoiler merely prompted tireless efforts by Western diplomats to bring him into the fold and to officialize him as a leading player in Libya.

Outlook

We conclude our comparative remarks and reflections on lessons for IR with a very short retrospect and prospect on armed groups. Earlier works on rebels and revolutionaries had always emphasized that the history of armed rebellion is a history of social contradictions, socio-economic change, and globalized conditions. Scholars like Hobsbawm (1959), Wolf (1969), or Migdal (1974) had stressed this.[1] Our book confirms these older insights. Looking back at the historical times covered in this volume and at our case studies, we can see that the struggle of armed groups and their fight for legitimacy has always been part of a much broader process of social change. The historical times that we identified are replete with contested politics anchored in social contradictions. This applies to the decolonization movements as to the revolutionary guerrillas of the Cold War or to more present-day rebel groups that fight against human rights violations or for an ideal Islamic order.

Much has seemingly changed in the politics of rebellion. Current processes of politicization might no longer be based on rural life-worlds, nor do they necessarily aim at social transformation. Today, cities rather than the countryside are the hotbeds of massive protest and rebellion (Beissinger 2022). And the dislocations of capitalist modernization are no longer confined to national populations as they have evolved into an unresolved global social question (Breman et al. 2019). The sovereigntism of our times and the competition between self-assertive states might distract attention away from the underlying origins of contemporary civil wars and intra-state warfare. But it will, lest the findings about global social inequality are entirely wrong, not last long until old and new contradictions will dynamize domestic politics, international relations and armed rebellions again. The ongoing effects of social change might be reason enough to assume that the causes for organized violence will not disappear. Armed groups are here to stay, and so is their legitimation.

References

Bakonyi, Jutta, Stephan Hensell, and Jens Siegelberg, eds. 2006. *Gewaltordnungen bewaffneter Gruppen. Ökonomie und Herrschaft nichtstaatlicher Akteure in den Kriegen der Gegenwart.* Baden-Baden: Nomos.

Beissinger, Mark. 2022. *The Revolutionary City: Urbanization and the Global Transformation of Rebellion.* Princeton, NJ: Princeton University Press.

Bhambra, Gurminder. 2007. *Rethinking Modernity: Postcolonialism and the Sociological Imagination.* London: Palgrave.

Breman, Jan, Kevan Harris, Lee Ching Kwan, and Marcel van der Linden, eds. 2019. *The Social Question in the Twenty-First Century: A Global View.* Berkeley, CA: University of California Press.

Byrne, Jeffrey J. 2016. *Mecca of Revolution: Algeria, Decolonization, and the Third World Order.* Oxford: Oxford University Press.

Coggins, Bridget. 2014. *Power Politics and State Formation in the Twentieth Century: The Dynamics of Recognition.* Cambridge: Cambridge University Press.

[1] For an adaption of this perspective in German and French speaking research see Jung et al. (2003), Marchal and Messiant (2003), and Bakonyi et al. (2006).

Connelly, Matthew. 2002. *A Diplomatic Revolution: Algeria's Fight for Independence and the Origins of the Post-Cold War Era.* Oxford: Oxford University Press.

Duursma, Allard and Feike Fliervoet. 2021. 'Fueling Factionalism? The Impact of Peace Processes on Rebel Group Fragmentation in Civil Wars'. *Journal of Conflict Resolution* 65(4): pp. 788–812.

Emirbayr, Mustafa. 1997. 'Manifesto for a Relational Sociology'. *American Journal of Sociology* 103(2): pp. 281–317.

Geis, Anna, Maéva Clément, and Hanna Pfeifer, eds. 2021. *Armed Non-state Actors and the Politics of Recognition.* Manchester: Manchester University Press.

Hobsbawm, Eric. 1959. *Primitive Rebels: Studies in Archaic Forms of Social Movement in the 19th and 20th Centuries.* Manchester: Manchester University Press.

Jung, Dietrich, Klaus Schlichte, and Jens Siegelberg. 2003. *Kriege in der Weltgesellschaft: Strukturgeschichtliche Erklärung kriegerischer Gewalt (1945–2000).* Opladen: VS Verlag für Sozialwissenschaften.

King, Charles. 1997. *Ending Civil Wars*, Adelphi Paper No. 308, Oxford: OUP.

Lewis, Su Lin. 2023. 'Decolonising the History of Internationalism: Transnational Activism across the South'. *Transactions of the Royal Historical Society* 2: pp. 345–369.

Lovell, Julia. 2019. *Maoism: A Global History.* New York: Alfred A. Knopf, Inc.

Marchal, Roland and Christine Messiant. 2003. 'Les guerres civiles à l'ère de la globalisation: Nouvelles réalités et nouveaux paradigmes'. *Critiques Internationale* 18: pp. 91–112.

Migdal, Joel S. 1974. *Peasants, Politics, and Revolution: Pressures toward Political and Social Change in the Third World.* Princeton, NJ: Princeton University Press.

Osterhammel, Jürgen. 2016. 'Global History and Historical Sociology'. In *The Prospect of Global History*, edited by James Belich, John Darwin, Margret Frenz and Chris Wickham, pp. 23–42. Oxford: Oxford University Press.

Roy, Olivier. 2004. *Globalized Islam: The Search for a New Ummah.* New York: Columbia University Press.

Schlichte, Klaus. 2009. *In the Shadow of Violence: The Politics of Armed Groups.* Frankfurt am Main: Campus; Chicago: University of Chicago Press.

Wallerstein, Immanuel. 2022. *The Global Left: Yesterday, Today, Tomorrow.* New York: Routledge.

Wendt, Alexander. 1992 'Anarchy is What States Make of It: The Social Construction of Power Politics'. *International Organization* 46(2): pp. 391–425.

Wolf, Eric J. 1969. *Peasant Wars of the Twentieth Century.* London: Harper & Row.

Some Further Thoughts

Friedrich Kratochwil

Introduction

This book is a timely book in several ways. For one, it is carefully researched and provides us with abundant information about some important contemporary problems that are usually not accepted as they are taboo. They are lumped together under issues of terrorism or civil strife from which we recoil as they violate our expectations of how our social world should be. Not long ago we were told that we had reached the *End of History* (Fukuyama 1992) and ours was a rosy future, but subsequent reality has been different.

In this situation, it is, for me, a glimmer of hope that I have come across this book that actually takes seriously the challenges which we have to face today. I read the book with some relief and profit, even though what one reads is not necessarily to one's liking.

Although the book deals mainly with local issues, they have the universality of the particular that Hegel noted without committing us to his philosophy of history. This is the second reason that made the book noteworthy for me. A third reason is the timing; this book could not have been published at a 'better' time, not because it offers quick fixes, but because it calls into question some of the most widely shared myopias of social theory, when one deals with issues on the ground rather than abstracts from them, because that would disturb the 'elegance' of theories (Kratochwil 2018). Obviously, this does not 'sum up' the insights I have gained from reading this book ranging from law to war, from theory to the practice of traditional statecraft, or from interdependence to social networks.

Instead, I shall catch some of the balls and run with them although the borders of the playing field, the rules which govern our moves, the study of the gains and losses, are far from clear or uncontested. Instead of submitting to the traditional notion that progress in knowledge is 'coming nearer and nearer to the goal or the limit', I shall begin with challenging this wrong analogy. Coming nearer to the goal in a play, or coming nearer to the limit, presupposes *direct perception* of all knowledge (in the case of Popper's argument) or of the goal posts to which I must get as otherwise I cannot say whether I am now nearer to it or not. All I could say is that after running around I am now somewhere different.

Indeed, nothing is easier and more difficult to do when faced with such perplexities than changing the metaphor and seeing what that does! This is what Wittgenstein

Friedrich Kratochwil, *Some Further Thoughts*. In: *Armed Groups and the Politics of International Legitimation*. Edited by: Stephan Hensell and Klaus Schlichte, Oxford University Press. © Friedrich Kratochwil (2025).
DOI: 10.1093/9780198974529.003.0013

did when he shifted from the word and its place in a sentence to the 'way we use it' (Wittgenstein 2009), from the task of making the concepts as sharp-edged and clear as possible, to calling our attention to the fact that words have meaning not only in larger structures, but that their extensions have heuristic value because the overlaps created thereby provide new useful ways for *going on*, instead of using ladders of abstractions, emptying concepts of content. In dealing with this problem, I once proposed the metaphor of a three-dimensional Scrabble game (Kratochwil 2007). It goes on in many directions, not without constraints, but without a central axis, allowing thereby new words and combinations to arise, that can branch out widely or get compressed in a particular path until someone finds a connection to a word in a different segment. It thus expands our knowledge by making 'new' connections possible, without however having to 'think from the end' (teleology) or from the limit (as in a circle that gets exhausted by finer and finer polygons, which it supposes to represents the growth of knowledge).

One of its implications then is that the issue of individual and social order is ill-served by starting with 'designs' and pretending that creating order is like following a preconceived plan or design. The *proprium* of praxis is that of '*ordering*' (*ordo ordinans*) rather than of visualizing *an* '*order*' (*ordo ordinatus*) and producing it step by step, as if it were a functional system or a product, that emerge from a process of making, rather than *choosing*.[1] Orienting ourselves is not like executing a plan. In following this thought, I want to elaborate it further by focusing on action in historical time and in a lifeworld, which is not 'natural' but 'artificial'.

Given these (revisable) prejudices, I think the authors of this book are perhaps a bit too optimistic that detailed case studies will help 'grand theorizing'—even though this view is common. The dream of grand theory, of the 'full view', has mesmerized the Western tradition since Plato, and supposedly returned again with Descartes. The latter, however, seemed to realize—when one gives a closer reading to his *Discourse* and *Meditations*[2]—that a turn to unshakable foundations might be for the understanding of practice the fastest way to nowhere. At least his remarks on opting for a 'preliminary morality' relying on customs, rules-of-thumb, and shared ideas suggest this.

If one takes historical time seriously, then the assumption that the accumulation of knowledge ought to lead to some grand theory might then be no more than a function of the reflexes of social scientists trained to believe that 'theory' is the key to everything and that there exists only one view. The idea of a 'perspective', however, is that it is always a *view from* '*somewhere*' which militates against the 'view from nowhere'.

Similarly, Kant is surprisingly inconsistent when he observes actual history and the havoc which results from pretending to know the end of history. We need not get that deep into the Kantian oeuvre to see that his 'What is Orientation?' and his *Rechtslehre*, in the 'Metaphysics of Morals' [1785], are rather at odds with some positions he

[1] For an excellent treatment of this problem in the study of (international) law see Lindahl (2019).
[2] For both texts see Cress (1980).

exposed in his 'The Idea of a Universal History with a Cosmopolitan Intent"' [1784] or 'The Contest among the Faculties' [1798].[3]

It is indeed one of the strengths of the present book that it gives those philosophical and sociological speculations fascinating empirical content, as it examines in detail the different arenas and grouping in movements and self-appointed representatives/warlords that have entered the game of recognition. To understand its different configurations—perhaps a 'post-modern' type of struggle for the exercise of rule—it is useful to compare it with some other historical forms of *dominium* (*Herrschaft*).

To see the specifics of *politics*, as distinguished from other modes of rule, and to understand our present attempts to find ways of reconstituting our individual and collective choices, faces considerable difficulties. Both the technological revolutions compressing time and space, as well as our patterns of opposition and collaboration are in flux. We can see this by the growing contemporary opposition of the market and of the extension of its 'logic' to other domains, but equally evident are the transformation of markets themselves, which are devoted less to the provision of goods but increasingly provide platforms for investment and speculation, dealing with 'assets' rather than goods.

Consequently, a type of 'bricolage politics' emerges, where images of the family have survived—when we refer to the 'other' as 'brother'—but where the 'citizen' is absent; where, under the name of 'égalité', welfare has made the political agenda, despite the ever-widening gap of unequal distribution of wealth across the globe. It also differs from framing the issue of politics in early modernity—centering on compliance with the commands of the sovereign- that preoccupied Hobbes (1974). Perhaps the most important change in Hobbes's construct is today, however, mostly passed over, as it is so familiar. It is the notion of *individual subjects* forming the body politic as individuals—as the title page of the *Leviathan* shows—and not as members of an estate or 'house', as in the old feudal regimes. All the 'houses' or corporations of the former politics in Bodin's work (Bodin 1955) have disappeared—Hobbes mentions only the family as a social formation. Although he draws on a variety of metaphors and sources—beast, machine, religion and geometry, rationality, and terror—the unencumbered individual serves to 'filet out' the state from all other social relations.

Finally, our present 'mapping of politics' no longer submits to some drift of history towards state-making, or some 'functionalist' or equilibrium form of analysis—as we do find plenty of dysfunctional set-ups in both the domestic and international arena. It is no secret that institutions often work at cross purposes, as 'dilemmas' arise that lead to socially sub-optimal outcomes, or they have multiple equilibria and thus cannot show a necessity for choosing a particular 'solution', or worse, they simply have none.

The book tries not only to direct our attention to new issues and new organizational forms that constitute the politics of our time, it also necessitates a critical

[3] All texts can be found in Kant (1991).

examination of the stated or silent assumptions underlying our 'politics'. Nevertheless, we should not only be interested in how 'politics', as the struggle for recognition and representation, is reconfiguring at present the various 'publics"' emerging first from the organizations that sit above the state—although they remain clients of the states that provide them with a charter—and second, from private or semi-private organizations, both giving voice to newly, transnationally emerging interest. We also have to understand why and how statehood still figures prominently in the calculus of an, at first sight, 'anti-systemic' opposition. As the book argues, this politics is definitely against the status quo but not against the 'game' as such (as e.g., anarchist programmes).

For explaining the emergence of supranational organizations two explanations exist: one, that under the subterfuge of the apolitical or technical character of issue areas, we create special regimes sharing supervisory and limited rule-making power with the state; and two, we enter the domestic political game and pursue through lobbying the adoption of special policies by governments. The latter is instantiated through the first humanitarian organizations such as the Red Cross, the former in the subsequent proliferation of public unions.

The new configuration of politics in post-modernity suggests that a different framework is needed if we want to understand why, with the growth of such international organizations, structures have become an important alternative for gaining recognition. It has become a 'third' way, similar to Tony Blair's attempt to find a third way between socialism and capitalism. In this way, official and market organization, as well as elements of civil society invoking 'humanity', interact and create norms and an ever-expanding catalogue of (subjective) human rights,[4] even if that catalogue is highly contested.

It is here that, for me, the book makes its greatest contribution. It does not submit that the realization of 'distinct epochs' all forming a single process of history, à la Hegel or Marx. Such an analogy is misleading because in practical choices we always have to deal with the paradox of the 'contemporaneity of the non-contemporaneous'[5] as Koselleck has (problematically) called it. But contrary to Koselleck, this is not just a 'developmental issue' when in contemporary social arrangements 'remnants' appear—when, e.g., 'nobility' has found a new place among the competing elites and tries its hand at selling real estate or overpriced cars. It has rather to do with the existential dimension of choice that bridges the three dimensions of time: the present, the past, and the future. We have a problem in the present which we understand by looking at how it came about, and we imagine a future which, however, cannot be the actual future but remains an imaginary project (what is called a 'future present') so that expectations and actuality usually differ significantly. Only in this way can

[4] The remain not only contested in regard to which substantive areas should be included in the catalogue, but much more importantly, whether they can be claimed against public authorities or whether they are standards that merely inform us of what is right (not what it means to 'have a right' which can also include to do the 'wrong thing' since it is *my* right).

[5] See the important collection of essays on this problem by Koselleck (2000, 2017).

we avoid a periodization that makes the contemporary 'liberal' empire simply one of 'humanity' (Barnett 2011).

By following those lessons of doubting 'theory' and a teleological conception of history, let's start with some *historical comparisons* rather than with a narrative of the historical process. Let's begin with the fundamental transformation of the world political system after WWI and the interwar period that resembles our present situation, as far as disorientation is concerned. Here, of course, together with the loss of the taken-for-granted life world, the acceleration of change has to be mentioned. For that reason, I shall make some preliminary remarks about information, meaning, and the *Lebenswelt* which we construct to be able to act; this leads me to a brief discussion of the role of media in 'making' but also destroying sense (*Sinn, Bedeutung*). For this the 'secularization' thesis and the surprising revival of religious zeal may serve as foils.

The problems of Zeitenwenden: 1918, the Interregnum, and the Universalist dream

That said, I think it is important to focus on the shift that occurred after WWI, as it dramatically changed the game and sets the stage for the similar 'turning points'after WWII and the present. Versailles narrowed the agenda of exercising rule through its insistence on a homogenized 'state' where a majority ruled, while at the same time extending the model increasingly to areas which had been kept in limbo by the empires, be it that of Great Britain, Russia, France, or even Belgium and Italy which had grown at the expense of similar empires at the periphery, such as the Ottoman Empire, China, or Japan.

One way of assessing the decisive break with the past is, however, that the formerly decisive topic of political practice, i.e. the 'just war', became a dead issue. War itself was now officially no longer a legitimate means to pursue a state's interests. Instead, 'the outlawing' of war together with modes of peaceful conflict resolution became the new preoccupations of lawyers and (at least some) statesmen. The formerly central issues of politics and law, of how war made states, how states made war, and how different forms of legitimacy shaped the international system became topics for historians. Except for the last ones, the 'old' problems were exorcized or conveniently 'forgotten' by assuming that 'democracy' could solve all problems. The bitter truth was that self-determination and the principle of 'one man one vote' pointed frequently in different directions—quite aside from the myopia that it excluded women and 'involuntary' or original residents (as in the US).

In reordering the world one invented (or let people) invent majorities and minorities, and tried to protect the latter by a feeble network of international norms; one drewnew conceptual boundaries distinguishing 'states' (in a nice equivocation of the term, as the word is also used for 'situation' or 'conditions', of rebellion and insurgency), and attached to them certain consequences. But one remained pretty helpless when social solidarity crossed the jurisdictional boundaries of a state, whose government was part of another population which would, or could not, call it 'their own'. The situation became worse when discontent could easily be used in neighbouring regions to create irredentist movements.

The blind spot of course of this Wilsonian vision was—as the authors of the book correctly identify—the issue of 'recognition', i.e. by what criterion you can make claims and have at least a presumption to be recognized in representing a group of subjects. Here the Wilsonian vision suggested also that 'illegitimate' violence—i.e. force used by non-state actors—has to be treated analogous to that in the domestic sphere. There, except in cases of self-defence, the use of force is prohibited and thus the whole attention in drafting was focused on the efforts to make war, as a primitive mode of self-help, 'a crime'. With it, we also 'forgot' the problems connected with civil wars and the often-tenuous claims that armed groups have used for suggesting that they actually provide at least some services of states, through their 'police' and the provision of basic infrastructures for commerce, education, and social welfare.

The authors have, in the tradition of Weber, chosen 'cases' for addressing these types of issues, by being interested in the challenge to the legitimacy presumption of governments as the unique international governance norm, but which paradoxically must also deal with those who use illegal violence in order to become the government or recognized participants in the international game. The authors—and the contributors with their added case studies—focus therefore on the problem of 'recognition'. It is no longer just a dichotomized category but shows—as the recent vote on Palestine in the UN, and as documented in the various cases—that recognition has become a multifaceted phenomenon, assembling a variety of practises that escape dichotomous choice: legal/illegal.

This has some methodological implications which violate not only the logic of classification but the taken- for-granted assumptions that since 'science' and theory have proven successful in knowing and manipulating the physical world, the same conceptual tools and methods are also applicable to the social world. The realization that concepts in the practical world are not like names for the elements of nature but are 'mind-dependent' has important implications for enabling us to orient ourselves and act in the social world.

First, we have to recognize that the role of language in the social world is not one of reference, but of constitution. That has surprising implications. Just think, aside from speech acts, such as promises, which may 'bind' me in my future actions but do not necessarily determine my eventual choices, as I might opportunistacally or 'with reasons" disregard them. Or think of on 'non-events' such as deterrence, which we try to make work. These examples suggest that language is not just pointing at 'things' by naming them, but leading us to what 'matters', although it is no 'matter' in a physical sense.

This raises a second issue, relevant for international relations: the usual disciplinary myopia of history. From the Westphalia moment[6] that established the sovereign states in the global order we take this development as the beginning of the

[6] Much of the myth of Westphalia, i.e. of the European system that has been 'extended' to the entire globe goes back to the article written by the international lawyer Gross (1948). Its historical accuracy has since severely criticized, see e.g. Osiander (1994).

expansion of the European state to the rest of the world. Unfortunately, this myth is, despite detailed studies to the contrary, due to the failure of social scientists who do not distinguish between a 'state' and an 'estate' so that a continuity is created, which actually is non-existent. Instead, it is a mixture of feudal remnants and of new social forces which determined the policies of the European powers until the end of 19th century.

With Versailles and the League of Nations such possibilities no longer offered themselves as viable templates for the future, as the elimination of Austria-Hungary showed, but which also sent a warning to other European powers with overseas possessions. Empires had to become a Commonwealth and the construction of 'overseas provinces' and the League mandate system—with different 'waiting periods' for becoming independent—were clear signs that times were changing.

Third, we need not review here the entire history of the beginning and the end of the Cold War to see how, in the *beginning*, issues of recognition become touchy not only in the case of membership of specialized organizations such as the UN. Right from the beginning, when China was still to play a leading role despite its weakness and civil war, the recognition issue led to preliminary exclusion. On the other hand, the *ending* of the Cold War occurred more by a series of surprises than by design. One surprise was the 'recognition' that the growth of a democratic groundswell in the Eastern bloc could no longer be ignored. As Gorbachev reminded Honecker, the East German leader: 'history punishes those who come too late'. It was then the unlikely understanding between Mikhail Gorbachev and President Reagan that created the possibility of a more constructive engagement; soon, however, it was undermined by both the question in the West of who 'won' the Cold War[7] and by the path Putin followed when he no longer felt 'recognized' as a great power.

Analysing these issues provides a more persuasive analysis than the 'progress and secularization' argument which most of us accept as the Enlightenment but which has left us wondering about the resurgence of political Islam as well as growing fundamentalism, even in some Western societies. Oddly enough, politics seems lost to rumour mongering, precisely at a time when we have more information than ever before, and the sectarian fever of political debates seems even more dumbfounding when one believes in 'enlightenment'.

Enlightenment, secularism, and the myth of progress

One of the most ingrained beliefs has been that 'religious conflicts', which devastated the world before modernity—were a thing of the past. We seemed to have learned the Hobbesian lesson to focus on our private affairs rather than engage in religious questions and to keep quiet, at least in public.

But there is something fishy about this argument. Two points should be considered here. First, contrary to such a sales-pitch, Enlightenment thinkers were not necessarily 'progressive' in the sense that we attribute this point of view to them, as the

[7] Out of the voluminous literature see just the discussions in Lebow and Stein (1994) and the collection of essays by Lebow and Risse-Kappen (1995).

example of Kant's warnings about the French Revolution show, which he at first endorsed. Voltaire was certainly no advocate of repression but wrote one of the most critical accounts debunking the common myths of his day in his *Candide*. Rousseau had a problem with liberalism's conception of the actor that is not only self-interested but self-concerned, without envy or mercy. It is, for him, an artificial man without morals or feelings. But even worse, against those tenets he pointed out that much of individual satisfaction and conception of the self comes from comparing oneself with others and receiving their recognition. In terms of natural rights, he was as suspicious of those constructs as was Hume, who considered the social contract a useless fiction. So, the Enlightenment was not and is not just the present standard liberalism that evolved from it.

But we can do even better when we remember Bentham—one of the icons of 'liberalism' —who considered the talk of natural rights 'nonsense' and of human rights 'nonsense upon stilts' (see his '*Anarchical Fallacies* treatise). It was also Bentham who conceived of the state as a surveillance institution, analogous to his plan of a 'rational' prison, where the wardens are invisible in the centre but can observe every prison cell. This reduces brute coercion by substituting 'transparency' and perhaps 'nudging' as social mechanisms, but whether this is already 'democratic' or 'civic' can be doubted not only by adherents of Foucault.

Second, even if we argue the Enlightenment fight was directed mainly against the Church which was struggling with the state for supremacy, we have to admit that even after the Reformation we were far from a 'secularized' political order but had to make our peace with state-churches. Finally, contrary[8] to the secularization trend in Europe, in the US we see other trends when we look beyond the established religious communities—as we notice a considerable growth of 'free churches' on the entire American continent, and the proliferation of types of private 'spiritual' encounters, seminars, and coachings—though they might differ from the classical teachings of the Christian churches, which left redemption to the afterlife and a final judgement at the end of time.

But it is surprising that what we accept nowadays, the notions of self-redemption and authenticity, as entirely 'secular ideas', but which are actually grafted on religious concepts as the debates from the gnostics to Nietzsche demonstrate. It also shows why much of the secularization thesis is confused, as it mistakes the changes in the 'artificial' social world as one of *displacing religion*—i.e. a particular notion that beliefs should find no public recognition. The latter is the ideology of 'secularism' and identified with 'progress', but has little to do with *secularization*. The latter 'means' actually the change in the use of 'holy' objects that are now transferred to other uses, such as making a hospital or a 'public' meeting place out of a church or vice versa (see the use of arenas for revival meetings!). To set this record straight might seem a mere quibble about words but is actually of fundamental importance for understanding changes in the social world.

[8] Here the largely quantitative analysis work of Pippa Norris and Ronald Inglehart (2004) comes to mind.

It explains on the one hand the general enthusiasm that even traditional rulers had in *all of Europe* to get their hands on rich monasteries and church property (secularization) not necessarily in secularism. Nor does the historical record exclude the possibility that the 'invisible' religion survives as Luckmann (2023), Berger (1969; 2014), and Cox (2019) argue. The US is a perfect example in which 'the nation under God' remains part of its foundational myth, while a strict anti-establishment regime is there also a constitutive element of the US 'civil religion'. In Germany we face this controversy at the moment in a new version: it concerns the vanishing magic of the formula of multicultural '*Willkommmenskultur*' which led to the slogan—invented by a journalist and sold to the German president—that 'Islam belongs to Germany' whereby the term 'belonging' remained somewhat opaque. It could mean that the belonging was of the kind as 'death belongs to life', or even comprise the radical demand that the West in general needs a Caliphate.

The most recent attempt of distinguishing between a 'political Islam' and Islam as a religious belief, shows that recognizing this difference is crucial. By not minding this difference, one is quickly seduced by an 'all-inclusive version' of politics, more akin to a cruise on one of the cruise ships with unlimited consumption offers, but with no envisaged destination (Fahrt ins Blaue). Not surprisingly, such wishful thinking leads to predictable confusions and self-deceptions among all concerned. To preach water while drinking wine is of course an age-old strategy of rulers who will seek to profit from the charismatic approval of their rule, while having to distance themselves because of the multiplicity of religions among their subjects or the possibility of other available 'defenders of the faith'. Thus, the recent Turkish regime has positioned itself as a 'protector"' of Islam, particularly in Western Europe, but defines itself also as an advocate of 'other minorities in Europe' in its campaign for a seat in the EU parliament.

Market, media, and the 'moral point of view'

The remarks on secularization as an ideology of consumption directs our attention to a new topic which in the last two decades has generated studies exploring the metaphor of the market and its role in the 'lifeworld.' It had, as its predecessor, the literature dealing with exchanges, such as Marcel Maus's treatise on gifts (Maus 1990), and of ritual exchanges, of matchmaking markets that were strictly regulated by clans, and Simmel's path-breaking analysis of money (Simmel 1991). There the latter focused on the transformation of 'debts' from a personal commitment to a commercial activity, where credits and risks and options can also be bought and sold. This provided a new conceptualization of a society as a network of debts (Schuldzusammenhang) but left open the exploration of the metaphors 'debt' and its religious overtones; it later kindled Weber's interest in world religions and resulted in his *Protestant Ethic and the Spirit of Capitalism* [1904].

Walter Benjamin's interest was less enticed by Weber's study in which the unintended consequence of predestination resulted in a capitalist social formation.

Rather, his curiosity was teased first by the 'Spirit of Ornaments', i.e. the depiction of sacred objects and persons on banknotes of different currencies. Thus, Fortuna is depicted on several banknotes across the world as goddess of fertility and fortune, but Benjamin found a quite impressive list of other ornaments. These allegories Benjamin found alluring since capitalism and religion address essentially the same 'worries, tortures and insecurities' for which 'formerly religion provided the answers' (Benjamin 1921).

His fragment (*Kapitalismus als Religion)* was only posthumously published and sparked a lively debate in 2018 when it became a topic for an anthology (Baecker 2019). Its focus on rituals of atonement and indebtedness became a heuristically fruitful programme, which I cannot possibly do justice to here. However, one of Benjamin's points needs to be mentioned: it is the role of the media which is crucial for restructuring social reality, making, for example, shopping centres new 'public places'—although they are legally speaking 'private'. There one can pay homage to consumption, meet others, and be on display. Similarly, sport arenas have become (rented) 'churches' where revivalist preachers practice 'healings' or produce miracles and provide opportunities for collective atonement.

But for understanding what these new forms of 'representation'—creating a presence for that which is absent—mean for individual and collective life, the role of the media needs to be examined further. While formerly only the 'voice'—that was heard—created a public, the book or leaflet, the telegraph, and electronic media generated an information-rich environment that organized people beyond this reach. It managed this by transmission and selection of information, presenting what was worth knowing—by distinguishing initially between 'confirmed' and 'non-confirmed' news. The 'non-confirmed' news were characterized by a sign "n.t." (*non-testatum*), which is supposed to be the ancestor of the German '*Zeitungsente*'—although the 'canard' in French must have a different pedigree. Benedict Anderson (2006) has shown how the emergence of a 'nation', and the growth of newspapers which generated these 'imagined communities', conditioned each other.

But there was also another crucial development for the creation of 'publics' which we can trace through an investigation of advertising in the papers. Originally the papers or broadsheets were financed through subscriptions by members of reading circles and salons of 'the better people' who represented the public, distinguishing them from the rough and ready fighters in parliament who acted in a partisan rather than 'patriotic' fashion. Furthermore, anyone familiar with the former political debates, such as when Lincoln debated Douglas and actually used arguments, will be shocked by the staged repetition of catchy phrases that characterize nowadays the presidential or election campaigns. Even the 'talk shows', whose only goal is to make a 'bella figura' where the colour coordination of your outfit may be more important than what you say, show the change produced by advertisement.

This change was long in the making but with the *inclusions of pictures* in product advertisement we see that the 'text' shrunk, the properties of the product became less interesting, as pictures appealed to unconscious desires. The 'brand' substitutes for values we hold and ensures public recognition because one could display a certain product, be it a handbag or a watch, that signals to others one's status (Esders 2017; Bolz 1999). The car will get you the girl you want to meet, the cigarette shows your sophistication and style, the cocktail puts you in a different class. But if that was too 'mondän' for you, one could also show one's disdain of the city-slickers by playing the lonely cowboy out West, who was riding into the sunset smoking a Marlborough.

So, the argument is that media are not only filters for information and for sorting them—which is the cognitive side—but that they also appeal to our desires, thereby affecting our 'will' to act. The market becomes less and less a mechanism for distributing goods efficiently, and more one for *creating* demand by getting your attention by focusing on the activation of impulses. Of course, sometimes the appeals are just clumsy, not to say stupid, as they appeal to gluttony or make you feel better since the abject behaviour you see is now treated as 'normal'. Perhaps you, as an outcast, can identify with some scatterbrain you 'meet' in the net and have now someone 'you can talk to'.

But seduction is best when it is not too obvious and here the 'influencers' come in. They show you how one can become famous, or be part of something big, while you 'loner' are sitting on a measly balcony of a tenement instead of lying on the beach in Copa Cabana. Your 'dream' will keep you busy. No need to waste time by sorting things out or making choices. Become authentic and spontaneous, by just making sure you 'follow' the right people with whom you click!

There is only one hitch. All the frantic activity and ritualistic trances cannot free you from the feeling of worry, even angst. Take the environment. Now, with the death of God, we are suddenly *responsible for everything* which can overwhelm the best of us. So there remain Kant's 'big questions'. Aside from 'What can we know?' and 'What shall we do?' the third one is the big one :'What can we hope for?' But now this question seems somehow to have morphed into one of: 'What must we fear?' And the one of 'What shall we do?' has become 'What must we avoid?' which becomes all the more taxing as we run out of time when we dare to ask the first one: 'What we can know'?

Everything is now important and urgent, and playing God is, after all, not the fun the snake in paradise promised us (*eritis sicut deus*: you will be like God). We should have had an inkling of the future after the 'apple' had become a logo for a corporation that runs more and more of our life. As a prankster so aptly said about the many changes in meaning we have to witness: the SOS signal first meant 'save our souls', then it mutated into the 'save ourselves', and now it just means 'save our sales'. The promise of salvation (and not only of salvage) has now become the responsibility to ourselves; and because only a few, or only one can win, the only safety we get is that of becoming at least a recognized victim, as suffering is now, not accidentally, the idiom of identity. Ironically the better our lives have become—when compared to other times—the more historical stigmata matter.

On the other hand, this hurt is an even better generator of income than previously an insurance policy. In the latter case you have to pay in, so that you can collect the premium; but in the case of the stigmatas of the past you as victim is entitled to make claims on the offspring of the former perpetrators. It is not quite clear whether or when the debt will be paid up and thus, aside from the victims, the others will most likely remain, associations of 'perpetual guilt'. With that, what we have is then not the end of history, but perhaps of politics.

References

Anderson, Benedict. 2006. *Imagined Communities: Reflections on the Origin and Spread of Nationalism.* New enlarged edition, London: Verso.

Baecker, Dirk, ed., 2019. *Kapitalismus als Religion.* 4th edn. Berlin: Kadmos.

Barnett, Michael. 2011. *Empire of Humanity: A History of Humanitarianism.* Ithaca, NY: Cornell University Press.

Benjamin, Walter. [1921]. *Kapitalismus als Religion*, in 2019 Dirk Becker (Edited.) *Kapitalismus als Religion*, 4th edn. Berlin: Kadmos.

Berger, Peter. 1969. *The Rumor of Angels.* New York: Anchor.

Berger, Peter. 2014. *The Many Altars of Modernity.* Boston-Berlin: de Gruyter.

Bodin, Jean. 1955. *Six Books on the Commonwealth.* Translated and selected by M. J. Tooley. Oxford: Basil Blackwell.

Bolz, Norbert. 1999. *Die Wirtschaft des Unsichtbaren.* München: Econ.

Cox, Harvey. 2019. *The Market as God.* Cambridge: Harvard University Press.

Cress, Donald, ed. 1980. *Rene Descartes: Discourse on Method and First Philosophy.* Indianapolis, IN: Hackett.

Esders, Michael. 2017. *Alphabetisches Kapital: Über die Ökonomie der Bedeutungen.* Bielefeld: Aisthesis.

Fukuyama, Francis. 1992. *The End of History and the Last Man.* New York: Free Press.

Gross, Leo. 1948. 'The Peace of Westphalia 1648–1948'. *American Journal of International Law* 42(1): pp. 20–41.

Hobbes, Thomas. 1974. *Leviathan.* Edited by C. B. S. McPherson. Harmondsworth: Penguin.

Kant, Immanuel. 1991. *Political Writings.* Edited by Hans Reiss. Cambridge: Cambridge University Press.

Koselleck, Rainer. 2000 *Zeitschichten.* Frankfurt: Suhrkamp.

Koselleck, Rainer. 2017. *Vergangene Zukunft.* 10th edn. Frankfurt: Suhrkamp.

Kratochwil, Friedrich. 2007. 'Evidence, Inference and Truth in Problems of Theory-Building in the Social Sciences'. In *Theory and Evidence in Comparative Politics and International Relations*, edited by Richard Ned Lebow and Mark I. Lichbach, pp. 25–54. New York: Palgrave Macmillan.

Kratochwil, Friedrich. 2018. *Praxis: On Acting and Knowing.* Cambridge: Cambridge University Press.

Lebow, Richard and Thomas Risse-Kappen, eds. 1995. *International Theory and the End of the Cold War.* New York: Columbia University Press.

Lebow, Richard Ned and Janice Stein. 1994. *We all Lost the Cold War.* Princeton, NJ: Princeton University Press.

Lindahl, Hans. 2019. *Authority and Globalization.* Cambridge: Cambridge University Press.

Luckmann, Thomas. 2023. *The Invisible Religion.* [1967] reissued. Amsterdam: Amsterdam University Press.

Maus, Marcel. 1990. *The Gift.* London/New York: Routledge.

Norris, Pippa and Ronald Inglehart. 2004. *Sacred and Secular Religion and Politics.* New York: Cambridge University Press.

Osiander, Andreas. 1994. *The State System of Europe.* New York: Oxford University Press.

Simmel. Georg. 1991 [1900]. *Die Philosophie des Geldes.* Edited by David Frisby. London/New York: Routledge.

Wittgenstein, Ludwig. 2009. *Philosophical Investigations.* Translated by G. E. M. Anscombe, rev. 4th edn. Chichester: Blackwell.

Index

For the benefit of digital users, indexed terms that span two pages (e.g., 52–53) may, on occasion, appear on only one of those pages.

The manufacturer's authorised representative in the EU for product safety is Oxford University Press España S.A. of el Parque Empresarial San Fernando de Henares, Avenida de Castilla, 2 – 28830 Madrid (www.oup.es/en or product.safety@oup.com). OUP España S.A. also acts as importer into Spain of products made by the manufacturer.

www.ingramcontent.com/pod-product-compliance
Lightning Source LLC
Chambersburg PA
CBHW060220310126
39050CB00034B/572/J
* 9 7 8 0 1 9 8 9 7 4 5 1 2 *